# Plant
# Growth
# *and*
# Development

## A Molecular Approach

# Plant Growth and *Development*

## A Molecular Approach

**Donald E. Fosket**
Department of Developmental and Cell Biology
School of Biological Sciences
University of California, Irvine
Irvine, California

**Academic Press**
A Division of Harcourt Brace & Company

San Diego   New York   Boston   London   Sydney   Tokyo   Toronto

*Cover Photograph:* Staghorn sumac (*Rhus typhina*)  (Photograph by Tom Pantages.)

This book is printed on acid-free paper. ∞

Academic Press, Inc.
525 B Street, Suite 1900, San Diego, California  92101-4495

*United Kingdom Edition published by*
Academic Press Limited
24–28 Oval Road, London NW1 7DX

Library of Congress Cataloging-in-Publication Data

Fosket, Donald E.
    Plant growth and development  :  a molecular approach  /  Donald E.
  Fosket.
      p.    cm.
    Includes bibliographical references and index.
    ISBN  0-12-262430-0
    1.  Growth  (Plants) --Molecular aspects.  2.  Plants--Development.
  3.  Plant molecular biology.    I.  Title.
  QK731.F69     1994
  581.3--dc20                                              93-46734
                                                              CIP

PRINTED IN THE UNITED STATES OF AMERICA
94  95  96  97  98  99     MM     9  8  7  6  5  4  3  2  1

# Contents

## 1    Introduction

# 2

## The Genetic Basis of Plant Development

# 3    The Size and Complexity of Plant Genomes

# 4    Regulation of Gene Expression

# 5    Characteristics of Plant Cells That Are Important in Development

# 6

## Light, Hormones, and Cell Signaling Pathways

# 7 Cell Division, Polarity, and Growth in Plant Development

# 8

## Embryogenesis, Seed Development, and Germination

# 9

# Apical Meristems and the Formation of the Plant Body

# 10

## Biotic Factors Regulate Some Aspects of Plant Development

# Preface

I have taught a course on the molecular analysis of plant growth and development to junior and senior undergraduates for more than 10 years. At the inception of this course, an understanding of plant development from a molecular–genetic perspective seemed unlikely in my lifetime. Still, it was obvious that this approach was necessary if we were to gain an understanding of the mechanisms responsible for angiosperm development, and that in some time frame we would understand plant development at the molecular level. Additionally, the course gave me the opportunity to present the many intriguing questions about how plants carry out their unique development to students who knew nothing about plants, but had a good background in molecular and cellular biology. It was an opportunity to proselytize, and possibly get young people excited about the opportunity to apply molecular–genetic techniques to these important questions.

This book came about because of the difficulty I had in finding an appropriate text for my course. Although there are some excellent books on plant development, none of them approached the subject from the perspective of molecular biology and genetics, and the treatment of plant development in molecular biology or developmental biology texts was too rudimentary, if the subject was presented at all, to be useful in a course devoted to plant development. Texts on plant physiology usually include chapters on plant development, but they also contain a great deal of information that is not directly related to development. Additionally, their perspective is physiological and biochemical rather than molecular and genetic. As a result, I began to provide the students with my typewritten summaries of different topics in plant development so that they would have something to read to reinforce what they had heard in lecture. When Dr. Phyllis B. Moses at Academic Press approached me about writing a text in plant development, it seemed like an opportunity to provide an important service to this rapidly developing field, as a text certainly was needed. Additionally, since I already had the summaries I provided to my students, I naïvely assumed that it wouldn't require much additional work to turn these and my lecture notes into a text. The book is now complete, five years later. It is primarily an advanced undergraduate textbook that is intended to be used in courses in plant development which follow an introductory biology course. No special knowledge of plant biology is assumed or necessary as a prerequisite since sufficient basic plant biology is provided in introductory chapters to give students the background necessary to understand how plants develop. Additionally, the book can be used as a text for courses in plant molecular biology at the advanced undergraduate level. I believe that graduate students and postdocs who are just beginning to work in this field also will find much valuable information in this book.

The field of plant development has moved exceptionally fast in the past five years, largely as a result of the increasing exploitation of the remarkable plant, *Arabidopsis*. It could have been designed by biologists seeking a model angiosperm for the molecular and genetic analysis of plant development. The small genome, rapid life cycle, and ease of transformation of *Arabidopsis*, as well as the relatively large number of laboratories that are utilizing this plant for their research, have led to an exponential increase in information about plant developmental mechanisms, and they will continue to do so. It has been a challenge to keep up with this flood of information and to incorporate it into this text. It is a very exciting time for plant developmental biologists and it is my hope that some of this excitement will be conveyed to students through this book.

Donald E. Fosket

# Acknowledgments

I thank my teachers who first made me aware of this fascinating subject, particularly Dr. Lorin W. Roberts and Dr. John G. Torrey. In addition, I thank Dr. Russell Jones and Dr. Richard Cyr for their careful reading of the manuscript and for their many helpful comments and corrections.

# 1

# Introduction

Development is the process that builds the organism. Development is the sum of all of those events during the life of a plant or animal that produce the body of the organism and give it the capacity to obtain food, to reproduce, and to exploit the opportunities and deal with the hazards of its environment. Development is a process. When we study plant or animal development, we not only describe what has changed, but, more importantly, we analyze the process by which the change has come about. We ask: How does a plant grow? How does it produce flowers? How are seeds formed? How can seeds survive without water and then germinate when water is provided? Developmental biologists also try to explain how an organism acquires its unique abilities. For example, How do humans develop the ability to communicate through language? How do photosynthetic plants acquire the ability to use light energy to drive chemical reactions? A child attains the ability to speak not only because she learns a language, but also because developmental changes occur in the structure of her brain that make language possible. Similarly, a seedling acquires the ability to carry out photosynthesis because chloroplasts, the photosynthetic organelles, are formed within specialized leaf mesophyll cells. The leaf also acquires many other adaptations and structures in the course of its development that enhance photosynthesis, such as stomatal openings permitting gas exchange with the atmosphere. Many aspects of leaf development, including the formation of the chloroplasts, are initiated and regulated by light, and they occur because light regulates the expression of specific genes. These are examples of developmental changes and this brief presentation suggests what we must investigate to identify the mechanisms responsible for these changes. A developmental biologist determines not only what changes take place, but also the mechanisms that bring change about.

Biologists use many different methods and approaches to study development. These include (1) analysis of the molecular–genetic mechanisms that underlie all developmental processes, (2) characterization of the biochemical reactions that actually carry out development, (3) investigations of the structures of cells and how these structures help bring about developmental changes, and (4) investigations of the structures and integrated functions of tissues and organ systems. Plant development differs from animal development most radically at the tissue and organ levels, but there is a strong cellular basis for these differences. Although plant cells share many of the major structural features found in other eukaryotic cells, they also have unique properties, such as the presence of a rigid cell wall and the lack of motility systems. These and other unique properties of plant cells make plant development rather different from animal development.

Development occurs in response to instructions contained in the genetic information the organism inherits from its parents. An acorn becomes an oak as a result of an orderly readout of information stored in its genes. It will come as no surprise to anyone with even the briefest exposure to biology that the sequence of bases in DNA encodes the amino acid sequence of cellular proteins. It is also true that genes control both quantitative characteristics, such as plant height, and qualitative characteristics, such as leaf shape and organ identity. Genes determine both the rate and the extent of growth. They determine where growth can occur as well as where it cannot occur. As we shall see, differential growth is one of the major mechanisms by which plant organs acquire their shape. Plants are very responsive to their environment and environmental factors determine whether or not growth occurs; however, when environmental factors permit growth, it is the genetic makeup of the plant that determines where and how rapidly this growth occurs. Nevertheless, the precise manner in which genes control growth or complex characteristics, such as the arrangement of leaves on the stem, is not completely understood at this time.

On the molecular level, plant development and animal development do not appear to be very different. The genetic code is the same in higher plants and animals and the mechanism responsible for regulating the expression of genes appears to be remarkably similar in all eukaryotic organisms. Plants differ from animals because of the specific nature of their genes, not because their genetic information is organized in a different fashion. Still, some important molecular–genetic differences between plants and animals may emerge when we learn more about the molecular mechanisms responsible for determination.

Organisms are constructed of cells. The structure and functions of a particular tissue or organ are determined by the collective shapes, functions, and interactions of its component cells. As a result, it is essential that we understand how individual cells function and how they change their characteristics. All eukaryotic cells share the same general cellular architecture. That is, all cells are constructed along the same lines, using largely similar molecules that interact to form organelles with similar functions. Plant cells, however, possess several unique structures, principally cell walls and plastids, that animal cells do not. These unique structures place constraints on plant development, as well as present the plant with developmental potentials, that do not exist in animals. To understand plant development, it is necessary to understand the structure and potential of plant cells.

This book deals primarily with flowering plant development, although occasionally we examine development in other organisms, such as ferns, yeasts, and algae, when an important aspect of development has been particularly well studied in these organisms. These organisms, however, are examined only insofar as their study helps us understand aspects of flowering plant development. Flowering plants, or **angiosperms,** are the most abundant, successful, and diverse group of land plants. They are the dominant plants in tropical and most temperate regions of the world. They include almost all of the plants that have been domesticated for agriculture. Wheat, maize, beans, rice, oats, tomatoes, potatoes, and soybeans are only a few of the important angiosperm crop plants. Most ornamental plants and many trees, including orchids, petunias, zinnias, asters, oak trees, maples, and lawn grasses, also are examples of angiosperms. Wheat, maize, and orchids belong to a group of angiosperms known as **monocots,** or plants that have only a single cotyledon. The **cotyledon** is a seed leaf that is formed during embryogenesis. In contrast, soybeans, oak trees, maples, petunias, and tomatoes are examples of **dicots,** or plants that have two seed leaves. Monocots and dicots differ in many other ways as well. Their stems and roots have different structures and they have consistent physiological differences as well, such as their response to herbicides. Despite these many differences, dicots and monocots share similar developmental mechanisms.

Some knowledge of plant structure is essential for the study of plant development. Many individuals begin their study of plant development after completing a course in general botany or plant morphology. A brief presentation of plant structure may be unnecessary for such individuals. In other cases, plant development is the student's first exposure to plant biology. The following synopsis of the essential features of plant structure is presented for the benefit of those individuals who intend to study plant development without previous exposure to plant morphology and anatomy and for individuals who require a quick review of the subject. The reader should, however, be aware that this summary is highly abbreviated. Plants present a rich diversity of anatomical features, with many variations on the structures portrayed here. Each of these divergent structures has ecological and evolutionary significance and represents the culmination of a particular developmental pathway. Further study of plant structure, beyond that presented here, would greatly enhance the student's appreciation and understanding of plant development.

## The Plant Body

A typical flowering plant has the general form shown in Fig. 1.1. It is highly elongate and polarized, with its axes oriented at right angles to the earth's surface. The force of gravity plays an important role in determining and maintaining the form of the plant. The above-ground portion of the plant is the system of integrated organs known as the **shoot;** the **root system** usually is underground. The root system may be highly branched, but it has no other lateral organs or appendages attached to it. In contrast, the shoot system may consist of a number of complex organs, including stems, leaves, thorns, tendrils, flowers, and fruits. The main axis of the shoot is the **stem,** which consists of alternating **nodes,** the points where leaves are attached, and **internodes,** the regions between nodes. Leaves, bracts, buds, flowers, and fruits typically are attached to the stem at the nodes and they occur in some regular, ordered pattern. Actually, although very few plants resemble the "typical" plant closely, most have many or all of the structures listed in some form. For example, the stem is not readily seen in the aquatic monocot often called water lily (e.g., *Nuphar lutea*, Fig. 1.2A). Instead, the only parts of the plant readily visible are its floating leaves and flowers. For most angiosperms, a **petiole** or stalk attaches each leaf to the stem at a node. In the water lily, the petiole is very long and flexible, connecting the leaf to an underground stem (known as a **rhizome**) that grows horizontally through the mud at the bottom of the pond. The leathery leaves are supported by the water and present a large surface area for photosynthesis. At the opposite extreme, the leaves of the cactus known as prickly pear (*Opuntia polycantha*) have been reduced to thorns and are not photosynthetic at all (Fig. 1.2B). The photosynthetic organs of the cactus are its stems, which are thick and flattened. These thickened stems also store water, which helps the prickly pear survive in areas with infrequent rainfall; the thorns protect the plant against foraging animals. The prickly pear and the water lily are only two examples of plants with very different morphologies, each of which has evolved a unique set of adaptations to its environment.

The xylem and phloem together constitute the vascular tissue of the plant. Water and minerals are transported from roots and into leaves, as well as all other parts of the plant, through specialized conducting cells of the xylem. The food produced during photosynthesis is transported out of the leaf and throughout the plant by means of specialized cells in the phloem.

The **xylem** supplies a stream of water and dissolved minerals from the roots to the leaves and to all other tissues and organs of the plant. The xylem tissue consists of four different kinds of cells: (1) parenchyma cells, (2) fibers, (3) tracheids, and (4) vessels (Fig. 1.3). A **parenchyma cell** is any kind of mature, nondividing cell with a large vacuole, a primary cell wall (usually), and a relatively unspecialized function. Many different plant tissues are composed in whole or part of parenchyma cells. Xylem parenchyma cells may be somewhat atypical in that they often have thickened cell walls, but they remain living at maturity and they do not participate directly in water transport. Xylem **fibers** are highly elongate, often pointed cells, with thick secondary walls. Fibers also do not participate in water transport. They provide mechanical support for the stem.

The conductive cells of the xylem, known as **tracheary elements,** are dead at maturity and have reinforced secondary cell walls (Fig. 1.3). Tracheary elements include tracheids and vessels. The **vessel** is a linear file of cells, each of which is called a **vessel element.** A file of vessel elements differentiates as a unit to form a single vessel. Vessel element differentiation involves the synthesis and deposition of a secondary cell wall, which reinforces the primary cell wall. A **primary cell wall** is the wall formed during cell growth and it consists of cellulose, noncellulosic polysaccharides, and some unique proteins. The **secondary cell wall** is formed after growth is complete. It also is composed of cellulose and noncellulosic polysaccharides, but it lacks protein and contains a complex polymer known as lignin. The secondary wall may be laid down in an annular, spiral, or reticulate pattern so that some portions of the wall receive no reinforcement. Some vessel elements also may have uniformly thickened lateral walls, but even then there is no secondary wall deposition on parts of the end walls. During their differentiation, portions of the unreinforced primary cell wall separating vessel elements are dissolved away, forming **perforation plates** between adjacent cells. The formation of perforation plates eliminates the physical barrier between vessel elements. After secondary wall deposition and perforation plate formation are complete, the vessel elements die and the protoplasts disintegrate. As a result, the vessel is a more or less continuous tube through which water may flow (Fig. 1.3).**Tracheids** differ from vessels in that each tracheid is a single cell, but usually it is much longer than an individual vessel element. Tracheids also undergo secondary wall deposition during their differentiation. The wall is usually more or less evenly thickened, except where connections occur between tracheids, called **pits,** which permit the passage of water. These openings are much smaller than those connecting the vessel elements. As a result, more water can move through xylem containing vessels than through an equal amount of xylem containing only tracheids. Gymnosperms, whose vascular tissue contains only tra-

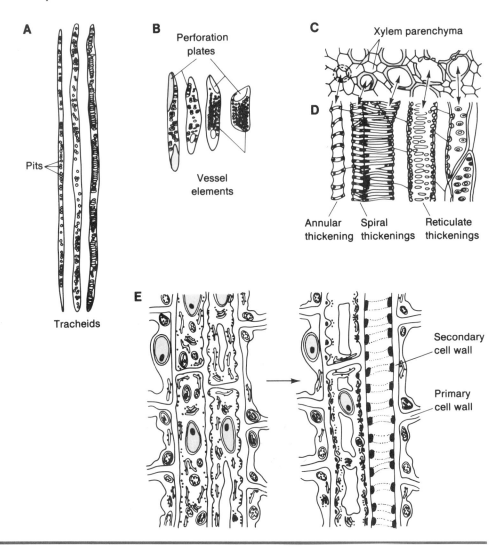

**Figure 1.3  Tracheary elements and other cell types of the xylem**
The water-conducting tracheary elements are tracheids (A) and vessels (B). The location of tracheary elements within the xylem tissue is shown in cross section (C) and longitudinal section (D). The tracheary elements shown here are vessels which have annular, spiral, or reticulate secondary wall thickenings. The cells surrounding the tracheary elements are xylem parenchyma. Vessel elements are formed when a file of parenchyma cells undergoes secondary wall deposition, followed by lysis of the cytoplasm and the cross walls separating them (E). Redrawn with permission from Jensen and Salisbury (1984).

cheids, are much less efficient in transporting water to their leaves than are angiosperms, whose vascular tissue contains vessels as well as tracheids.

The **phloem** component of the vascular tissue conducts food materials synthesized during photosynthesis; in most plants this food is sucrose. Phloem also consists of four different cell types: (1) parenchyma cells, (2) fibers, (3) sieve elements, and (4) companion cells (Fig. 1.4). Phloem parenchyma cells and fibers are similar to the cell types with the same names found in the xylem, although phloem parenchyma cells usually have primary cell walls. The principal food-conducting structures of the phloem are multicellular structures known as sieve tubes. Like vessels, sieve tubes are constructed from a linear file of cells, each component of which is known as a **sieve element.** Unlike the vessel, however, the sieve element remains alive at maturity, although its nucleus disintegrates and its cytoplasm and end walls are extensively modified. Sieve elements undergo some secondary wall deposition, but they do not become lignified, as do the tracheary elements. A file of cells, all of which usually arise from a common meristematic precursor, differentiate coordinately, with the construction of specialized end walls known as **sieve plates,** during the formation of the sieve tube. There are cytoplasmic connections between the sieve elements, through the pores in the sieve plate. Sieve elements and companion cells share a common origin: They are daughter cells derived from the same cell division. One daughter becomes a companion cell, whose function is largely unknown, whereas the other becomes the sieve element.

## Meristems

One of the most unusual aspects of higher plant development is the fact that some regions, called **meristems,** may remain embryonic throughout the life of the plant. These regions of perpetual embryogenesis generate the plant body by producing the cells that will become the leaves, stem, roots, and flowers of the mature plant (Fig. 1.5). Meristems contain small cells known as initials that continue to divide indefinitely and do not differentiate. Meristematic activity is regulated by physiological and environmental signals, so meristems are not active at times when the climate is unfavorable for growth, but they retain the potential for growth. The pattern in which cells divided within the meristem will determine the placement of leaves and the organization of the tissues within the organs. Plant development is mostly postembryonic. Embryogenesis establishes only a rudimentary plant axis, with the shoot and root apical meristems at either end. With the exceptions of the cotyledons and the first leaves in some plants, none of the organs of the mature plant are formed during embryogenesis. Rather, the plant body is constructed by the meristems, which do not begin activity until after embryogenesis is complete and seed germination has begun. For these reasons the process of embryogenesis does not have the same central importance for plant development as it does for animal development.

**A**

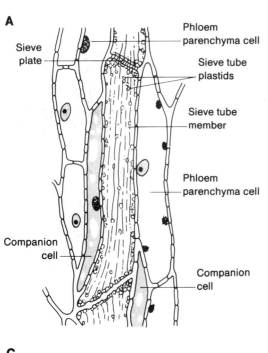

Sieve
plate

Phloem
parenchyma cell

Sieve tube
plastids

Sieve tube
member

Phloem
parenchyma cell

Companion
cell

Companion
cell

**B**

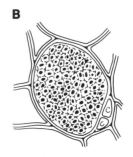

**C**

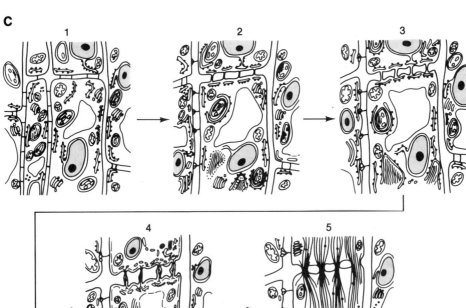

1   2   3

4   5

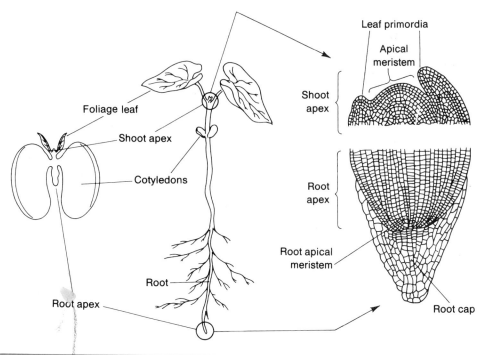

**Figure 1.5    Primary meristems**

Primary meristems are formed during embryogenesis and include the root and shoot apical meristems. This diagram demonstrates both the origin and the position of the apical meristems in a bean seedling. Modified with permission from Sussex (1989).

**Figure 1.4    Phloem cell types**

(A) A mature sieve tube is shown as it would appear in longitudinal section. To the right of the sieve element is a companion cell; a phloem parenchyma cell lies to the right of the sieve element. A sieve plate is shown in face view in (B). The black areas represent cytoplasmic passages through the wall between adjacent sieve elements. In (C) the progressive stages in the formation of a sieve tube are depicted in the series of drawings 1 to 5. Unlike tracheary elements, the sieve tube retains its cytoplasm at maturity, although it is highly modified and these cells lack a nucleus. Phloem fibers are depicted in (E). Redrawn with permission from Jensen and Salisbury (1984).

Two meristems, the **shoot and root apical meristems,** are formed during embryonic development, whereas additional or **secondary meristems** may develop from mature cells later in development. The root and shoot apical meristems form the primary plant body; secondary meristems are responsible for producing secondary tissues, such as wood and bark. Herbaceous plants, such as beans and petunias, may lack secondary meristems entirely, or these meristems may be poorly developed. In contrast, woody plants, such as trees, have well-developed secondary meristems that persist for many years. In either case, meristems frequently persist throughout the life of the plant, exhibiting periods of activity followed by quiescence.

The shoot apical meristem is at the extreme tip of the shoot, although it may be enclosed by the most recently formed young leaves. Cell divisions within the flanks of the shoot apical meristem lead to the formation of protruding mounds of cells, which will grow and differentiate to become mature leaves. These are known as **leaf primordia.** Cell divisions that occur within a region just below the apical meristem generate the cells that differentiate to become the tissues of the stem. The vegetative shoot apical meristem is very repetitive in its activity. It produces the same structures (leaves, lateral buds, and stem tissues) over and over again. Its activity is indeterminate, but also usually periodic. That is, vegetative shoot apical meristems may be active in the spring and early summer, but then become dormant in late summer or fall, after producing special overwintering structures such as bud scales. Growth then is resumed the following spring. Vegetative shoot apical meristems can be transformed into floral meristems. A floral meristem usually exhibits **determinant growth;** that is, it produces a set number of organs and grows no more. Floral meristems are progressive rather than repetitive. After all of the floral organs are produced, all meristematic capital is used up and there can be no more growth.

In contrast to the shoot apical meristem, the root apical meristem is not terminal, but rather is subterminal. It is covered by another tissue, known as the **root cap.** The root cap protects the meristem as the root grows through the soil. The root apical meristem also differs from the shoot apical meristem in that it does not produce any lateral organs. The root apical meristem generates cells that become the root cap and the cells that make up the primary tissues of the root axis. Lateral roots are formed by adventitious meristems that appear later in root development in mature regions of the primary root.

### Roots

Water and minerals are essential nutrients for all forms of life. Animals obtain essential minerals such as iron, calcium, potassium, and nitrogen from their diet, whereas plants obtain water and these essential minerals from the soil solution. Roots are specialized organs for water and mineral absorption and transportation. They also anchor the plant in the soil.

Roots increase in length only at their tips (Fig. 1.6). It is here that new cells

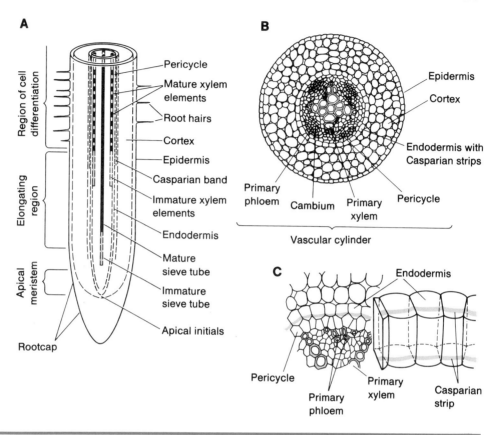

**Figure 1.6** **Some important aspects of root structure**
(A) This longitudinal section through a growing root illustrates the root cap, apical meristem, region of elongation, and region of differentiation. It is within these regions that the primary body of the root is formed and all primary growth and differentiation occur. (B) In this cross section of a young dicot root, note all the vascular tissue occupies the center of the root. (C) The staructure of the endodermis and its position within the root. (A) Redrawn with permission from Esau, K. (1965). (B) and (C) Redrawn with permission from Esau, K. (1977). *Anatomy of Seed Plants,* 2nd ed. Wiley, New York.

are added to the root cylinder by the meristem. These new cells subsequently elongate and then differentiate to assume specialized tasks within the root. After they differentiate, most cells neither divide nor elongate again. As these different developmental events tend to be spatially separated, the root is said to consist of four zones: (1) the root cap, (2) the meristem, (3) the region of elongation, and (4) the region of differentiation. As growth occurs in the region of elongation, three types of tissues can be recognized: **protoderm, ground meristem,** and **procambium.** The protoderm will differentiate as the epidermis, the procambium becomes the vascular tissue, and the ground meristem differentiates as cortical tissue.

The vascular tissue forms a cylindrical central core of the root. It is surrounded by four concentric rings of tissue. Proceeding from the inside out, these are the **pericycle, endodermis, cortex,** and **epidermis.** The pericycle actually is considered to be the outermost layer of the vascular cylinder, because it was derived from the procambium, but it is a region of the vascular cylinder with a unique function. The pericycle is only one cell layer thick. Its cells are thin-walled and undifferentiated, but they retain a capacity for growth and play an important role in root development. The pericycle gives rise both to the vascular cambium of the root and to branch roots later in development. Cortical tissue is composed mainly of parenchyma cells, and its primary function usually is carbohydrate storage; however, the innermost layer of the cortex, the endodermis, is specialized and functions in water transport through the root. Endodermal cells lack secondary walls, but lignin and a waxy material called suberin are deposited in their radial and transverse walls. These deposits are known as the **Casparian strip.**

Many epidermal cells develop **root hairs.** These are formed within the region of differentiation and they are thin-walled, linear outgrowths that remain part of the epidermal cell. Root hairs greatly increase the surface area of the root that is exposed to the soil solution, making the uptake of water and minerals much more rapid. It is within the region of the root containing root hairs that minerals and water are taken up and loaded into the tracheary elements for transport into the shoot system. To reach the vascular tissue, water and minerals must cross the cortex. Primary cell walls are very hydrophilic. As a result, water can flow through the matrix of the cell walls as it crosses the cortex, without entering the cytoplasm of the cortical cells. When water reaches the endodermis, however, the suberin-containing Casparian strip of the epidermal wall prevents further movement of water or minerals through the cell walls. Water and dissolved minerals reaching the xylem must, as a result, pass through the cytoplasm of the endodermis. The plasma membrane of the endodermal cells then acts as a selective barrier, allowing some minerals to pass while excluding others. All dissolved minerals that will be transported by the xylem must first pass through the selective filter of the endodermal cytoplasm.

Not all species develop root hairs or rely exclusively on root hairs for water and mineral adsorption. Many plants have developed a symbiotic relationship with soil-dwelling fungi in which the fungi serve the function of the root hairs. These fungi are known as **mycorrhizae** (singular, mycorrhiza). Although laboratory-germinated seedlings lack mycorrhizae, most seeds growing in soil in nature have mycorrhizae associated with their roots.

The arrangement of the xylem and phloem within the vascular tissue is characteristic of the different groups of angiosperms. In dicots, the strands of xylem alternate with the phloem strands, both of which initially form close to the pericycle. Tracheary elements first differentiate near the outside of the vascular cylinder, and then differentiation progresses inward. Often the whole central vascular cylinder contains a core of differentiated tracheary elements, from which arms radiate out toward the pericycle. The number of arms of xylem observed in a cross section of the root is a consistent characteristic of a given species, although it can be changed experimentally. The phloem continues to be present as separate bundles which lie between the xylem arms (see Fig. 1.6). In monocots, the cells in the center of the vascular cylinder do not differentiate into tracheary elements, but remain as undifferentiated parenchyma and are collectively called **pith.** The xylem and phloem of the root are continuous with the vascular tissues of the stem, although the arrangement of the vascular tissues in the shoot is quite different from that of roots.

## Stems

Cell division and development within the shoot apical meristem form leaf primordia, as well as cells that grow and differentiate to become the tissues of the stem. Primary stem growth and tissue differentiation occur only in the terminal several centimeters of the shoot tip, but the elongating and differentiating zones are not spatially separated, as they are in the root. Growth and development of the shoot apical meristem initially form three different tissues—protoderm, ground meristem, and procambium—as in the root (Fig. 1.7). These three tissues further differentiate into epidermal, ground, and vascular tissues, respectively; however, where the root had a solid core of procambium that produced a central cylinder of vascular tissue, the procambium of the stem consists of several separate strands within the ground tissue. Each of the procambial strands then differentiates into a separate vascular bundle. In monocots, the vascular bundles are more or less randomly scattered throughout the ground tissue, but in dicots the vascular bundles are arranged in a ring and the central part of the stem lacks vascular bundles. Each vascular bundle contains both xylem and phloem in both monocots and dicots. Usually, the phloem is toward the surface of the stem whereas the xylem is more internal, but they are aligned on the same radius. Undifferentiated procambial tissue separates the xylem and phloem during primary growth, but the vascular cambium will develop from the remaining procambial tissue later in development as secondary growth is initiated. Pith parenchyma occupies the central core of the dicot stem, while the vascular tissues are separated from the epidermis by a layer of cortical parenchyma. Both cortex and pith are derived from the ground meristem in the stem. The cortical parenchyma of the stem usually contains chloroplasts and is active in photosynthesis.

## The Vascular Cambium and Secondary Growth

The vascular cambium and cork cambium are secondary meristems that are formed in stems and roots after the tissues of the primary plant body have differentiated. The vascular cambium is responsible for increasing the diameter of stems and roots and for forming woody tissue. The cork cambium produces

**A**

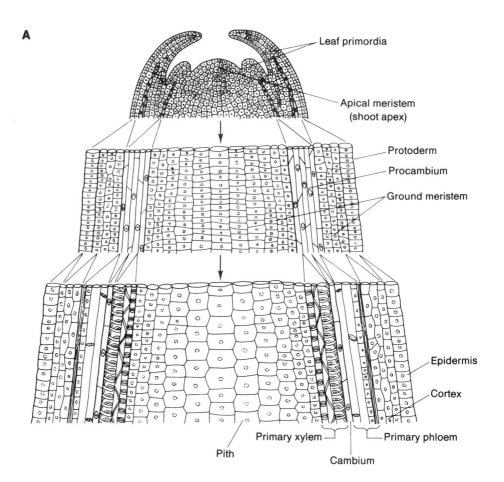

Leaf primordia

Apical meristem
(shoot apex)

Protoderm

Procambium

Ground meristem

Epidermis

Cortex

Primary xylem

Cambium

Primary phloem

Pith

**B**

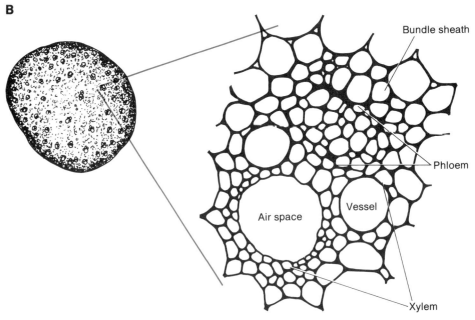

Bundle sheath

Phloem

Air space

Vessel

Xylem

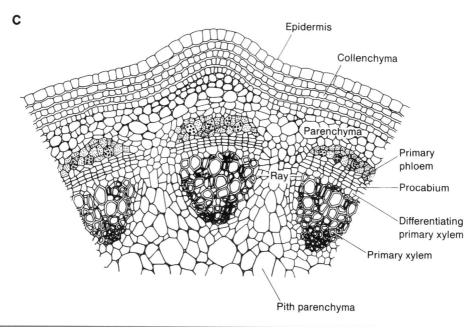

C

Epidermis

Collenchyma

Parenchyma

Primary phloem

Procabium

Differentiating primary xylem

Primary xylem

Ray

Pith parenchyma

**Figure 1.7** **Some important aspects of stem structure**
The shoot apical meristem produces not only leaf primordia, but also the tissues of the stem: protoderm, procambium, and ground meristem. The procambium is present as strands that run through the ground meristem. Procambial cells differentiate to become the vascular bundles (A). In contrast to the root, a young stem will have several vascular bundles either scattered throughout the stem in the monocots (B) or arranged in a ring in dicot stems (C). (A) Redrawn with permission from Wilson, C. L., and Loomis, W. E. (1967). *Botany.* Holt, New York. (B) Redrawn with permission from Esau, K. (1977). *Anatomy of Seed Plants,* 2nd ed. Wiley, New York. (C) Redrawn with permission from Weier, T. E., Stocking, C. R., and Barbour, M. G. (1970). *Botany: An Introduction to Plant Biology,* 4th ed. Wiley, New York.

some of the bark. In dicot stems, the vascular cambium initially differentiates from procambial cells within the vascular bundles (Fig. 1.8A). This **fascicular cambium** may contribute additional cells to both the xylem and the phloem of the bundle. At some point the cambium expands into the ground tissue between the vascular bundles, forming an **interfascicular cambium,** completing the ring of vascular cambium (Fig. 1.8B). Cell division by the cambium produces cells that become secondary xylem and phloem. As secondary phloem and xylem tissue accumulates, it both increases the girth of the stem and forms wood and bark. Because cambial activity is seasonal in temperate zone plants, the wood and bark are laid down in distinct annual rings (Fig. 1.8C). Monocots do not have a vascular cambium, even though some of them, such as palms and the Joshua tree, exhibit secondary growth. Instead, they have a thickening meristem that produces secondary ground tissue. This increases the girth of the stem and additional vascular bundles differentiate within the secondary ground tissue.

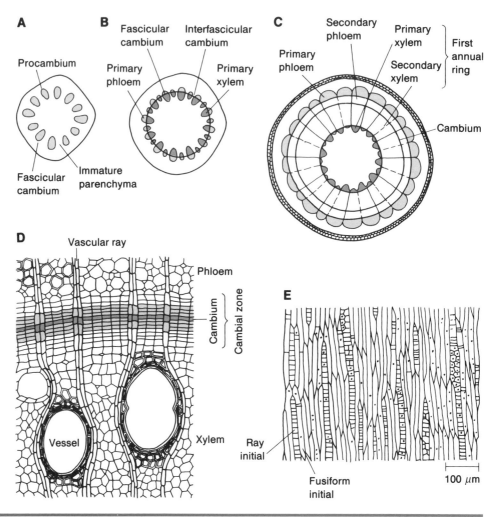

**Figure 1.8   Secondary growth: the origin and structure of vascular cambium in thc stem**

The vascular cambium is formed in mature dicot stems after stem elongation stops. (A) Primary xylem and phloem differentiate from procambial tissue in the vascular bundles, and a fascicular cambium is formed from procambial tissue separating these tissues. (B) Later, an interfascicular cambium appears between the vascular bundles that is continuous with the fascicular cambium. (C) The further development of the cambium results in the formation of a cylinder of vascular tissue. (D) The vascular cambium is a layer of pluripotent dividing cells whose derivatives differentiate as either xylem elements (vessel members, tracheids, fibers, or xylem parenchyma) or phloem elements (sieve tube members, companion cells, fibers, or parenchyma). (E) The dividing cells of the vascular cambium consist of long, narrow fusiform initials, from which the tracheary elements are derived, and ray initials, from which ray parenchyma is formed. Based on Wilson, C. L., and Loomis, W. E. (1967). *Botany.* Holt, New York.

The vascular cambium is composed of two kinds of cells, **ray initials** and **fusiform initials.** In cross section these look very similar. Both are small, flattened cells with thin walls. When viewed in tangential section, however, ray initials can be seen to be relatively short, small cells, whereas fusiform initials are very long and narrow (Fig. 1.8D). In gymnosperms the fusiform initials often are several millimeters in length. Dicot fusiform initials are much shorter, but some still are up to 0.5 mm in length. Cell division in the fusiform initials usually is tangential and the cell is partitioned down its long axis, forming two equally long, narrow cells. Some of the cells produced by the cambial initials continue to divide, whereas others differentiate. Tracheary elements or sieve elements differentiate from derivatives of the fusiform initials, and derivatives of the ray initials differentiate as ray parenchyma. The ray parenchyma permits transport of water from the xylem into the cambium and the tissues of phloem, as well as transport of photosynthate from the phloem into the cambium and the living cells of the xylem.

The cork cambium also is a secondary meristem, containing meristematic cells. The cork cambium forms a major portion of the bark of woody plants. The secondary phloem also is part of the bark, but of course phloem is produced by the vascular cambium. The cork cambium first arises within the cortex as a concentric layer forming a cylinder of dividing cells (Fig. 1.9). The derivatives of this meristematic cell layer differentiate as cork, or **phellem,** toward the outside of the stem, whereas derivatives produced toward the inner part of the stem differentiate as **phelloderm.** Suberin is deposited in the cell walls of the phellem and they are dead at maturity. They protect the stem from water loss and from mechanical damage. As the tree increases in girth, the outer layers of bark are sloughed off. Additional cork cambia arise within the secondary phloem as the plant develops.

### Leaves

A typical angiosperm leaf is a thin, flattened structure that may be only a few cells thick. Photosynthesis occurs in the chloroplasts, which are abundant in the mesophyll cells of the leaf. Photosynthesis consists of two separate processes: (1) the fixation of carbon from carbon dioxide and (2) the reduction of carbon using energy obtained from light. Technically, the fixation of $CO_2$ does not require light energy and can occur in the dark. In carbon fixation, an intermediate reacts enzymatically with carbon dioxide to form an organic acid. Two different mechanisms have evolved to fix carbon for photosynthetic reduction. These are known as the C3 and C4 pathways, and plants exhibiting these are said to be C3 and C4 plants (Fig. 1.10). The first stable product of $CO_2$ fixation in C3 plants is a three-carbon organic acid. The enzyme **ribulose-1,5-bisphosphate carboxylase–oxygenase (RUBISCO)** combines $CO_2$ with the phosphosugar ribulose 1,5-bisphosphate to form two molecules of **3-phosphoglyceric acid**. In contrast, carbon fixation in C4 plants results in the formation of four-carbon organic acids, **aspartate** and **malate,** as the first stable product of carbon fixation.

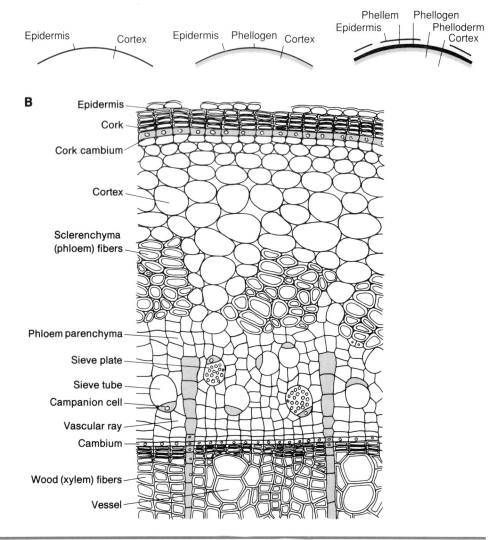

**Figure 1.9  Cross section through the stem of a woody dicot showing the develompent of a cork cambium**

(A) Based on Raven, P. H., and Curtis, H. (1970). *Biology of Plants*. Worth Publishing Company, New York. (B) Redrawn with permission from Wilson, C. L. and Loomis, W. E. (1967). *Botany*. Holt, New York.

In C3 plants, the 3-phosphoglyceric acid subsequently is reduced by a series of reactions known as the **C3 photosynthetic carbon reduction (PCR) cycle** to form a six-carbon sugar as its ultimate product and to regenerate the intermediate $CO_2$ acceptor. Both the carbon fixation and the photosynthetic reduction cycle occur within the leaf mesophyll cells of the C3 plants. The structure of a leaf of a typical C3 plant is shown in Fig. 1.11A. It contains two kinds of mesophyll

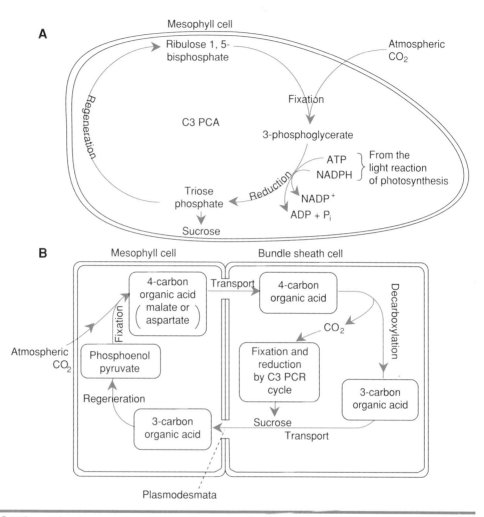

**Figure 1.10 The photosynthetic carbon reduction (PCR) cycle in C3 and C4 plants**

(A) In C3 plants, the fixation of atmospheric $CO_2$, carbon reduction, and the synthesis of sugars all occur in chloroplasts of the same mesophyll cell. Atmospheric $CO_2$ reacts with the five-carbon sugar phosphate ribulose 1,5 diphosphate, forming two molecules of 3-phosphoglycerate. This is then reduced, using ATP and NADPH formed in the light reactions of photosynthesis, to produce triose sugar phosphate. Some of the triose phosphate is used to regenerate ribulose 1,5-bisphosphate; the rest is used to synthesize sucrose. (B) In C4 plants, the initial fixation of atmospheric $CO_2$ fixations occurs in mesophyll cells, but it is used to make organic acids which are transported into bundle sheath cells. The organic acids are decarboxylated in the bundle sheath cells and the $CO_2$ released is fixed again and reduced by the C3-type PCR cycle in bundle sheath chloroplasts. There are several variants of the C4 photosynthetic carbon assimilation cycle which differ as to which organic acid is used to transport carbon between the mesophyll and bundle sheath cells.

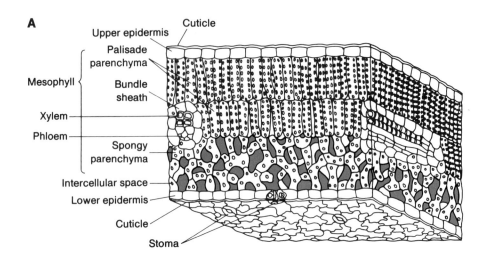

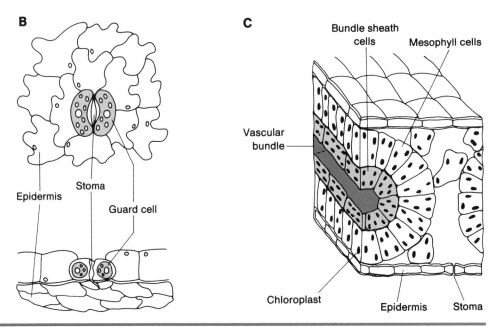

**Figure 1.11 Leaf structure**

(A) Diagram of the structure of the leaf of a C3 plant. The mesophyll is organized into distinct palisade layers which overlie a more disorganized spongy mesophyll. The bundle sheath cells surrounding the vascular bundles lack chloroplasts. (B) Surface view of the epidermis of a leaf illustrating the structure of guard cells and their position around stomata. (C) Structure of the leaf of a C4 plant. The major difference between C4 and a C3 leaf are that the bundle sheath cells contain chloroplasts and play a major role in photosynthesis in C4 leaves. In addition, the mesophyll may not be organized into spongy and palisade layers. Redrawn with permission from (A,B) Wilson, C. L., and Loomis, W. E. (1967). *Botany.* Holt, New York and (C) Alberts (1989).

tissue, spongy and palisade, which differ in their organization, although the cells in both layers contain many well-developed chloroplasts and both fix $CO_2$ and participate in the photosynthetic carbon reduction cycle. The sugar produced by the mesophyll cells is transported into the bundle sheath cells, which surround the vascular tissue, and then into the phloem for transport out of the leaf. The bundle sheath cells of C3 plants lack chloroplasts.

The structure of a C4 leaf is diagrammed in Fig. 1.11C. In C4 plants, the mesophyll cells have chloroplasts, but they lack RUBISCO and do not participate directly in the photosynthetic carbon reduction cycle. The four-carbon organic acids formed as a result of carbon fixation are transported out of the mesophyll cells and into the bundle sheath cells. Bundle sheath cells of C4 plants contain chloroplasts, although their structure differs from that of the mesophyll cells, and RUBISCO is present in these chloroplasts. After entering the bundle sheath cells, the organic acids are decarboxylated and the $CO_2$ is fixed again as 3-phosphoglyceric acid, which is then reduced via the C3 PCR pathway (see Fig. 1.10B). This concentrates the $CO_2$ within the bundle sheath cells and makes the RUBISCO reaction more efficient. It also eliminates an inefficient process known as photooxidation which competes with photosynthesis. The C4 pathway was first discovered in the tropical grass sugarcane. It is also found in maize, crab grass, and many other monocots and dicots scattered among 18 different families of angiosperms. Apparently, this photosynthetic pathway has evolved independently several times. Although it makes photosynthesis slightly less efficient under conditions of high $CO_2$ concentration, when $CO_2$ is limiting for photosynthesis, which it nearly always is under natural conditions, C4 plants are considerably more productive photosynthetically.

In both C3 and C4 plants, the mesophyll tissue is surrounded by a layer of epidermal cells that lack chloroplasts and secrete a waxy material that coats the outer surface of the leaf and prevents its desiccation. As photosynthesis requires atmospheric $CO_2$, gas exchange must take place between the air and the inner portions of the leaf. This can occur because there are small openings through the epidermis called **stomata** (singular, **stoma**) (see Fig. 1.11B). Stomata can be opened or closed by the action of specialized epidermal cells known as **guard cells.** Unlike the other epidermal cells, stomatal guard cells contain chloroplasts, as well as special wall thickenings.

The vascular bundles in the leaf are known as **veins.** The leaf veins are part of the vascular system of the plant that forms an interconnected network running throughout the plant. There are two main types of venation patterns in flowering plants: reticulate and parallel. In **reticulate venation,** the veins occur in a netted pattern; in **parallel venation,** the large veins run side by side with only small lateral interconnections. The pattern of leaf venation is carefully controlled and it may represent one of the unique characteristics by which plants are identified. For example, grasses and other monocots usually have parallel venation, whereas many dicots have reticulate venation. In either case the veins are interconnected and form a system that serves all parts of the leaf.

# Differences between Plants and Animals as Organisms

Plants and animals are radically different kinds of organisms. Plants are not simply green animals. A soybean plant and a human, for example, are very similar in the nature of their genetic material and in the mechanism by which their genetic information is read and used. There are, however, few similarities between them on the level of the whole organism. This is because plants and animals have evolved very different mechanisms to cope with the challenges and opportunities of their environments—an inevitable result of these organisms' having to use different strategies to cope with the main problems all living organisms face:

1. Obtaining the nutrients necessary to sustain life
2. Reproducing
3. Surviving under adverse environmental conditions and exploiting favorable conditions
4. Protecting themselves against harmful organisms

All organisms have structures or behaviors that help them meet these challenges, but plants and animals meet these in radically different ways, each of which has important developmental consequences.

## Mechanisms for Obtaining Food and Other Nutrients

Plants are predominantly **autotrophic;** that is, they make their own food. Plants use atmospheric carbon dioxide, which is a very dilute resource in the environment, as well as water and the energy of sunlight to manufacture their own food by the complex photochemical reactions of **photosynthesis.** In most flowering plants, photosynthesis takes place primarily, if not exclusively, in leaves, which are highly adapted for photosynthesis. Other organs, such as roots, may be heterotrophic; they depend on the food made by photosynthesis in leaves to supply their needs. The plant as a whole, however, is autotrophic. In contrast, animals are **heterotrophic;** that is, they obtain their nutrition by eating already synthesized food. Some animals eat other animals whereas others eat plants, but, directly or indirectly, animals use the food that plants have manufactured through photosynthesis.

## Reproductive Mechanisms

**Plants do not establish a germ line**   Plants do not make a distinction between cells that form the body and reproductive cells, whereas animals usually set aside special cells as the germ line. The germ line cells do not participate in the formation of the animal body; they are reserved early in embryogenesis as the cells that will reproduce the organism. Eggs and sperm (gametes) are formed only by descendants of the germ line cells that divide relatively infrequently during the life of the organism. In contrast, all plant cells retain the potential for reproduction and the probability that any given cell will give rise to gametes is simply a function of its position within the organism. This can have important consequences for plant genetics, as it has been shown that long-lived plants such as trees accumulate mutations and pass these onto their offspring. Mutations can occur during the replication of DNA during cell division; so, the more cell divisions that occur in the life of the organism, the more mutations its descendant will inherit. Reproductive structures most frequently are produced by the youngest tissues on the plant and these will have undergone the greatest number of cell divisions in the life of the organism. The longer a plant lives, the more cell divisions the tissue that will produce the gametes will have completed and the more likely that these gametes will contain mutations.

Plants reproduce asexually as well as sexually     Another major difference in the strategy for reproduction is the fact that most plants reproduce both sexually and asexually, whereas, with a few exceptions, higher, multicellular animals reproduce only sexually. Asexual reproduction results in offspring that are genetically the same as the parent. They are clones, and plants that reproduce by asexual mechanisms are said to be *vegetatively propagated.* In contrast, sexual reproduction produces offspring that differ genetically from either parent. Many plants can be propagated vegetatively, and horticulturists have developed a variety of techniques to exploit this common plant trait. In fact, some important crop plants, such as seedless grapes, and many horticultural varieties can be propagated only vegetatively because they are sterile. A sterile plant would have difficulty surviving in nature over evolutionary time because it most certainly would at some point encounter a major environmental change to which it could not adapt. Over the space of even a few thousand years, however, asexual reproduction can make it possible for a species to expand rapidly into a particular environment to which it is particularly well adapted.

Many plants produce an underground stem, known as a **rhizome,** from which new shoots can arise. A rhizome is similar to a stem and quite different from a root in both its external morphology and internal anatomy. The rhizome's response to gravity is different from that of either roots or shoots. A rhizome usually grows parallel to the surface of the ground instead of vertically. Many grasses, including bamboo, reproduce asexually by producing rhizomes. Bamboo reproduces sexually very infrequently, but it reproduces asexually and can spread rapidly by producing large numbers of genetically identical individuals from rhizomes. When bamboo does flower, which may not happen for many years, all members of the clone flower at the same time and then die. The swamp plant known as cattail also reproduces asexually from rhizomes (Fig. 1.12).

Sexuality is not as well defined in plants as it is in animals     Most animals are either male or female and they have well-defined genetic mechanisms

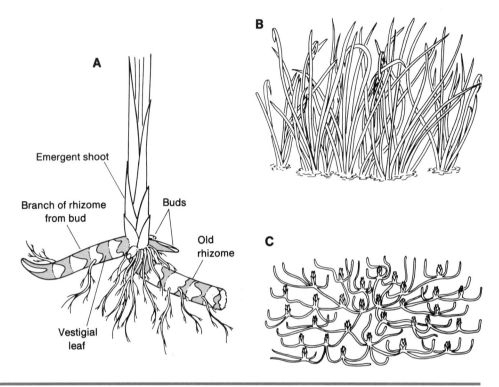

**Figure 1.12 Rhizomes and asexual reproduction**
Rhizomes are underground stems that often play an important role in asexual
reproduction. (A) In the cattail, the rhizome grows horizontally, giving rise to buds
from which young plants develop. (B) The many cattails shown are clones derived
from rhizomes produced by a single individual within a 6-month period. (C) When the
leaves and roots are cut off and the soil washed away, the individual plants can be seen
to be interconnected by the rhizomes, from which they emerged. Redrawn with
permission from Wilson, C. L., and Loomis, W. E. (1967). *Botany.* Holt, New York.

for sex determination. Of course, there are some exceptions. The garden snail,
for example, has both male and female sexual organs and is said to be **hermaphroditic.** All mammals have a well-defined mechanism for sex determination, and
the developmental differences between males and females extend beyond reproductive differences into many other characteristics of the animal, such as body
build and distribution of hair. In contrast, approximately 90% of all angiosperms
produce both male and female gametes. They are hermaphroditic, like the snail.
In a typical angiosperm flower (Fig. 1.13), sperm are formed inside pollen grains
produced by the anthers, whereas the egg develops within an embyo sac, inside
the ovary. Sexual differentiation in animals serves to ensure genetic outcrossing.
Plants achieve this by a different mechanism. Although some plants are self-
fertile, many mechanisms have evolved to ensure that the eggs of one individual

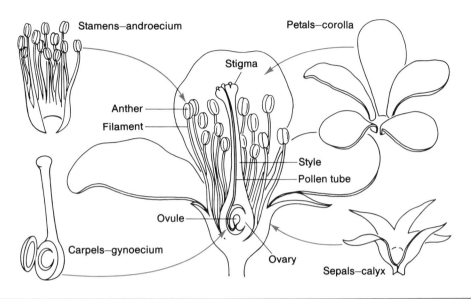

**Figure 1.13 Major parts of an angiosperm flower**
Based on Jensen and Salisbury, 1984.

are not fertilized by its own sperm. These self-incompatibility mechanisms not only are important genetically, but also are examples of self-recognition and cell–cell interactions in plant development, which are presented in detail later in this book.

The plant life cycle is more elaborate than that of animals   Sexual reproduction in most plants involves an alternation of dramatically different generations, whereas in animals reproduction involves the direct production of sperm and eggs as a result of meiosis in the germ cell line. The life cycle of a flowering plant is shown in Fig. 1.14. Plants have two alternating generations: the dominant **sporophytic generation** and a much reduced, parasitic **gametophytic generation.** The sporophytic generation begins with the fertilization of the egg, which undergoes embryogenesis and ultimately develops to become the body of the plant as we usually think of it. The gametophytic generation begins in the flower when cells within the anther or the ovule undergo meiosis to produce haploid spores. There are actually two gametophytic generations within a typical flower, one male and the other female. The male gametophyte generation develops into the **pollen grain,** and the female gametophyte generation develops into the **embryo sac.** Before fertilization a pollen grain contains two nonmotile, but highly differentiated sperm and a tube nucleus, which controls the growth of the pollen tube. A typical angiosperm embryo sac is a polarized structure con-

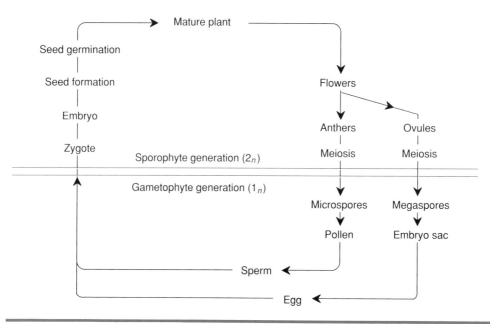

**Figure 1.14 Alternation of generations in angiosperms**

taining eight nuclei in seven cells. Three of these, the antipodal cells, appear to have relatively little to do with embryogenesis. Fertilization begins when the pollen tube ruptures and the two sperm enter the embryo sac.

Double fertilization occurs in the angiosperms. One sperm unites with the egg to form the diploid **zygote.** This marks the end of the gametophyte generation and the beginning of the sporophyte generation. The zygote undergoes **embryogenesis;** it grows and develops into the embryo. The second sperm unites with both of the polar nuclei to form the endosperm nucleus. The endosperm nucleus will divide to give rise to **endosperm** tissue. Endosperm tissue is composed of triploid cells with a genetic composition different from that of the embryo. It forms around the developing embryo and plays an important role in its nutrition, either during embryogenesis or later during seed germination. Embryogenesis is accompanied by **seed formation.** Typically, embryogenesis and seed formation culminate in the formation of a dormant seed. The mature seed has the capacity not only to develop into a new plant, but also to withstand drought, heat, cold, and other environmental extremes that would kill a growing plant. The seed is a complex structure designed to contain, nourish, and protect the embryo. Most seeds are surrounded by a **seed coat,** which is a protective layer derived from the mother plant, and they may contain endosperm tissue in which food reserves have been stored to be used by the embryo during germination.

### Survival Mechanisms

Plants and animals use different strategies to protect against adverse environmental conditions and invading microorganisms  Plants usually are rooted in the soil and are not able to move about in their environment. In contrast, most animals are motile. Animals can simply leave or hibernate when environmental conditions become too difficult and threaten their survival. Animals are well adapted to their environment through evolutionary mechanisms, and the normal seasonal change in climate rarely brings about a major developmental change. The onset of winter may trigger a change in fat deposition or coat thickness and color in some mammals, but the overall shape and form of the animal are unaltered. In contrast, plants cannot

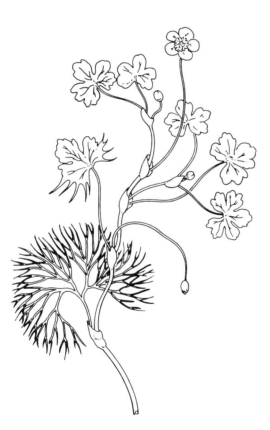

**Figure 1.15 Strong influence of environmental factors on plant development**
Leaf morphology differs radically (heterophylly) between the aerial and submerged leaves of the water buttercup, *Ranunculus aquatilis*. Redrawn with permission from Jensen and Salisbury (1984).

escape changing or adverse conditions and frequently respond to climatic challenges with developmental changes that enable them to adapt to the new conditions. The importance of the environment in plant develoment can be illustrated by the somewhat extreme example of amphibious plants. Consider the water buttercup, *Ranunculus aquatilis*. This plant can grow both in the water as a submerged aquatic and in the air above a pond. The submerged leaves are very finely dissected and little mesophyll tissue surrounds the veins, whereas the aerial leaves are broad with well-developed mesophyll tissue. These two leaf forms are so different from each other that they seem to belong to completely different plants (Fig. 1.15). This phenomenon, known as **heterophylly,** is quite common and is only one example of how a plant responds to its environment with adaptive developmental changes.

Plants may alter their development or use chemical warfare to protect against pathogens and predators. Higher animals have a well-developed immune system to protect them from invading microorganisms. Plants also must protect themselves from invading microorganisms and they have evolved many mechanisms to do so, but these mechanisms are quite different from the immune system of animals. In many cases, the plant's response to invading microorganisms is developmental. These developmental changes include alterations in the structure of the cell wall and localized cell death around the site of microorganism invasion. In other cases the response is a biochemical change leading to the synthesis of special poisonous substances to kill the invading pathogen. Plants mount a chemical defense against predators as well as pathogens. They synthesize special organic molecules that repel, sicken, or kill animals or microorganisms that might harm them.

## Differences in the Developmental Mechanisms of Plants and Animals

### Postembryonic versus Embryonic Development

The majority of animal development takes place between conception and birth in the process known as **embryogenesis.** In fact, animal development is almost synonymous with embryology. Of course, animals continue to develop after birth. The average person is most familiar with the developmental changes that occur between birth and adulthood. Also, as we grow older we become acutely aware of the relatively slow developmental changes that occur during aging; however, these more readily observable changes are relatively minor when compared with the changes during embryogenesis that shape the fertilized egg into a functioning multicellular organism. Animal embryogenesis usually forms most of the major organs of the mature organism. Even though these organs continue to grow and develop after birth, the organs nevertheless are recognizable, and usually functional, at birth.

Angiosperms also undergo embryogenesis and the seed contains an embryonic plant; however, the embryonic plant lacks most of the major organ and tissue systems of the mature plant. The tissues and organs of the mature plant, instead of developing during embryogenesis, are formed after germination through the activity of the meristems.

## Nature of Cell Commitment for Differentiation

In the course of development animal cells become irreversibly committed to a particular developmental pathway, whereas with plant cells the commitment

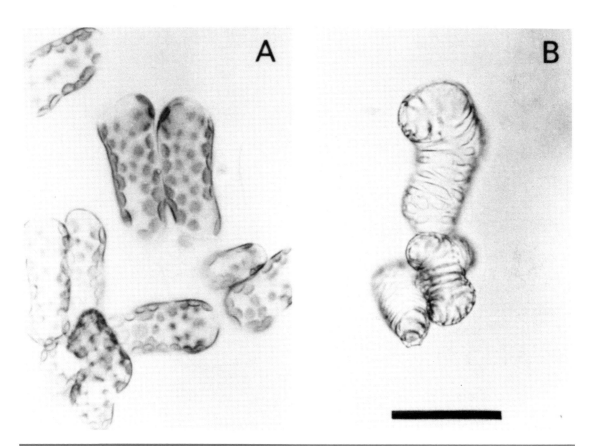

## Figure 1.16 Differentiation of tracheary elements from leaf mesophyll cells

(A) *Zinnia elegans* mesophyll cells were removed from leaves of seedling plants and placed in tissue culture. (B) Four days later many of these cells had differentiated as treacheary elements, with thickened secondary walls. Reprinted with permission from Sugiyama and Komamine (1990).

rarely is irreversible. Virtually any plant cell that retains its nucleus at maturity can, in some circumstances, be induced to divide, grow, and perhaps even regenerate the whole plant. Furthermore, nucleated, but highly differentiated, plant cells can change into a radically different cell type. For example, green photosynthetic leaf mesophyll cells can be made to differentiate into tracheary elements (Fig. 1.16). In contrast, it can be very difficult, if not impossible, to induce a differentiated animal cell, such as a neuron, to become a different cell type, such as a muscle cell.

### Fate of a Plant Cell as Determined by Its Position in the Organism

Cell migration is an important part of animal development. Animal cells are motile. They change their neighbors during development and migrate in well-defined directions according to information they obtain as a result of contact with other cells. The developmental fate of a plant cell is determined by its position within the body of the plant. Because plant cells do not move, the position of a given cell is determined by the plane in which its parent cell divided. Although cell lineages are established in some cases, cell differentiation is not dependent on cell lineage. Plant cells can and do communicate with each other and the information obtained in this fashion plays a role in initiating and maintaining development. Although cell–cell interactions play a role in development in both animals and plants, clearly the nature of the information exchanged and what the cells are able to do with this information will be more limited in plants than in animals.

## Brief Developmental History of a Typical Flowering Plant

The fertilization of an egg gives rise to a zygote that will grow and develop into an embryo. Typically the zygote exhibits intracellular differentiation. Its nucleus and most of the cytoplasmic organelles are near one end, whereas a large vacuole occupies the opposite half of the cell (Fig. 1.17). The unfertilized egg is already polarized within the embryo sac. When the zygote divides, this polarity is maintained and persists through the life of the plant. The structure containing the embryo is known as an **ovule.** Ovules are formed inside the ovary of a flower. The embryo and the ovule together develop into a seed, while the ovary may develop into a fruit. Most plant embryological development has been completed by the time the seed matures. At maturity, further development of the embryo is

**Figure 1.17 Egg and zygote of an angiosperm**
(A) Electron micrograph of the egg of *Capsella bursa-pastoris*, illustrating its polarity and its relationship with one of the synergid cells. (B) Micrograph of the zygote of the same species, before the first cell division. Reprinted with permission from Schulz and Jensen (1968).

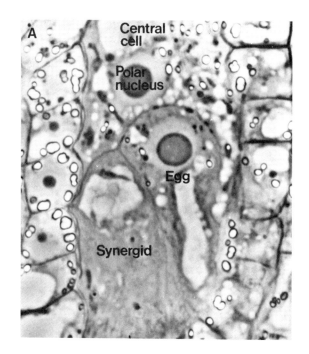

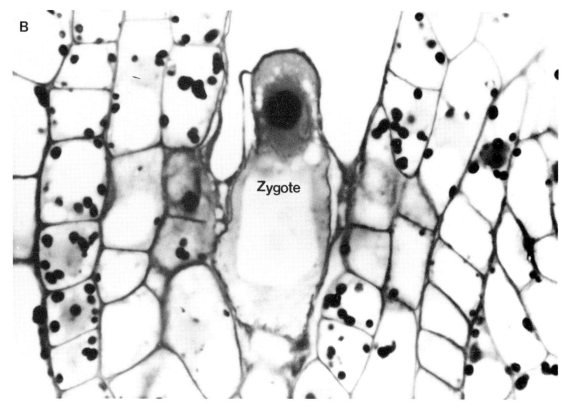

halted and the seed enters a dormant phase remarkably resistant to environmental adversity. The seed, or the fruit containing the seed, is released from the parent plant and can germinate and reinitiate growth and development if environmental factors permit, but it also could remain dormant for a considerable time.

Development resumes when the seed germinates. Germination may be initiated by water uptake, although there may be additional requirements for the resumption of growth, such as exposure to light or low temperature. Germination marks the beginning of a phase of rapid vegetative growth. The dormant embryo inside the seed contains a recognizable root and shoot axis and may contain several immature leaves as well. At the histological level, the dormant embryo contains the beginnings of recognizable specialized tissue, but these tissues do not mature in the embryo. On germination, not only does the plant increase in size, but internal cell differentiation occurs to form the tissues of the shoot and root. As mentioned earlier, plants contain meristems that are regions of persistent embryology. The apical meristems at the end of each growing root and shoot are primary meristems. They are self-perpetuating, growing regions that form the cells that make up the main root and stem axis of the plant. The shoot apical meristems also form lateral appendages, including leaves, flowers, and lateral meristems. Many plants also develop secondary meristems, which give rise to additional tissues and organs. The vascular cambium is a secondary meristem that consists of a cylinder of proliferating cells; some of the cambial derivative cells differentiate as water-conducting xylem elements, whereas others become food-conducting phloem elements. The vascular cambium increases the girth of the stem.

Reproductive development may begin only after a distinct juvenile phase that is marked by rapid vegetative growth. Vegetative meristems are transformed into floral meristems during the transition to reproductive maturity. Specialized cells are formed within the flowers that will undergo meiosis. The haploid cells that result from these reductive divisions will grow and differentiate to form a second generation, known as the male and female gametophytes, which are parasitic on the vegetative plant. The female gametophyte is a structure known as the embryo sac, which contains the egg; the male gametophyte is a pollen grain, which forms two sperm cells. The sporophytic generation is initiated with union of the egg and sperm, beginning the cycle again.

## Model Systems for the Analysis of Plant Development

Biologists have made rapid progress by focusing their studies on a single organism that not only is representative of a much broader group of organisms, but also is easily manipulated in the laboratory. This approach has been followed more or less unconsciously throughout modern biology, but occasionally groups of biologists with similar interests have quasi-formally agreed to concentrate their efforts on a particular organism or group of similar organisms. In the most

**Figure 1.18 Arabidopsis**

*Arabidopsis thaliana,* a small weed in the same family as mustard and broccoli, is an exceptionally useful experimental plant. During the vegetative phase of *Arabidopsis* growth, the plant consists of a rosette of from 2 to 130 leaves, each of which may be no more than a few centimeters in diameter. When it flowers, it produces a cluster of elongated, flower-bearing stems. Each fruit, known as a **silique,** contains about 20 seeds. Redrawn with permission from Estelle and Somerville (1986).

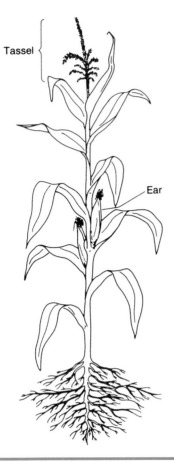

Tassel

Ear

**Figure 1.19 The mature maize plant**
The mature maize plant can stand over 1 m high and has few branches. It has
determinant growth, which means that it flowers after attaining a certain height, with
the stem ending in the male inflorescence (tassel). The female flowers, known as ears,
are formed at the internodes near the center of the stem. Redrawn with permission
from Kiesselbach, T. A. (1949). The structure and reproduction of corn. *Res. Bull.* **161,**
Univ. of Nebraska Press, Reprinted 1980.

notable example of this, the phage biologists, led by Max Delbruck, agreed to
concentrate their efforts on a group of bacteriophage known as the T-even phage.
The result was spectacularly rapid progress in understanding how these bacterial
viruses worked, and from these studies came important insights into the molecular
mechanisms of heredity. In genetics, many of the most important discoveries
have been made through studies of the fruit fly, *Drosophila melanogaster,* and
maize. *Drosophila* has been particularly important for investigations of the molecular–genetic basis of animal development. Plant molecular biologists in recent
years have had extensive discussions about the possibilities of finding the plant

equivalent of *Drosophila*. The weed *Arabidopsis thaliana* comes close to being an ideal laboratory plant because of its small stature, small genome size, the rapidity with which it completes its life cycle, and the ease with which it may be manipulated in the laboratory (Fig. 1.18). Because of these advantages, many plant molecular biologists are now concentrating their efforts on *Arabidopsis*, and there are regular meetings of molecular biologists working with *Arabidopsis*. In addition, *Arabidopsis* may be the only higher plant included in the Genome Project, which will completely sequence the genomes of humans and a few other important organisms that serve as model systems. Despite these considerations, it is unlikely that all the major problems of plant development will be solved through investigations conducted on a single species. As can be seen by comparing Figs. 1.1 and 1.18, the basic form of *Arabidopsis* is quite different from that of our "typical" higher plant. In fact, no single higher plant is truly representative of all higher plants. As a result, it would be unwise for molecular biologists to focus their attention completely on *Arabidopsis*, excluding all other species. Many important contributions to our understanding of higher plants have resulted from studies on maize, tomato, petunia, tobacco, soybean, and other plants. Maize also is an important model system because it is the best genetically characterized higher plant (Fig. 1.19). As a result, it is a particularly useful plant for many developmental studies. Furthermore, maize is both a monocot and an agronomically important plant. Information we obtain about maize could help improve the yield of this crop as well as give us a perspective on problems in other important grasses, such as wheat and rice.

## Summary

Development is the process by which an organism changes to acquire new structures and abilities. It occurs in response to instructions encoded in the nuclear genome. Development may be studied at many levels, from descriptive studies of the developmental process to mechanistic studies of the manner in which these changes are brought about. The most profound insights into the mechanism by which developmental change occurs have come from molecular–genetic analysis of the nature of the genetic instructions for developmental change and the process by which these instructions are used to bring about the changes.

A typical angiosperm is a highly polarized terrestrial organism with two main organ systems, the root and the shoot. The root system may be highly branched, but it contains no other organs. In contrast, the shoot system consists of several organs and appendages, principally leaves, stems, and flowers. An interconnected vascular system runs throughout the plant, con-

*continues*

necting the root system with the shoot and its appendages. The vascular tissue consists of two tissues, xylem and phloem. Xylem conducts water and dissolved minerals through tracheary elements. The latter, made up of cells known as tracheids and vessel elements, are dead at maturity, with reinforced secondary walls and specialized openings through the cell walls that permit the passage of water. Xylem also contains fibers and xylem parenchyma. Phloem is made up of sieve elements, highly specialized living cells that transport food assimilated in photosynthesis, as well as companion cells, phloem parenchyma, and fibers.

The plant body is constructed through the activity of meristems. Meristems are regions that retain embryonic characteristics throughout the life of the plant. They contain small cells known as initials that retain the potential for cell division long after most embryonic cells have differentiated and stopped dividing. Cell divisions within meristems produce the cells that differentiate to become the leaves, stem, roots, and flowers of the mature plant. Most plant development is postembryonic and the plant body is constructed almost entirely by the activity of meristems. There are three kinds of meristems. The root and shoot apical meristems generate the primary tissues of the root and shoot. In addition, secondary meristems produce secondary tissues such as wood and bark. Vegetative meristems usually are indeterminate and highly repetitive in their activity. They produce the same structures over and over again. In contrast, some meristems are determinant and are capable of only a limited, predetermined amount of growth. Roots and stems grow by the addition of new cells at their ends through the activity of the primary meristems, as well as by the subsequent expansion and differentiation of these cells. The primary vegetative meristems produce cells that differentiate to form the three main tissue systems: vascular, ground, and epidermal tissues. The shoot apical meristem also produces leaf primordia and lateral buds. Leaf primordia are determinant meristems. Further cell division and differentiation from the leaf primordia produce the leaf, an organ specialized for photosynthesis.

Vegetative meristems can be transformed into floral meristems. The floral meristem is highly determinant in its activity. It produces four whorls of organs—sepals, petals, stamens, and pistils—but is completely consumed by this activity, and no meristematic cells remain once all the floral organs have been initiated. The pistil is the female reproductive organ and contains one or more ovules in which an egg cell will be formed. The anthers are the male reproductive organs. They produce pollen grains in which sperm cells differentiate. The floral organs do not produce the sex cells (gametes) directly. The haploid cells formed through meiotic divisions in the ovule or anther develop further before the gametes are differentiated in the haploid tissue. The union of the sperm with an egg results in the formation of a zygote which will undergo embryogenesis. Embryogenesis is accompanied by seed development and the mature embryo is contained within a seed, which is both a protective and a nutritive structure. The angiosperm embryo is relatively simple and contains only the axis of the plant body, in which the tissue pattern has formed, and the root and shoot apical meristems, from which the rest of the plant will develop during postembryonic development, which begins with seed germination.

## Questions for Study and Review

1. How do the two plants depicted in Fig. 1.2 differ from the "typical" plant shown in Fig. 1.1? Where are the leaves in the cactus? Where is photosynthesis conducted in the cactus? Where is the stem of the water lily?

2. What is meant by the term *development* as it is used in biology?

3. What levels of analysis reveal the greatest similarity between plant and animal development? Why?

4. In what ways do plant cells differ from animal cells? How might these differences affect the ways in which plants develop?

5. Study the diagram shown in Fig. 1.10A. In which cells of the leaf does photosynthesis occur? What is the function of the stomata and the guard cells? What is the function of the epidermis?

6. What is the function of the specialized cells known as tracheary elements? How are tracheids different from vessel elements? What are the steps involved in the differentiation of vessel elements, as depicted in Fig. 1.3?

7. See Fig. 1.6. What is the function of the sieve tubes? How does a mature sieve tube differ from a vessel? How does a mature sieve element differ from a companion cell?

8. What is the function of the root hairs depicted in Fig. 1.6A? What is the nature and function of the Casparian strip of the endodermis? Why is this important in water and mineral uptake? Trace the pathway of water movement as it enters the root and flows into the shoot.

9. What is the main difference between the structures of the stem and the roots in a herbaceous dicot? What are the origin and function of the vascular cambium?

10. What is the function of the germ line cells of animals? What is the significance of the fact that plants do not set aside cells as a special germ line?

11. Examine the diagram shown in Fig. 1.13. List the organs of the flower and the functions of each of these different organs.

12. Diagram the life cycle of a typical animal. How does it differ from an angiosperm, such as the maize plant shown in Fig. 1.14? Why are plants said to have alternation of generations? What are the two generations that alternate in angiosperms? What do the pollen grains represent?

13. In what way has the plant shown in Fig. 1.15 responded to its environment? How does water affect its development?

14. Compare and contrast the role of embryogenesis in plant and animal development.

15. What are meristems? What role do meristems play in plant development?

## Further Reading

### An Introduction to Plants

Bell, P. R., and Woodcock, C. L. F. (1983). *The Diversity of Green Plants*, 3rd ed. Edward Arnold, London.

Bold, H. C., and La Claire, J. W., II (1987). *The Plant Kingdom*, 5th ed. Prentice–Hall, Englewood Cliffs, NJ.

Foster, A. S., and Gifford, E. M., Jr. (1974). *Comparative Morphology of Vascular Plants*. W. H. Freeman, San Francisco.

Jensen, W. A., and Salisbury, F. B. (1984). *Botany: An Ecological Approach*. Wadsworth, Belmont, CA.

Mauseth, D. (1991). *Botany: An Introduction to Plant Biology*. Saunders, Philadelphia.

Raven, P. H., Evert, R. F., and Curtis, H. (1992). *Biology of Plants*, 5th ed. Worth, New York.

Ray, P. M., Steves, T. A., and Fultz, S. A. (1983). *Botany*. Saunders, Philadelphia.

### Plant Anatomy and Physiology

Mauseth, J. (1988). *Plant Anatomy*. Benjamin/Cummings, Redwood City, CA.

Salisbury, F. B., and Ross, C. W. (1992). *Plant Physiology*, 4th ed. Wadsworth, Belmont, CA.

Taiz, L., and Zeiger, E. (1991). *Plant Physiology*. Benjamin/Cummings, Redwood City, CA.

Zimmerman, M. H. (1983). *Xylem Structure and the Ascent of Sap*. Springer-Verlag, New York.

### Other Books on Plant Development

Burgess, J. (1985). *An Introduction to Plant Cell Development*. Cambridge University Press, Cambridge.

Lyndon, R. F. (1990). *Plant Development, the Cellular Basis*. Unwin Hyman, London.

Roberts, L. W. (1976). *Cytodifferentiation in Plants: Xylogenesis as a Model System*. Cambridge University Press, Cambridge.

Sachs, T. (1991). *Pattern Formation in Plant Tissues*. Cambridge University Press, Cambridge.

Steeves, T. A., and Sussex, I. M. (1989). *Patterns in Plant Development*, 2nd ed. Cambridge University Press, Cambridge.

### Specific References

Alberts, B., Bray, D., Lewis, J., Raff, M., Roberts, K., and Watson, J. D. (1989). *The Molecular Biology of the Cell*, 2nd ed., Garland, New York.

Estelle, M. A., and Sommerville, C. R. (1986). The mutants of *Arabidopsis*. *Trends in Genetics*. **2,** 89–93.

Janzen, D. H. (1976). Why bamboos wait so long to flower. *Annu. Rev. Ecol. Syst.* **7,** 347–391.

Klekowski, E. J., and Godfrey, P. J. (1989). Aging and mutation in plants. *Nature* **340,** 389–391.

Meyerowitz, E. M. (1989). *Arabidopsis:* A useful weed. *Cell* **56,** 263–269.

Smyth, D. R. (1990). *Arabidopsis thaliana:* A model plant for studying the molecular basis of morphogenesis. *Aust. J. Plant Physiol.* **17,** 323–331.

Sugiyama, M., and Komamine, A. (1990). Transdifferentiation of quiescent parenchymatous cells into tracheary elements. *Cell Diff. Dev.* **31,** 77–87.

Sussex, I. M. (1989). Developmental programming of the shoot meristem. *Cell* **56,** 225–229.

Walbot, V. (1985). On the life strategies of plants and animals. *Trends Genet.* **1,** 165–169.

# 2

# The Genetic Basis of Plant Development

This chapter examines some of the evidence that has allowed us to conclude that the developmental program is inherited, that it is found in the nucleus, and that it is encoded in the specific sequence of bases making up the DNA of many genes. In future chapters we will consider both environmental and internal factors that control or modify the developmental program.

## Mendel and the Laws of Heredity

The first systematic work showing that heritable factors control aspects of plant development was conducted by Gregor Mendel more than 100 years ago. In the course of this work he also uncovered the fundamental rules of heredity on which the science of genetics is based. The importance of these principles to genetics tends to obscure the fact that Mendel also made a significant contribution to plant development.

Mendel's most important studies were done with the garden pea, in which he examined the inheritance of several characteristics, such as plant height and seed color and shape (Fig. 2.1). These traits are all developmental characteristics. The flowers of the garden pea are self-fertile. By selecting plants carefully for these characteristics and making certain that they reproduced only by self-fertilization for many generations, Mendel was able to obtain true-breeding strains of peas for all of these characteristics. For example, one line of peas always produced smooth, round seed when it was self-fertilized, whereas a different line always produced wrinkled seed. Such traits are known as **phenotypic characteristics,** meaning that they are the characteristics we actually see when we examine the organism. Mendel's experiments showed that a given phenotypic characteristic is determined by heritable factors we now call **genes.**

One of the genes Mendel studied controlled the shape of the seed. By crossing true-breeding plants bearing wrinkled seeds with true-breeding plants bearing smooth seeds, he showed that this aspect of seed morphology in garden peas was determined by a pair of closely related genes. The gene resulting in smooth seeds has been designated $R$, and the gene resulting in wrinkled seeds, $r$. The $R$ and $r$ genes are **alleles,** and plants inherit two of these alleles in some combination, $RR$, $Rr$, or $rr$. Plants that contain two identical alleles ($RR$ or $rr$) are said to be **homozygous;** those with different alleles ($Rr$) are called **heterozygous** (Box 2.1). This is the **genotype** of these individuals with respect to these genes. Note that we do not necessarily know the genotype of the plant from observing its phenotypic characteristics. In experiments to examine the inheritance of a particular character, the two plants that are crossed are called the parental generation (P). The seeds resulting from such a cross and the plants that can be grown from these seeds constitute the first filial generation (F1).

In the cross between a smooth-seed plant and a wrinkled-seed plant, all the plants of the F1 generation produced smooth seeds. This occurs because the gene for smooth seeds is **dominant** over the gene for wrinkled seeds. The plants of the F1 generation can be self-fertilized to give a second filial generation (F2). In Mendel's work, the wrinkled-seed characteristic reemerged in some of the F2 plants, showing that the factor that led to wrinkled seeds was not lost in the F1 generation; it was simply not expressed. The $r$ gene is **recessive** with respect to its dominant $R$ allele and its effect is observed only when the plant is homozygous

**Figure 2.1    Seven phenotypic traits of garden pea whose inheritance Mendel studied**

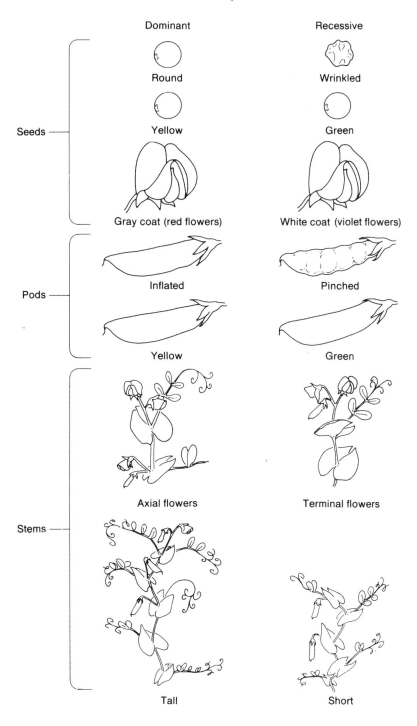

Dominant          Recessive

Seeds

Round          Wrinkled

Yellow          Green

Gray coat (red flowers)      White coat (violet flowers)

Pods

Inflated          Pinched

Yellow          Green

Stems

Axial flowers          Terminal flowers

Tall          Short

---

### Smooth Seed versus Wrinkled Seed

True-breeding pea plants that produce only smooth seeds when self-fertilized are crossed with true-breeding plants that always form wrinkled seeds.

$$\text{Smooth} \times \text{Wrinkled} \qquad \text{(phenotypic characteristics)}$$

$$RR \times rr \qquad \text{(genotypic composition)}$$

F1 generation:  Phenotypically, all seeds are smooth.
Genotypically, all are $Rr$.

The F1 seeds can be germinated and will grow to become the plants of the F1 generation. If the flowers of the F1 plants are allowed to self-fertilize, they will produce seed which begin the F2 generation.

$$Rr \times rR$$

F2 generation:  Phenotypically, three-fourths of the F2 seed are smooth and one-fourth are wrinkled.
Genotypically, plants are $RR$, $Rr$, or $rr$, and these occur with a frequency of 1:2:1.

---

for this gene. A dominant phenotypic characteristic would be observed in plants both heterozygous and homozygous for that allele.

## Hammerling's Work on *Acetabularia*

Although Mendel's work clearly established that heritable factors, or genes, controlled developmental characteristics, nothing was known about the nature of the genes or where they might reside within the cell. One of the clearest demonstrations that factors residing in the nucleus determine the developmental form of an organism can be found in the work of Hammerling and his students on the cenocytic alga, *Acetabularia* (Fig. 2.2). Hammerling first recognized the value of *Acetabularia* for experiments on development. *Acetabularia* are algae that grow in shallow waters in tropical and subtropical regions. Although an individual

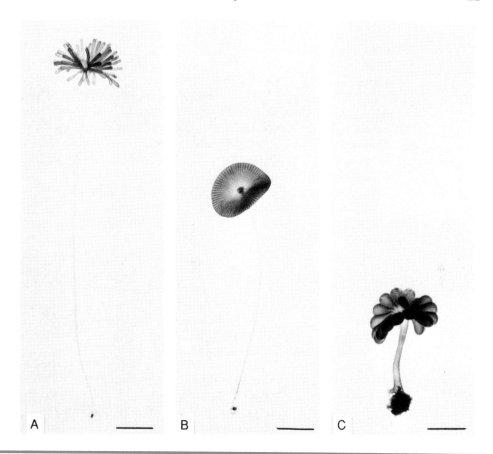

**Figure 2.2  Three different species of _Acetabularia_**
(A) _Acetabularia major._ Bar = 1 cm. (B) _Acetabularia mediterranea._ Bar = 0.7 cm.
(C) _Acetabularia peniculus._ Bar = 1 cm. Reprinted with permission from Berger _et al._
(1987)

_Acetabularia_ has only one nucleus throughout much of its life cycle, it is a surprisingly large and complex organism (Fig. 2.3). Some species of _Acetabularia_ are nearly 20 cm long, and the single nucleus, which increases in volume as the organism grows and develops, can reach 100 to 150 $\mu$m in diameter. At maturity the nucleus resides in the basal end of the cell, in a specialized, rootlike structure known as the **rhizoid,** whose function is to anchor the plant to a rock at the bottom of the pond. A structure known as the **cap** is formed at the opposite end of the cell and contains many chloroplasts. The shape of the cap is different in different _Acetabularia_ species and the characteristics of the cap are inherited. For example, the cap of _A. mediterranea_ is disk-shaped (Fig. 2.4A), whereas the cap of _A. crenulata_ is crenulate, or lobed (Fig. 2.4B).

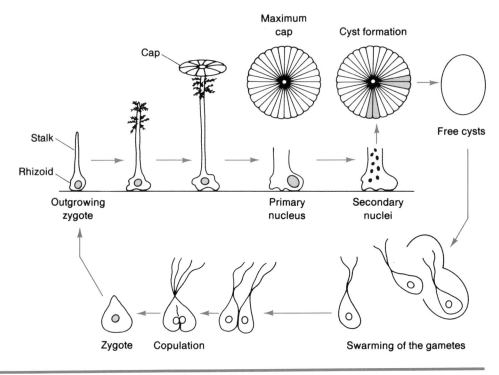

**Figure 2.3** **Life cycle of *Acetabularia***
The diploid zygote develops into a large, differentiated single cell with a rhizoid, stalk, and cap structure. The nucleus remains in the rhizoid until the cap is completed, whereupon it divides repeatedly, without cellular division. The secondary nuclei migrate up the stock and into the cap where they become encysted. Subsequently the cysts release the flagellated, haploid gametes. Redrawn with permission from Berger *et al.* (1987).

### Regeneration after Wounding

An unusual feature of *Acetabularia* is its ability to withstand and survive wounding. Because it is not partitioned into cells, this capacity probably is essential for its survival. This property also makes this organism very useful for experiments that involve surgical manipulation. For example, the cap can be cut off an *Acetabularia* and the stalk not only will continue to live, but also will regenerate a new cap. Furthermore, the regenerated cap will have the same appearance as the first cap. This can be repeated many times, and each time the regenerated cap will be the same. A stem segment lacking both a cap and a rhizoid also will regenerate a new cap, even though it lacks a nucleus, provided the stem segment is given light so that photosynthesis can continue. If, however, the regenerated cap is cut from a plant that lacks a nucleus, there will be no further cap regeneration.

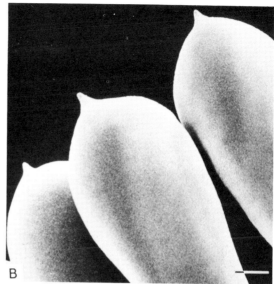

**Figure 2.4    Structure of the cap in two *Acetabularia* species**
(A) The rays of *A. mediterranea* closely associate with each other, forming a disklike cap.
(B) The rays of *A. crenulata* are nipplelike at their ends and separate, forming a
crenulate or lobe-shaped cap. Reprinted with permission from Berger *et al.* (1987).

## Grafting of Parts of Different *Acetabularia* to Each Other

When an enucleated stem segment is grafted onto the nucleated basal segment
of another species, it regains the ability to regenerate another cap, but this time
the regenerated cap will have the characteristics expected for the species from
which the nucleus was derived. In the example shown in Fig. 2.5, an enucleated
*A. crenulata* stem was grafted onto a nucleated *A. mediterranea* rhizoid. The regenerated cap has hybrid characteristics, indicating that the *A. mediterranea* nucleus
influenced the morphology of the regenerated cap. If this cap is now removed, the
stem will regenerate a second cap, but this cap no longer has hybrid characteristics.
Instead, it has the morphological characteristics of *A. mediterranea*, the species
that contributed the nucleus. Although these experiments strongly suggest that
the nucleus contains the instructions for forming a new cap, they do not prove
it conclusively because the rhizoid fragment contains other subcellular structures
in addition to the nucleus; however, the specific contribution of the nucleus to
the developmental program of this organism was unambiguously demonstrated
by nuclear transplantation experiments. The nucleus can be isolated from the
rhizoid by means of a steady hand and an instrument known as a micromanipula-
tor. When an isolated nucleus is placed into an *Acetabularia* stalk fragment with

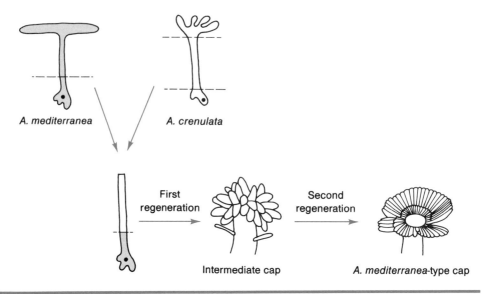

**Figure 2.5**   **Cap regeneration in *Acetabularia***
A graft between two *Acetabularia* species initially will regenerate a cap with
intermediate characteristics, but if the hybrid is decapitated, it regenerates a cap with
the characteristics of the species donating the nucleus. Redrawn with permission from
Grant (1978).

both the rhizoid and cap cut off, the fragment will regenerate both a rhizoid and
cap with the characteristics of the species from which the nucleus was taken.

### Information Flows from the Nucleus to the Cytoplasm via Messenger RNAs

These experiments clearly show that the *Acetabularia* developmental program
resides in the nucleus. You may, however, wonder why an enucleate stem fragment would form a new cap, and why, in the grafting experiment, the first cap
regenerated was hybrid in its morphology. These observations seem to suggest
that the cytoplasm also contains cap-forming instructions, but that these instructions do not persist as long as those found in the nucleus. The answer to these
questions is that the cytoplasm does indeed contain information that can be
used to construct cellular structures. This information resides in macromolecules
known as messenger RNAs (mRNAs). These mRNA molecules contain the instructions necessary for making specific cellular proteins, according to the directions provided by the nuclear DNA. They are transcribed from DNA templates
in the nucleus and contain a sequence of RNA bases complementary to one
strand of the DNA, as described in Chapter 3. All proteins synthesized in the
cytoplasm are assembled according to instructions in the DNA, which are carried

to the cytoplasm by RNA molecules. Each mRNA directs the synthesis of a specific protein. In most organisms mRNA molecules are short-lived. The gene that produced a particular mRNA must be repeatedly transcribed to maintain a stable population of mRNAs with instructions for the synthesis of a particular protein. Some of these mRNA molecules are particularly long-lived in *Acetabularia*, however. These long-lived mRNAs include transcripts of genes that control cap formation and they provide the information necessary to form a new cap. Because they are long-lived, they continue to direct the synthesis of the required proteins even after the nucleus is not immediately present. After they have decayed away, they must be replenished before another cap can be produced, and they can only be replenished with information contained in the nucleus.

## Gene Linkage

One of the conclusions Mendel reached was that the genes he studied were independently distributed to the progeny. This is often called the **law of independent assortment.** Although this "law" is certainly valid for many genes, including the genes governing the seven characteristics shown in Fig. 2.1, most genes are linked to other genes and are inherited together. This is because every gene occupies a particular position on a chromosome within the nucleus. Each chromosome is composed of one very long DNA molecule, to which certain proteins are associated. Every organism has a characteristic number of chromosomes. For example, maize has 10 different chromosomes, whereas *Arabidopsis* has 5. Each vegetative cell of the plant body contains at least two sets of chromosomes, one of which is derived from the maternal parent, the other from the paternal parent. The vegetative cells are said to be **diploid.** For example, the vegetative cells of an *Arabidopsis* plant have 10 chromosomes: two each of the five kinds of chromosomes that are the basic genetic inheritance of this species. Every chromosome in a vegetative cell has a homolog. Each pair of homologous chromosomes contains the same kinds of genetic information, but the exact nature of the genes may be different. Homologous chromosomes contain allelic genes. In contrast, reproductive cells (sperm cells and eggs, as well as other gametophytic cells) are **haploid.** The reproductive cells contain only one of each of the homolog chromosomes. Union of the sperm and egg restores the diploid condition found in each cell of the plant body.

### Chromosome Structure

Chromosomes are easily seen when dividing cells are examined by light microscopy. Each homologous pair of chromosomes has a characteristic structure, and in most cases, the individual chromosomes of an organism can be identified and distinguished from each other by size, specific staining properties, and other structural features. Each homologous chromosome contains a single unique DNA molecule, if it has not undergone replication. Chromosomal DNA molecules typically are very long, often many times longer than the diameter of the cell. DNA is very fragile and easily broken and the consequences of chromosome breakage to the organism can be severe. Chromosome fragments can be lost during cell division, which means that the information these fragments contain also is lost. For these reasons, the chromosomal DNA must be protected to prevent its breakage and to ensure that it can be packed into the nucleus, accurately transcribed into mRNAs, and replicated at the appropriate times. This packaging is accomplished by the interaction of the DNA with specific chromosomal proteins, most importantly the histones. Chromosomes consist not only of DNA, but also of proteins that interact with the DNA in specific ways to package and protect it from breakage. Histones bind to DNA, folding and packing it into the nucleus (see Chapter 3).

## Distribution of Chromosomes to Daughter Cells in Cell Division

A duplicate of each chromosome is constructed between divisions (interphase) in meristematic cells as a result of the replication of each chromosomal DNA molecule and the synthesis and assembly of the chromosomal proteins. The two daughter chromosomes remain in close association after chromosome replication is completed, and at this time each chromosome is said to consist of two **chromatids.** The chromatids will not be called chromosomes until after they are separated during **mitosis.** Mitosis accurately distributes the chromatids to daughter cells as a cell divides (Fig. 2.6). In mitosis, each daughter nucleus receives one complete set of the chromosomes, so mitosis does not change the number of chromosomes or the genetic information of the dividing parental cell. In contrast, **meiosis,** which is the division process that leads to the formation of the reproductive cells, reduces the number of chromosomes by half and alters the genetic composition of the daughter cells. The chromosomes initially are replicated in cells undergoing meiosis, as in preparation for a mitotic division, but then the chromosomes are divided twice so that four daughter nuclei are produced, each containing the haploid chromosome number.

## Linkage Maps

If genetic recombination did not occur during meiosis, breeding studies would show that the genes controlling certain traits are always inherited together,

**Figure 2.6  Mitosis**

The different stages of mitosis in *Hemanthus* endosperm cells are viewed with differential interference contrast microscopy. These cells are not typical of dividing plant cells in that they lack a cell wall, but the events of mitosis are particularly easily seen in these large cells. Micrographs courtesy of Andrew Bajer.

*continues*

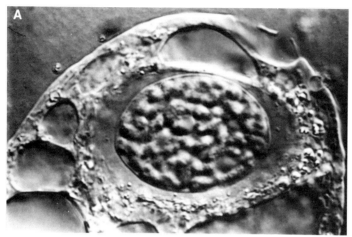

(A) **Prophase:** Chromosomes are not readily visible during interphase because the chromatin fibers are relaxed. As cells enter prophase, the chromatin fibers begin to condense so the individual chromosomes become visible. Each chromosome was replicated during the S phase of interphase, and as they condense, they can be seen to consist of two chromatids, held together at their centromeres. The mitotic spindle is initiated here as polar determinants initiate spindle microtubule formation, but the nuclear envelope is intact throughout this period and the chromosomes are not attached to the spindle fibers.

(B) **Prometaphase:** This phase is initiated with the dissolution of the nuclear envelope. Some of the spindle microtubules are captured by a special structure, known as the kinetochore, that is attached to the chromosomal centromeres. Each chromatid has a kinetochore attached to its centromere and the kinetochores of the two chromatids face in opposite directions. The chromosomes move dramatically around the cell as they are pulled first one way and then another by the spindle fibers.

*continues*

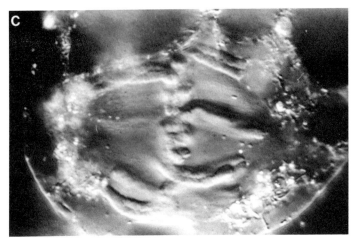

(C) **Metaphase:** Once spindle microtubules are attached at the kinetochores of both chromatids of each chromosome, they come to rest in the center of the mitotic spindle, held there by a dynamic tension as the two chromatids are pulled toward the opposite poles.

(D) **Anaphase A:** Anaphase is initiated suddenly when the two chromatids separate and begin to move toward the pole. As soon as they separate, the chromatids are called chromosomes. The microtubules attached to the kinetochores shorten as the chromosomes move toward the poles.

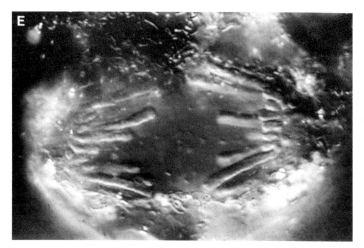

(E) **Anaphase B:** The two sets of chromosomes are further separated by an increase in the length of the mitotic spindle. This is brought about by an increase in the length of the spindle microtubules which extend from each pole into the equatorial plane where they overlap and interact.

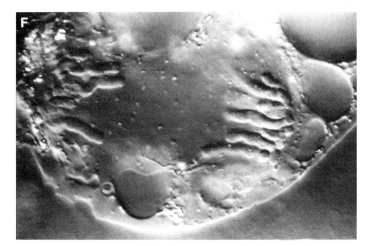

(F) **Telophase:** Telophase is initiated when the chromosomes arrive at the poles and the kinetochore microtubules disappear. The nuclear envelope will be re-formed around each chromosome aggregation, and the individual chromosomes begin to decondense and assume an interphase configuration.

*continues*

(G) **Cytokinesis:** A phragmoplast forms between the two daughter nuclei and begins to construct the partition, known as the cell plate, between them. The phragmoplast consists of two overlapping sets of microtubules that are attached to the cell periphery by microfilaments. Vesicles containing cell wall precursors are transported by the microtubules to the growing cell plate.

whereas others always are inherited independently; however, **genetic recombination,** or crossing over, results in the exchange of portions of homologous chromosomes so that the linkage between genes may be broken. As a result, the degree of linkage of two genes can be expressed as a frequency. Genes that are widely separated on a chromosome have a low frequency of linkage; genes that are very close together exhibit a high frequency of linkage (Fig. 2.7B). Geneticists have constructed **linkage maps** on which the positions of genes are marked on chromosomes according to how frequently they are inherited together. Linkage maps have been constructed for *Arabidopsis* and maize, as well as many other crop plants (Fig. 2.8). Each gene occupies a particular place on one linkage group in relation to the other genes on the same linkage group. It is only the different linkage groups that are independently assorted to the progeny, not the individual genes. In the case of maize, the linkage map has been shown to be very similar to the physical map of the chromosomes.

## Genetic Control of Developmental Characteristics

As an example of one of the ways a gene might control a developmental characteristic, let us consider the mechanism by which the *R* gene studied by Mendel controls the shape of the pea seed. The *R* gene is known to affect starch deposition in the seed. During their development the cotyledons of the pea seed synthesize

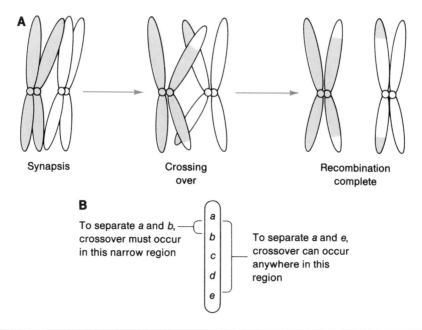

**Figure 2.7  Crossing over during meiosis may break the linkage between genes located on the same chromosome**
(A) Progressive stages in the genetic recombination that occurs as a result of crossing over during meiotic prophase I. (B) Genes closer together will be linked more often than those further apart on the chromosome. As a result, the frequency with which genes are linked is a measure of their position on the chromosome. This is the basis for the construction of gene linkage maps. (A,B) Redrawn with permission from Darnell *et al.* (1990).

and store large amounts of starch, which will be used as a food reserve for the germinating seedling. To explain how the *R* gene controls seed shape, we must first know something about the nature of starch and how it is synthesized. Starch is a polymer composed of glucose subunits. There are two kinds of starch, amylose and amylopectin. Amylose consists of about 300 glucose units, joined by ester bonds between the No. 1 carbon of one glucose molecule and the No. 4 carbon of the next glucose molecule. Amylose molecules are unbranched. In contrast, amylopectin is branched starch. It contains about 1000 glucose units, most of which are joined with the same $1 \rightarrow 4$ linkages between glucose molecules found in amylose; however, about once in every 25 glucose subunits, another glucose molecule is attached with a $1 \rightarrow 6$ linkage to initiate a branch (Fig. 2.9).

Starch is synthesized from the disaccharide sugar sucrose. Sucrose is transported from the photosynthetic leaves into the pea pods during seed growth, where it is converted to starch in the developing cotyledons of the pea embryo. The synthesis of starch in the developing seed requires the expression of a group

**A**

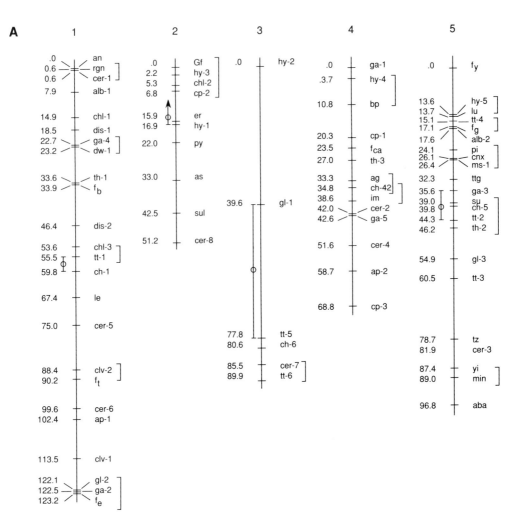

**B**

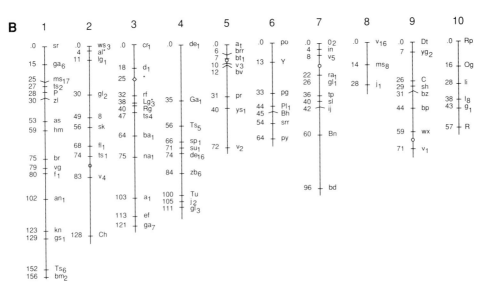

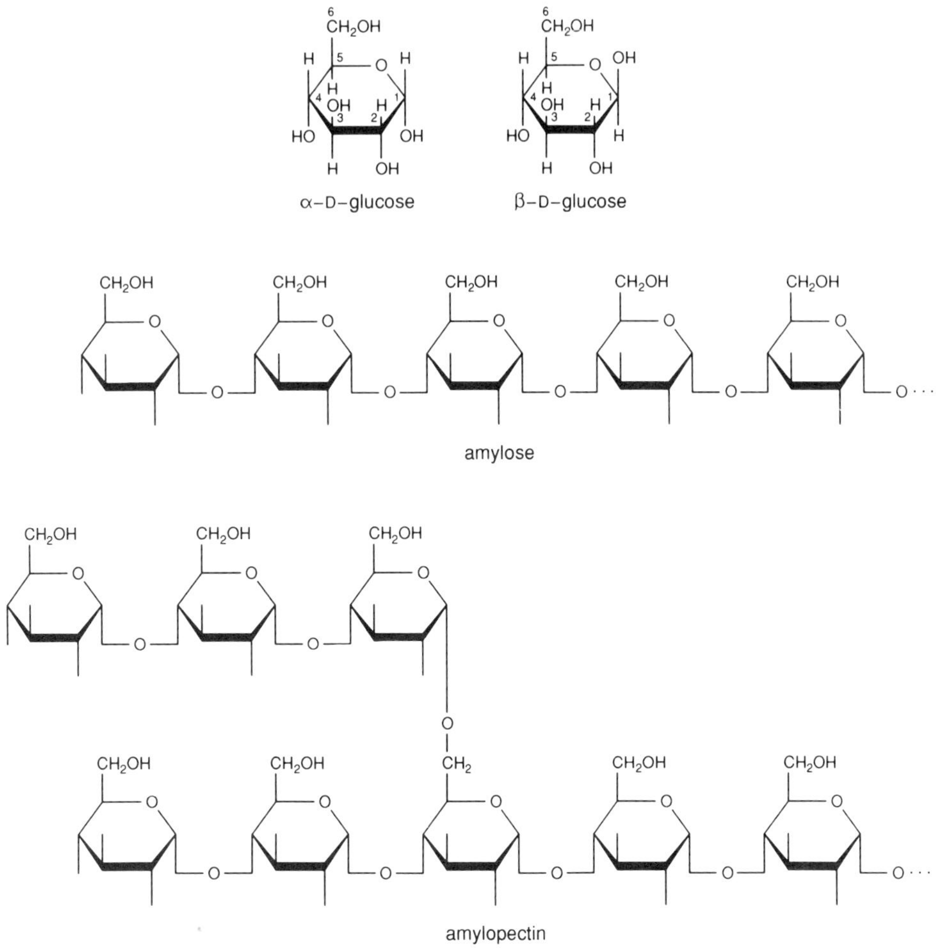

Figure 2.9  **Structures of glucose, amylose, and amylpectin**

Figure 2.8  **Linkage maps for the *Arabidopsis* (A) and maize (B) genomes**
The numbers refer to the relative positions of specific marker genes on each linkage group; the initials are abbreviations for those genes. Redrawn with permission from (A) Koornneef *et al.* (1983) and (B) Rhoades (1950).

of genes encoding enzymes required for the conversion of sucrose into starch. The enzyme sucrose synthase, despite its name, is involved primarily in converting sucrose into fructose and the nucleotide diphosphate derivative of glucose, according to the reaction

$$\text{ADP} + \text{sucrose} \xrightleftharpoons{\text{sucrose synthase}} \text{ADP-D-glucose} + \text{fructose} + P_i$$

The enzyme starch synthase then uses the ADP-D-glucose to add glucose units to the ends of growing amylose chains, forming $1 \rightarrow 4$ linkages. Starch branching enzymes convert amylose to amylopectin by inserting glucose units with $1 \rightarrow 6$ linkages along the amylose backbone. This initiates new growing points to which more glucose subunits can be attached with $1 \rightarrow 4$ linkages.

The seeds of pea plants that are homozygous $rr$ are relatively inefficient in converting sucrose to starch, as compared with the heterozygous $Rr$ or homozygous $RR$ seeds. As a result, the cotyledonary cells contain a higher concentration of sucrose and are under greater osmotic pressure. The increased osmotic pressure causes the $rr$ seeds to become much larger as a result of osmotic water uptake during the active phase of cell expansion. Although the $rr$ seeds grow larger than the $Rr$ or $RR$ seeds, most of their increased volume is due to water. The final phase of seed development is the loss of water as the seeds mature and become quiescent. Because the $rr$ seeds contain a larger percentage of water than the $Rr$ or $RR$ seeds, they lose a greater proportion of their volume as they go dormant and this leads to the wrinkling of these seeds.

Scientists at the John Innes Institute in England have shown that the decreased starch deposition in seeds with the wrinkled phenotype is the result of a deficiency in one of the starch branching enzymes. With fewer growing ends in the wrinkled phenotype, starch chains grow more slowly and sucrose accumulates within the cells. When the $r$ gene was isolated and characterized, the primary difference between it and the $R$ gene proved to be the inactivation of the gene encoding a starch branching enzyme as a result of the insertion of a genetic element known as a *transposon* into the structural gene. The starch branching enzyme gene is not able to function as a result of this insertion. The $R$ gene is the fully functional gene encoding a starch branching enzyme, whereas its $r$ allele is a defective gene. This example illustrates one of the ways genes control developmental characteristics. Most genes, such as the $R$ gene, contain instructions for the synthesis of a protein. If the protein is an enzyme, it can carry out a chemical reaction in the cells in which the gene encoding the enzyme is expressed, and this chemical reaction has an effect on development. By controlling the activity of their genes, cells can come to have different characteristics.

## Phenotypic Consequences of a Change in the Expression of a Single Gene

Pea plants that differ genetically only in being either $RR$ or $rr$ are indistinguishable, except for the characteristics of their seeds. That is, these alleles do not affect the vegetative growth or the development of the pea plant; however, the seeds of the *rr* plant differ from those of the *RR* plant in properties other than starch content and shape. Most notably, seeds of the *rr* plant have markedly smaller amounts of the seed storage protein legumin and higher levels of lipid than do seeds of the *RR* plant. These differences are **pleiotropic** effects of the *r* gene. A single change or mutation in a gene can have several phenotypic effects on the organism, and these multiple phenotypic effects are said to be pleiotropic. In an effort to assess the magnitude of the pleiotropic effects of the *r* gene in peas, Gottlieb and co-workers at the University of California, Davis, compared the proteins present in the seeds and vegetative organs of pea plants that differed genetically only at the *R* locus. They used a technique known as two-dimensional gel electrophoresis, which enabled them to visualize approximately 700 of the most abundant proteins in these plants (Fig. 2.10). They could detect no differences in the proteins present in the *RR* and *rr* plants when vegetative organs such as leaves and roots were compared. When the mature seeds were compared, however, there were differences in about 10% of the proteins visualized by this technique. Approximately 70 proteins were either increased or decreased in the mature seed as a result of the lack of the starch branching enzyme in the plants homozygous for the *r* gene. These pleiotropic differences are a consequence of the increased sucrose content of the cotyledons of the *rr* seeds which results from the deficiency in starch formation. The increased osmotic concentration of the cytoplasm of the *rr* seeds affects protein synthesis, but the synthesis of some proteins is affected more dramatically than that of others. The increased osmotic concentration of the *rr* seeds inhibits the synthesis of some proteins, such as the legumin storage protein, whereas it stimulates the synthesis of other proteins.

## Identification of Genes through Mutations

Any change in the ability of a gene to be expressed or in the character of the protein that gene encodes is called a **mutation.** Mutations are very useful for the identification of genes that play an important role in development. Mutations arise for a number of reasons, including relatively rare mistakes that occur during the replication of the genetic material, effects of various agents that interact with and change DNA, such as x-rays and other forms of ionizing radiation, and effects of a variety of chemicals. For example, it is possible to induce mutations in *Arabidopsis* and other plants by soaking the seeds in a chemical mutagen, such as ethyl methanesulfonate, or by treating the seeds with ionizing radiation, such as x-rays. X-rays frequently break chromosomes or cause other rearrangements of the chromosomal DNA. Sometimes breakage and repair result in the deletion of a portion of the gene. In contrast, chemical mutagens usually cause small changes in DNA, although the consequences of even these small changes can be severe. The plants developing from these treated seeds, designated the **M1 generation**, are allowed to self-fertilize, and the seeds thus formed are sown to produce the **M2 generation**. Mutations occur at random. If a mutation occurs

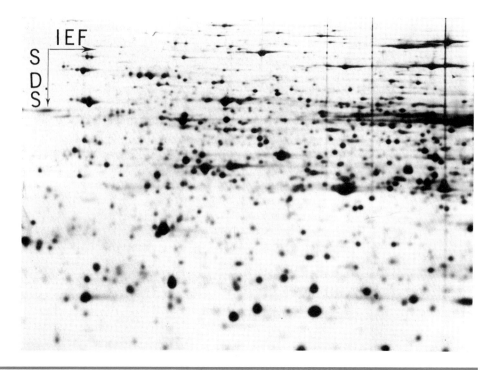

**Figure 2.10** **The abundant proteins of pea stipules after their separation by two-dimensional gel electrophoresis**
Proteins extracted from the stipules of an *RR* plant were separated according to their isoelectric points on a polyacrylamide gel in one dimension (IEF). The proteins were then denatured with the detergent sodium dodecyl sulfate (SDS) and separated according to their molecular weight on a slab acrylamide gel in the second dimension. After their electrophoretic separation, the proteins were stained. Each spot represents one protein, whereas the streaks represent many proteins with similar properties. Reprinted with permission from Gottlieb and de Vienne (1988).

in a particular gene, it is highly unlikely that its allele also is mutated, so the M1 generation will be heterozygous for the mutation. Most mutations result in a loss of function of the gene. The character controlled by the gene would be normal in the heterozygote as a result of the function of the unmutated wild-type allele of the mutant gene. In the M2 generation, 25% of the progeny will be homozygous for the mutant allele, so the effects of the change in that gene can be determined.

Many of the mutations that have been identified using this approach have affected the development of the plant and they can easily be identified by visual inspection of the M2 plants. These include mutations that affect the shape of leaves, the number and/or placement of floral organs, growth habit, and shape of the silique. Examples of mutations affecting trichome morphology in *Arabidopsis* are shown in Fig. 2.11. In this case the mutation occurred as a result of the insertion of another DNA fragment into a gene whose function it is to regulate

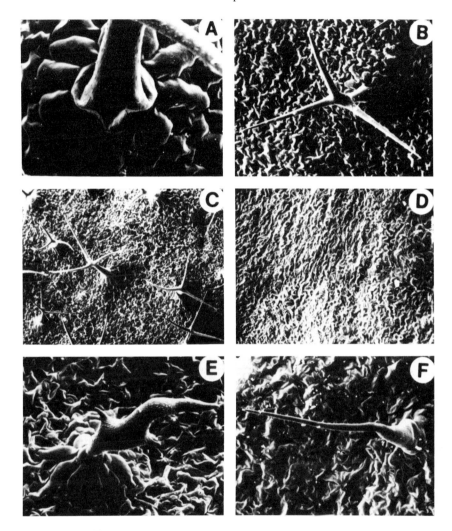

**Figure 2.11 Mutations affecting trichome morphology in *Arabidopsis***
Trichomes are multicellular appendages that project from the epidermis of leaves and stems. Wild-type trichomes of *Arabidopsis* (A–C) usually have three branches. In the mutant Gl1, no trichomes are formed (D), whereas the trichomes of the mutants Dis1 (E) and Gl3 (F) are unbranched. Reprinted with permission from Haughn and Somerville (1988).

the expression of other genes involved in trichome differentiation. Although it is necessary to be cautious in interpreting such results, this approach can be an extremely powerful tool for identifying genes that play a central role in a given developmental pathway. In other cases, the mechanism by which a mutation in a given gene has led to a change in some aspect of plant development has not been determined. Nevertheless, the molecular characterization of genes identified through mutagenesis will tell us a great deal about how development occurs. The power of this approach is most clearly seen in the analysis of a group of mutations affecting flower development, which is discussed later.

# Genetic Control of Developmental Processes

Developmental programs are the result of the expression of at least two different classes of genes, regulatory genes and effector genes. **Effector genes** encode proteins that carry out some task in the cell. They modify the structure or metabolism of the cell, leading to a developmental change. Many effector genes encode enzymes that are required for a particular biochemical pathway. In contrast, **regulatory genes** encode proteins whose job it is to regulate the expression of other genes. The difference between these two classes of genes can be illustrated by examining the control of anthocyanin synthesis in plants.

## Synthesis of Anthocyanin Pigments

Many plant organs and tissues are brightly colored as a result of the synthesis and accumulation of pigments, such as anthocyanins and carotenoids. Anthocyanins are responsible for many of the red and purple colors of flower petals, leaves, roots, and seeds. The red color of beet roots and the blue color of delphinium flowers are due to anthocyanins. Anthocyanins play many diverse roles in plants. The anthocyanin pigments in flower petals attract pollinators, and anthocyanins also are synthesized in response to stresses such as cold, exposure to damaging ultraviolet light, mechanical damage, and attack by pathogens.

Anthocyanins are members of a large class of compounds known as flavonoids and over 1000 different anthocyanins have been identified. Flavonoids, which occur very commonly in plants, have the basic 15-carbon skeleton shown in Fig. 2.12A. The different flavonoids are formed through chemical reactions that add methyl or hydroxyl groups at specific positions on this skeleton. The precursor for the 15-carbon flavonoid skeleton is the amino acid phenylalanine. Phenylalanine is converted to a coenzyme A (CoA) derivative known as $p$-coumaroyl CoA by means of a three-step enzyme-catalyzed reaction (Fig. 2.12B). This series of enzyme-catalyzed reactions, which has been termed **general phenylpropanoid metabolism,** is used by plants to synthesize many different secondary products, and is important in lignification and in the response to disease organisms, as well as in the synthesis of flavonoids. The first reaction in this pathway is catalyzed by the enzyme **phenylalanine ammonia lyase,** which removes the amino group from this amino acid, forming cinnamic acid. A hydroxyl group then is added to the carbon in the 4-position by the enzyme **cinnamic acid hydroxylase** to give $p$-coumaric acid. The enzyme $p$-coumaroyl CoA ligase then catalyzes the synthesis of the CoA ester of $p$-coumaric acid. The C-15 flavonoid skeleton is then

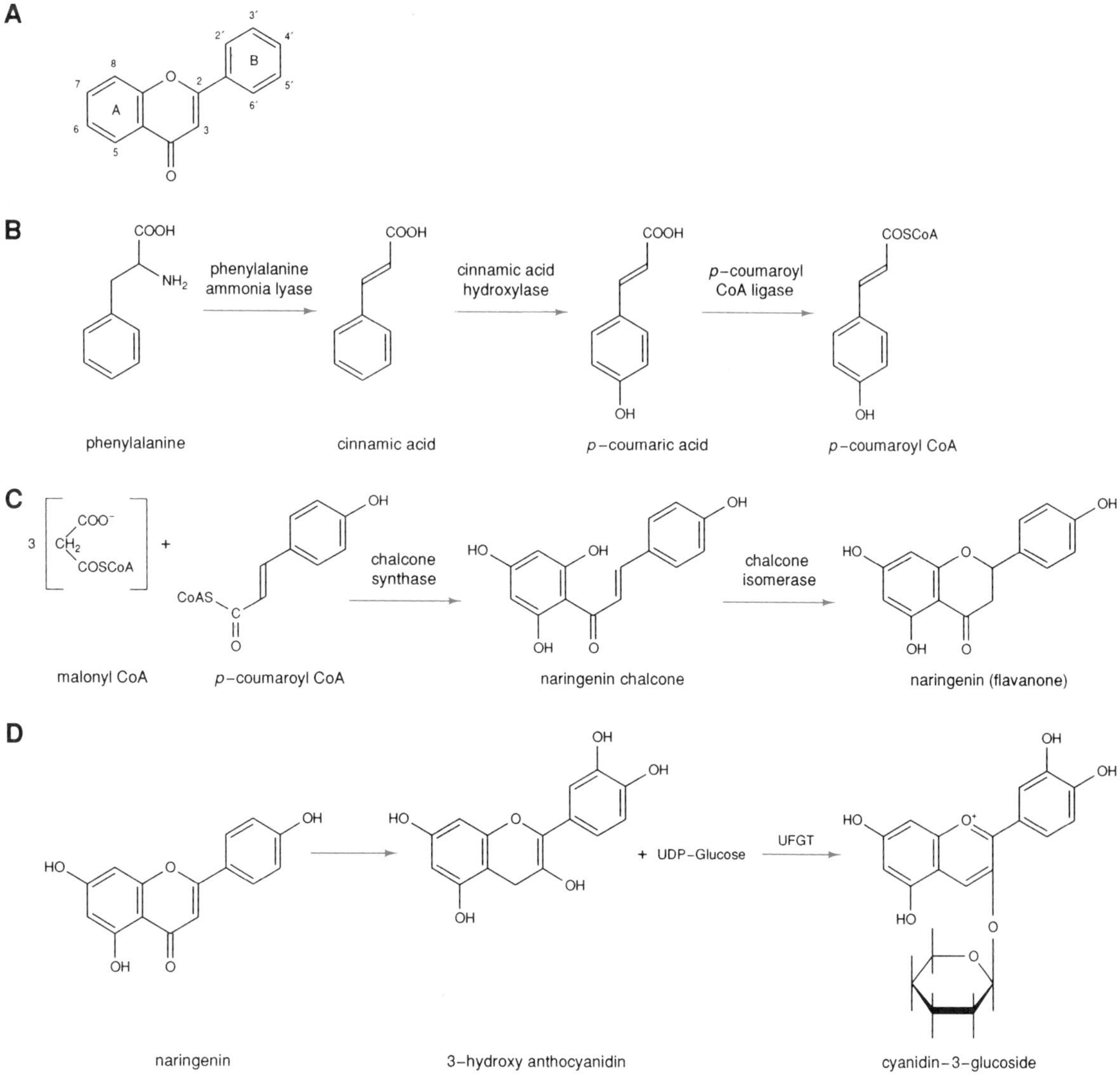

**A**

**B**

phenylalanine     cinnamic acid     p-coumaric acid     p-coumaroyl CoA

**C**

malonyl CoA     p-coumaroyl CoA     naringenin chalcone     naringenin (flavanone)

**D**

naringenin     3-hydroxy anthocyanidin     cyanidin-3-glucoside

**Figure 2.12 Anthocyanin synthesis**

(A) The basic flavonoid skeleton is shown, with the numbering scheme for the carbon atoms. (B) The chemical reactions of phenylpropanoid metabolism which forms the principal precursor of the flavonoids, an ester of *p*-coumaric acid and coenzymeA (*p*-coumaroyl CoA). (C) Synthesis of the flavonoid skeleton occurs as a result of the condensation of three molecules of malonyl CoA with one molecule of *p*-coumaryl CoA, a reaction catalyzed by the enzyme **chalcone synthase.** The product, a chalcone, is converted to a flavanone by the enzyme **chalcone isomerase.** (D) The stable anthocyanins are glycosides, so the final step in anthocyanin biosynthesis is the transfer of a sugar residue from the intermediate uridine diphosphate glucose to the hydroxyl group on the No. 3 carbon of the flavonoid. The enzyme that accomplishes this step is known as UDP-glucose:flavonoid 3-*O*-glucosyl transferase, or **UFGT.** Reprinted from Reddy and Coe (1986).

synthesized from *p*-coumaroyl CoA and malonyl CoA by the enzyme **chalcone synthase** (Fig. 2.12C). The anthocyanin pigments found in higher plants are complexed with sugars as glycosides. Thus, one of the last steps in anthocyanin biosynthesis is the formation of an ester linkage between a sugar, often glucose, and the No. 3 hydroxyl group of the flavonoid. This reaction is catalyzed by an enzyme known as **UDP glucose:flavonol 3-*O*-glucosyltransferase** (UFGT) (Fig. 2.12D).

### Developmental Regulation of Anthocyanin Synthesis

Anthocyanin synthesis is part of a developmental pathway. That is, the synthesis of anthocyanins occurs in a particular tissue and organ as that organ develops; other tissues and organs do not synthesize these pigments. For example, in the plant *Antirrhinum* (snapdragon), anthocyanins are synthesized by petals during their development; the sepals, bracts, and leaves usually do not synthesize these pigments. Furthermore, anthocyanin synthesis occurs only in petal epidermal cells; the cells of other petal tissues do not make the pigment. Thus, at least some of the genes regulating anthocyanin synthesis show very strong organ, cell, and tissue specificity. Anthocyanin synthesis also is affected by stress and environmental factors such as temperature and light. In *Antirrhinum,* for example, anthocyanin synthesis may be greatly stimulated by cold, whereas ultraviolet light can initiate anthocyanin synthesis in many plants.

The genetic analysis of anthocyanin synthesis in maize has been particularly informative, resulting in the identification of regulatory genes and also of genetic elements known as transposons, discussed later. Anthocyanin synthesis occurs in the seed of the type of maize known as "Indian corn," although most commercial maize cultivars do not synthesize any significant quantities of anthocyanins. Geneticists have confined their attention largely to an analysis of the genetic factors responsible for anthocyanin synthesis in the seed tissue known as the **aleurone.** At least 20 genetic loci are involved in anthocyanin biosynthesis in maize. Many of these are known to encode the enzymes required to catalyze specific chemical reactions in the pathway leading to anthocyanin synthesis, such as those described earlier. The genes encoding structural enzymes can be considered to be the effector genes for this biochemical pathway. The most important of these are the *C2* gene, which encodes the enzyme **chalcone synthase,** the *A1* gene, which encodes the enzyme **dihydroquercertin reductase,** and the *Bronze* gene (*Bz1*), which encodes **UFGT enzyme**, which transfers glucose to the anthocyanin. Both the C2 gene and the A1 gene are essential for the formation of colored anthocyanins. In the absence of the *Bronze* gene, the anthocyanin is still synthesized, but is not glucosylated. As a result, it is unstable and does not accumulate in the cells that make it; they therefore are brownish instead of purple. *C2, A1,* and *Bz1* are effector genes whose expression can be turned on by specific regulatory genes.

Any living part of a maize plant can form anthocyanins and become pigmented. Whether or not it does so depends on the expression of one or more of the regulatory genes. There are several different but related genes whose function is to regulate the activity of the *C2, A1,* and *Bz1* genes. For example, the gene *Pl* triggers the formation of anthocyanins in many tissues and organs of the maize plant, including leaf blade and husks. In contrast, the *C1* and *B* genes control the formation of anthocyanin in the aleurone layer of the seed, but not in the vegetative parts of the plant. When these regulatory genes are expressed in a particular maize tissue, anthocyanins are synthesized. The regulatory genes and the proteins they encode do not themselves function catalytically in the synthesis of anthocyanins. Rather, these genes are positive regulators of genes encoding the anthocyanin biosynthetic enzymes, namely, the *C2, A1,* and *Bz1* genes. Clearly the regulatory genes are themselves regulated in their expression.

The regulatory genes *C1* and *B* normally are expressed in the aleurone tissue of the seed, whereas the *Pl* gene is expressed in vegetative tissues. This is because the factors controlling the expression of *Pl* and *C1* are different; however, the proteins encoded by the *C1* and *Pl* genes are very similar. They belong to a class of proteins known as **transcription factors.** Transcription factors are proteins that bring about the expression of other genes by binding to specific DNA sequences that act as controlling elements within genes. Binding of the protein product of the regulatory genes to these controlling elements initiates the transcription of the gene and the formation of the mRNA from which the protein encoded by the gene will be synthesized. The protein products of the *C1* and *Pl* genes bind to similar controlling elements in the *C2, A1,* and *Bz1* genes, so either of these genes can initiate anthocyanin synthesis, if they are expressed.

A number of the regulatory genes for anthocyanin synthesis have been cloned and characterized. The controlling elements within the cloned *C1* gene can be changed experimentally in such a way that the gene is expressed in most plant tissues instead of only in the maize aleurone. When the modified *C1* and *B* genes are introduced back into tobacco or *Arabidopsis* plants anthocyanins will be produced in many different organs, including some that normally produce no anthocyanins, such as roots. This demonstrates that the regulatory genes controlling anthocyanin synthesis encode common transcription factors that are able to recognize controlling elements present in the structural genes for anthocyanin biosynthesis in widely divergent plants. Whenever the regulatory genes are expressed, anthocyanins are synthesized. We examine the mechanism by which these regulatory genes control gene expression in Chapter 4.

## Transposons

**Transposons,** genetic elements that can change position within the genome, were discovered by Barbara McClintock and it was for this work that she received a Nobel prize. She was studying the genetic factors controlling the synthesis of

**A**

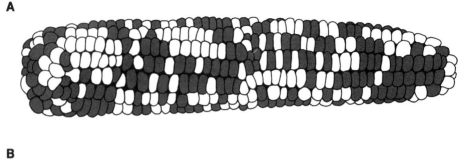

**B**

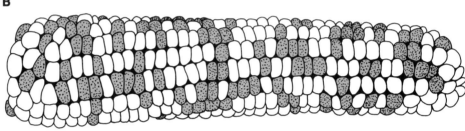

**Figure 2.13** **Different effects on pigmentation of maize kernels by stable and unstable mutations**
Each maize kernel is a seed that has developed from an independent fertilization event. Here we see maize ears in which some of the kernels are pigmented and others are colorless. (A) The pigmented kernels are purple (shown here as black) because cells in the aleurone layer have synthesized anthocyanins. No anthocyanin is formed in the colorless kernels as a result of a stable mutation that results from the insertion of a *Ds* transposon into the C gene encoding an enzyme essential for the synthesis of this pigment. (B) This ear also has both pigmented and colorless kernels, but the pigmented kernels are not uniformly colored. Instead, they show a pattern of pigmented spots in which some of aleurone cells have synthesized anthocyanin, whereas others in the same kernel have not and remain unpigmented. This is the result of an unstable mutation resulting from the insertion of an *Spm* transposable element in the C gene, inactivating it. The *Spm* element can, however, transpose out of the gene in some of the aleurone cells during kernel development, and movement of the transposon out of the gene results in the restoration of its activity, with the synthesis of anthocyanin in those cells.

anthocyanins in the aleurone cells of the maize seed. The pigmented kernels of maize that result from the unrestricted synthesis of anthocyanins are a deep purple (Fig. 2.13). Stable mutations that inactivate structural genes for enzymes required in anthocyanin synthesis result in yellowish-white kernels; however, not all mutations in these genes are stable. In some cases, the mutations are unstable at a high frequency and will revert to wild-type function many times during the development of a cob. As a result of this instability, individual kernels show a fine or coarse pattern of pigmented spots, depending on when in the development of the kernel the mutation was lost, as illustrated in Fig. 2.14. McClintock identified several genetic elements that were responsible for this instability. She named two of them *Ac* (for **Activator**) and *Ds* (for **Dissociation**). Both of these elements can move, or transpose, within the genome. For example, the insertion of *Ds* into the *C2* gene inactivates it, making it impossible for the cells to form anthocyanins. When it transposes back out of the *C2* gene, however, the function of the gene is restored and its expression once again results in anthocyanin synthesis (Fig. 2.15). The *Ac* element controls its own movement, but transposition of the *Ds* gene requires an *Ac* element.

Transposons were first discovered in maize, but they are present in the genomes of all organisms. A transposon such as *Ac* contains genes whose expression leads to its transposition or to the transposition of another element, such as *Ds*. *Ds* differs from *Ac* in that it lacks the genes necessary for transposition. The molecular details of the mechanism by which transposons move are discussed in Chapter 3. Because of their ability to inactivate genes and because their presence within a gene makes it possible to identify the gene, transposons are valuable tools for the molecular analysis of development. One of the most intriguing aspects of transposons is that in many cases their movement is developmentally regulated.

## Identification of Key Regulatory Genes through Homeotic Mutations

The class of mutations known as homeotic are particularly important and useful for understanding the genetic program controlling development. A homeotic mutation is one that changes organ identity so that recognizable structures are formed, but they appear in the wrong place or at the wrong time. For example, the *Drosophila* homeotic mutation *antennapedia* results in flies that have legs growing from their heads where they should have antennae. This homeotic mutation has changed the identity of structures that would have formed antennae so that other recognizable structures are produced, namely legs, but they are not the structures that should have been formed in this position. Mutations that change cell, tissue, or organ identity have been found in many other organisms, including angiosperms. Horticulturists have selected and propagated many mutations that cause "double flowers." The rose varieties grown in our gardens today usually have 20 or more petals as a result of mutations selected by rose breeders who found multipetal roses more attractive than those with the five petals of the wild type. Multipetal roses are the consequence of homeotic mutations that change the stamens of the wild-type rose into petals. Angiosperm flowers consist of successive whorls of organs that are formed in a specific sequence; bracts are

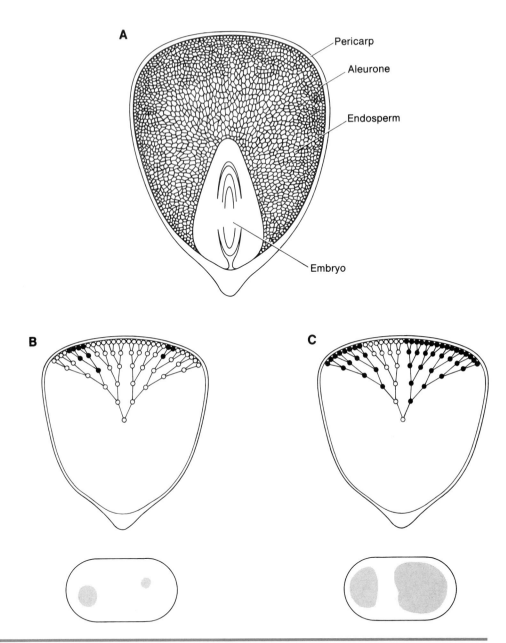

**Figure 2.14 Size of pigmented spots on the kernel depends on when in the development of the endosperm the transposon moves** (A) This diagram of the maize seed or kernel illustrates the position of the aleurone layer of the endosperm in which anthocyanins are produced. When a transposon moves out of a gene essential for anthocyanin synthesis, the function of the gene may be restored and anthocyanins formed. (B) If the transposon moves late in endosperm development, the clones of cells containing the revertant gene are small and scattered, (C) however, when movement of the transposon restores gene function early in development, the clones of pigmented cells are much larger. Redrawn with permission from Federoff (1984).

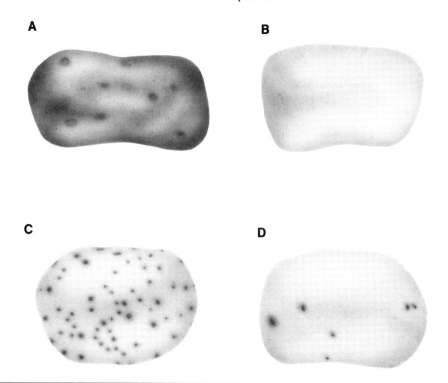

**Figure 2.15 Transposons alter the pattern of pigmentation of maize kernels**

In the wild-type kernel of purple-pigmented maize shown in (A), the kernel is uniformly pigmented as a result of anthocyanin synthesis in all the cells of the aleurone. The kernel in (B) is colorless as a result of the inactivation of the C gene, which encodes an enzyme essential for anthocyanin synthesis, by the insertion of the nonautonomous transposon *Ds*. The kernel shown in (C) has an autonomous *Spm* transposon inserted in a gene essential for anthocyanin synthesis, completely inactivating the gene and resulting in colorless aleurone cells; however, *Spm* has transposed out the gene in some of the aleurone cells, restoring anthocyanin synthesis in the clones where the transposon has moved and creating a pigmented spot. In the kernel shown in (D), a fully effective *Spm* transposon jumped out the anthocyanin gene later in development than in the kernel shown in (C), so the spots are smaller and there are fewer of them. Redrawn with permission from Federoff (1984).

formed fist, followed by sepals, petals, stamens, and finally carpels. Numerous homeotic mutations affecting floral development are known. These result in the conversion of sepals to bracts, or bracts to petals, or petals to stamens, and so forth (Fig. 2.16). Genes that result in homeotic transformations of cell or organ identity when they are mutated are known as **homeotic genes.** Homeotic genes

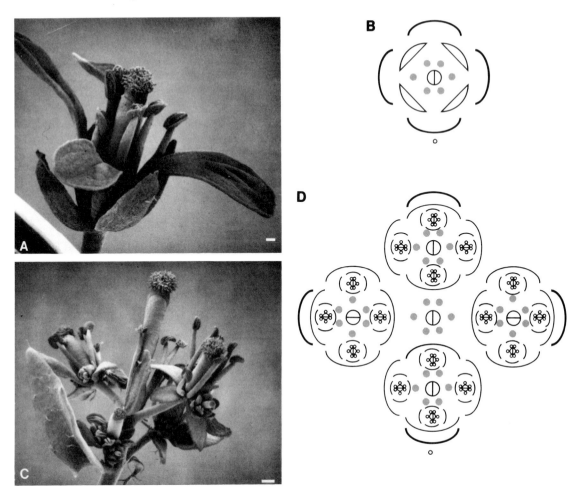

**Figure 2.16 Homeotic mutants of *Arabidopsis* change the structure of the flower**
(A) Scanning electron micrograph of a wild-type flower. (B) Diagram of the flower in cross section. (C) This flower was formed by an individual homozygous for a mutation known as *ap1-1* in which the sepals are converted to bracts and additional floral buds are formed in the axils of each transformed sepal. (D) Diagram of the *ap1-1* flower. Reprinted with permission from Irish and Sussex (1990). ©American Society of Plant Physiologists.

constitute a class of regulatory genes that control many important aspects of development, such as the time and place in which major cell types differentiate and the identity of tissues and organs that are to be formed.

Among the characters with which Mendel worked were genes that determined the position of the flowers on the pea plant. He found that some pea plants always had terminal flowers; that is, flowers were formed on the end of the branches. Other pea plants produced flowers only in the axils of the leaves and never at the ends of the branches. Plants with terminal flowers have a **determinant growth habit.** This is because the apical meristem is transformed from a vegetative to a floral meristem when the plant flowers. The transformation of the apical meristem into a floral meristem causes the plant to stop growing, as it no longer has vegetative apical meristems to produce additional leaves and internodes. In plants with axillary flowers, by contrast, only the lateral meristems in the axils of the leaves are transformed into floral meristems; the terminal apical meristems remain vegetative. As a result, plants with axial flowers continue to grow vegetatively from the terminal apical meristems after they have begun to flower. They have an **indeterminate growth habit.**

When Mendel crossed true-breeding plants bearing axillary flowers with true-breeding terminal-flower plants, all of the F1 offspring had axillary flowers. The axillary trait was dominant over the terminal trait. This means that the genes controlling flower position are alleles. Apparently the dominant allele can transform lateral meristems from vegetative to floral, but it cannot transform the apical meristems. The recessive allele, in contrast, transforms the apical meristem but not the laterals. Clearly this gene controls a major developmental event in plants: the transition between vegetative and reproductive growth. At present we do not know much more about how this gene acts to regulate this important developmental program in the pea plant, but it is likely to be a homeotic gene. As with other homeotic genes, it controls the identity and placement of plant structures; in this case it controls the formation of flowers by meristems.

Homeotic mutations of a gene known as *LEAFY* in *Arabidopsis* block the formation of flowers. Vegetative meristems do not directly initiate flowers in the wild-type *Arabidopsis*. Instead, as the plant begins to flower, the vegetative apical meristem is first transformed into an **inflorescence meristem.** The primary inflorescence meristem is indeterminate in its growth and it only forms stem tissue, bracts, and lateral meristems in the axils of the bracts. Some of the lateral meristems initiated by the inflorescence meristem go on to become the floral meristems from which the floral organs are differentiated. Mutations in the *LEAFY* gene are recessive. Plants homozygous for these mutations never form floral meristems, although they undergo the transition from vegetative to inflorescence meristems; however, the lateral meristems produced by the activity of the inflorescence meristem are themselves inflorescence meristems that cannot differentiate floral organs. The *LEAFY* gene normally is expressed in *Arabidopsis* floral meristems and the product of the *LEAFY* gene is required for floral meristem identity. Mutations that block the expression of the *LEAFY* gene in the lateral meristems prevent these meristems from assuring the identity of floral meristems and forming floral organs.

Plant organs such as leaves, petals, sepals, anthers, and carpels are initiated from meristems as small masses of cells known as **primordia** (singular, primordium). The primordium is relatively undifferentiated when it is first initiated from the shoot apex, but at some point during its growth the organ identity of the primordium is determined. It becomes a leaf, bract, sepal, petal, stamen, or carpel primordium. This determination event is the result of the expression of homeotic genes whose function it is to establish organ identity. Homeotic genes are regulatory genes whose function is to initiate the expression of other genes, including the effector genes whose protein products are necessary for the unique functions of this organ. For example, petals and leaves often have similar morphologies, but petals synthesize anthocyanins, or other pigments, and lack chloroplasts, whereas the mesophyll cells of leaves have well-developed chloroplasts. One of the major differences between leaves and petals is that genes encoding the anthocyanin biosynthetic enzymes, such as chalcone synthase, are turned on in the petals, or at least in the epidermal cells of the petals, whereas genes required for chloroplast development are turned on in leaves but not in petals. The expression of a unique pattern of homeotic genes in the primordium leads to the expression of the regulatory genes required for anthocyanin biosynthesis or for chloroplast development, respectively. In these cases, the correct pattern of expression of homeotic genes is necessary for the differentiation of the floral organs. The homeotic genes are expressed and this expression in turn controls the expression of other genes, including other regulatory genes. Ultimately the effector genes whose products characterize each of the floral organs are expressed. Homeotic genes encode proteins that regulate the transcription of other genes whose expression is necessary for cell, tissue, and organ differentiation. The mechanism by which the expression of these homeotic genes regulates the expression of other genes is discussed in Chapter 4.

## Role of Differential Growth in Plant Development

Plant organs achieve their shape as a result of differential growth. As the primordium grows, its different parts grow at different rates so that the original spherical shape is distorted (Fig. 2.17). For most plants, there is nothing spherical about a leaf. To develop the flat, expanded blade characteristic of many leaves, there must be rapid growth along the margins of the primordium, with little growth at right angles to this.

What is the evidence that differential growth rates are under genetic control? Once again, breeding experiments have identified genes whose expression plays a significant role in determining the shape of an organ. Many years ago, Edmund Sinnott investigated the inheritance of fruit shape in gourds. Some of these had flattened, disk-shaped fruit, and others, spherical fruit. He found that fruits produced by the F1 generation were all disk-shaped, whereas in the F2 generation, three-fourths of the plants produced disk-shaped fruit and one-fourth produced spherical fruit. Sinnott's work demonstrates that a single pair of alleles controls fruit shape in these cucurbits.

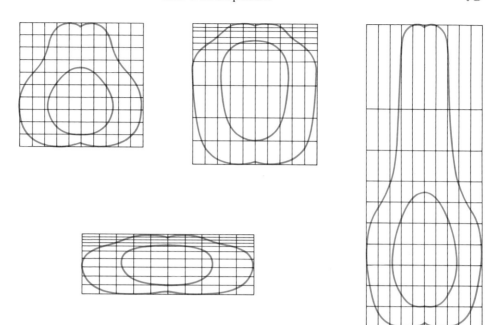

**Figure 2.17 Differential growth can produce many different shapes in a fruit**
A grid has been lain over the fruit shown at the upper left. The shape of this fruit then is modified by changing the relative sizes of the tissues in each of the grids. Redrawn with permission from Sinnott (1960).

Mendel also studied the inheritance of height in the garden pea. He studied two alleles, *T* and *t,* that had dramatic effects on plant height. This difference is due to the rate and extent of growth of the internodes. Plant stems consist of nodes and internodes. In most higher plants, the development of the stem is characterized by the periodic initiation of leaves by the apical meristem, followed by the formation of internodal tissue, without leaves. The leaves are attached to the stem at the nodes. When the stem elongates, it is the internodal tissue that elongates; there is relatively little elongation of nodal tissue.

Theoretically, it is possible to make a taller plant either by increasing the rate of node and internode formation or by increasing the length of the internodes. Short and tall pea plants produce leaves at the same rate, however. As a result, tall and short peas of the same developmental age have the same number of internodes, but internodes are longer in the tall plant. Such differences are, in some cases, due to the plant hormones that control the rate of stem elongation.

## Summary

In the course of his work with the garden pea which led to the discovery of many of the principles of genetics, Gregor Mendel also demonstrated that heritable factors controlled important aspects of plant development. The heritable factors, or genes, that determine the develomental program are in the nucleus and reside on chromosomes. This knowledge was established through grafting experiments between different species of the alga *Acetabularia,* in which the structural characteristics of the cap were shown to depend on information in the nucleus. Information flows from the genes in the nucleus into the cytoplasm in the form of mRNAs. An mRNA contains instructions for the synthesis of a specific protein in the cytoplasm. Each gene occupies a particular place on one of the chromosomes within the nucleus. Every organism has a characteristic number of chromosomes. Every chromosome in a vegetative cell has a very similar sister, and the pair is said to be homologous. In contrast, reproductive cells (sperm cells and eggs, as well as other gametophytic cells) are haploid. The reproductive cells contain only one of each of the homologous chromosomes. Each pair of homologous chromosomes contains the same kinds of genetic information, but the exact nature of the genes may be different. Homologous chromosomes contain allelic genes. *Arabidopsis* has a basic chromosome number of 5, so the diploid cells of the vegetative plant have a total of 10 chromosomes. If the alleles are the same at one genetic locus, or position, the organism is said to be homozygous for that gene, whereas it is heterozygous if they are different. Usually there are a number of different alleles for any given gene in most interbreeding populations. Typically the different alleles result in somewhat different characteristics when they are expressed. When two different alleles for the same character are present in the genome, frequently one is dominant and the other is recessive. The phenotypic characteristics of the organism would be determined by the dominant allele. We do not necessarily know the genotype of the plant from its phenotype. In the cross between a smooth-seed garden pea plant and a wrinkled-seed plant, all the plants of the F1 generation produced smooth seeds. This is because the gene for smooth seeds is dominant over the gene for wrinkled seeds. The plants of the F1 generation can be self-fertilized to give a second filial generation (F2). The wrinkled-seed characteristic will reemerge in approximately one-quarter of the F2 plants, showing that the factor that led to wrinkled seeds was not lost in the F1 generation; it was simply not expressed. The $r$ gene is recessive with respect to its dominant $R$ allele and its effect is observed only when the plant is homozygous for this gene.

A duplicate of each chromosome is constructed between divisions (interphase) in meristematic cells as a result of the replication of each chromosomal DNA molecule and the synthesis and assembly of the chromosomal proteins. The two daughter chromosomes remain in close association after chromosome replication is completed, and at this time each chromosome is said to consist of two chromatids. The chromatids will not be called chromosomes until they are separated by mitosis.

The genes controlling developmental traits can be identified by isolating mutations that alter or block that particular type of development. In some cases, mutations affecting development have been found to occur in structural genes encoding enzymes that are necessary for a specific function that characterizes a particular develomental pathway. This was the case for the mutation Mendel originally studied that led to wrinkled pea seeds. Mendel did not know why the *r* gene resulted in the wrinkled phenotype when homozygous; however, recently it was shown that the *r* allele is a mutation of the *R* gene encoding a starch branching enzyme. The wrinkled phenotype results from a reduced level of starch in seeds of plants homozygous for the *r* gene because of the inability of the mutant *r* gene to initiate branches in the growing starch molecules. A change in the expression of a single gene may have additional phenotypic consequences. Plants homozygous for the *r* gene also have reduced levels of seed storage proteins. These differences are secondary or pleiotropic effects of the expression of the gene. In addition to the identification of genes encoding enzymes, mutations can identify regulatory genes whose expression controls the expression of other genes. These regulatory genes are switches that control developmental and biochemical pathways by regulating the expression of genes encoding proteins that make these possible. For example, anthocyanins are pigments whose synthesis is regulated during plant development; the pigments are synthesized in the petals of flowers and in aleurone tissues of developing maize seeds, among other places. Anthocyanin synthesis is brought about by several enzymes, including phenylalanine ammonia lyase and chalcone synthase. Expression of the genes encoding these enzymes requires the expression of other regulatory genes encoding transcription factors. Transcription factors participate in the decoding of the genes for the enzymes. Tissues in which the regulatory genes are expressed will synthesize anthocyanin pigments; those that do not express the regulatory genes do not synthesize anthocyanins.

Some mutations are caused by transposons. Transposons are mobile genetic elements that hop in and out of genes. The insertion of a transposon into a gene frequently inactivates it, and activity can be restored when the transposable element jumps back out of the gene. Transposons were first identified by Barbara McClintock, who noted that some mutations altering

*continues*

the pattern of anthocyanin synthesis in maize seeds were highly unstable. Transposons were first discovered in maize, but they are found in the genomes of all organisms. Insertion of a transposon into a gene has, in some cases, made it possible to mark the gene so it can be isolated and characterized.

Homeotic mutations have identified key regulatory genes that act as master switches for tissue and organ identity. A homeotic mutation is one that changes tissue or organ identity. Homeotic genes constitute a class of regulatory genes that control important aspects of development, such as the time and place in which major cell types differentiate and the identity of tissues and organs. Many homeotic mutations affecting floral organ identity have been isolated.

## Questions for Study and Review

1. See Figs. 2.2 to 2.4 in answering both this question and Question 2. *Acetabularia mediterranea* and *A. crenulata* are different species belonging to the same genus. What are the morphological differences between them? How can you account for these differences and how do these differences come about?

2. When you graft an *A. mediterranea* rhizoid to an *A. crenulata* stalk, the graft will heal and the organism will regenerate a new cap. The first cap regenerated is intermediate in its morphology. Why?

3. You have crossed a tall strain of peas that have terminal flowers with a strain having axial flowers and short stature. Both strains are true breeding when they are self-fertilized. What are the phenotypic characteristics of the F1 generation resulting from this cross? What are the genotypic and phenotypic characteristics of the F2 progeny obtained by self-fertilizing the F1 plants?

4. Pea seeds that are homozygous *rr* differ from those homozygous *RR* in starch content, lipid content, and amount of legumin storage protein. Which of these is considered to be the primary effect of the gene and which is a pleiotropic effect? How does the genetic difference between these two plants bring this primary effect about? How can you account for these pleiotropic effects?

5. What is the ratio of pigmented to colorless kernels in the ear of maize shown in Fig. 2.13A? How can you account for this ratio? In answering this question, you should remember that the pigmented aleurone phenotype is dominant over the colorless aleurone and that the aleurone is part of the endosperm, which is formed as a result of the fusion of two maternal nuclei with one male nucleus.

6. How many different pigmented phenotypes can you detect in the kernels on the ear shown in Fig. 2.13B? How can you account for

these differences when each kernel differs by only one allele?

7. What is a homeotic gene? How is a homeotic gene different from a regulatory gene, such as those that regulate the anthocyanin biosynthetic pathway?

8. Explain how differential growth could turn a spherical structure into a banana. Would this differential growth occur by cell division or cell enlargement?

## Further Reading

### General References

Ayala, F. J., and Kiger, J. A. (1984). *Modern Genetics,* 2nd ed. Benjamin/Cummings, Menlo Park, CA.

Darnell, J., Lodish, H., and Baltimore, D. (1990). *Molecular Cell Biology.* Scientific American Books, New York.

Fristom, J., and Clegg, M. (1988). *Principles of Genetics.* 2nd Ed. Freeman, New York.

Grant, V. (1975). *Genetics of Flowering Plants.* Columbia University Press, New York.

Hawkins, J. D. (1991). *Gene Structure and Expression.* 2nd Ed. Cambridge University Press, New York.

Klekowski, E. J. (1988). *Mutation, Developmental Selection, and Plant Evolution.* Columbia University Press, New York.

Mendel, G. (1965). [*Experiments with Plant Hybridization.*] Translation of Mendel's work by the Royal Horticultural Society, London. Harvard University Press, Cambridge, MA.

Sinnott, E. W., Dunn, L. C., and Dobzansky, T. (1958). *Principles of Genetics,* 5th ed. McGraw-Hill, New York.

### Specific References

Berger, S., deGroot, E. J., Neuhaus, G., and Schweiger, M. (1987). *Acetabularia:* A giant single cell organism with valuable advantages for cell biology. *Eur. J. Cell Biol.* **44,** 349–370.

Bhattacharyya, M. K., Smith, A. M., Ellis, T. H. N., Hedley, C., and Martin, C. (1990) The wrinkled-seed character of pea described by Mendel is caused by a transposon-like insertion in a gene encoding starch-branching enzyme. *Cell* **60,** 115–122.

Biggin, M. D., and Tjian, R. (1989). Transcription factors and the control of *Drosophila* development. *Trends Genet.* **5,** 377–383.

Estelle, M. A., and Somerville, C. R. (1986). The mutants of *Arabidopsis. Trends Genet.* **2,** 89–93.

Federoff, N. V. (1984). Transposable genetic elements in maize. *Sci. Amer.* **250,** 84–90.

Federoff, N. V. (1989). About maize transposable elements and development. *Cell* **56,** 181–191.

Gehring, W. J., and Hiromi, Y. (1986). Homeotic genes and the homeobox. *Annu. Rev. Genet.* **20,** 147–173.

Gottlieb, L. D., and de Vienne, D. (1988). Assessment of pleiotropic effects of a gene substitution in pea by two-dimensional polyacrylamide gel electrophoresis. *Genetics* **119,** 705–710.

Grant, P. (1978). *Biology of Developing Systems.* Holt, New York.

Haughn, G. W., and Somerville, C. R. (1988). Genetic control of morphogenesis in *Arabidopsis. Dev. Genet.* **9,** 73–89.

Irish, V. F., and Sussex, I. M. (1990). Function of the *apetala-a* gene during *Arabidopsis* floral development. *Plant Cell* **2,** 741–753.

Koornneef, M., van Eden, J., Hanhart, C. J., Stam, P., Braaksma, F. J., and Feenstra, W. J. (1983). Linkage map of *Arabidopsis thaliana. J. Heredity* **74,** 265–272.

Lloyd, A. M., Walbot, V., and Davis, R. W. (1992). *Arabidopsis* and *Nicotiana* anthocyanin production activated by maize regulators *R* and *C1. Science* **258,** 1773–1775.

Ludwig, S. R., Bowen, B., Beach, L., and Wessler, S. R. (1990). A regulatory gene as a novel visible marker for maize transformation. *Science* **247,** 449–450.

Marks, M. D., and Feldmann, K. A. (1989). Trichome development in *Arabidopsis thaliana.* I. T-DNA tagging of the *glaberous 1* gene. *Plant Cell* **1,** 1043–1050.

Marx, G. A. (1987). A suite of mutants that modify pattern formation in pea leaves. *Plant Mol. Biol. Rep.* **5,** 311–335.

Meyerowitz, E. M. (1989). *Arabidopsis,* a useful weed. *Cell* **56,** 263–269.

Meyerowitz, E. M., Smyth, D. R., and Bowman, J. L. (1989). Abnormal flowers and pattern formation in floral development. *Development* **106,** 209–217.

Reddy, C. M., and Coe, E. H., Jr. (1986). Anthocyanin biosynthesis in maize: A model system in the study of gene action and gene regulation. In: *"Gene Structure and Function in Higher Plants,"* C. M. Reddy and E. H. Coe, Jr. (Eds). Oxford IBH Publishers, Pvt. Ltd., New Delhi.

Rhoades, M. M. (1950). Meiosis in maize. *J. Heredity* **41,** 59–67.

Roth, B. A., Goff, S. A., Klein, T. M., and Fromm, M. E. (1991). *C1-* and *R*-dependent expression of the maize *Bz1* gene requires sequences with homology to mammalian *myb* and *myc* binding sites. *Plant Cell* **3,** 317–325.

Schultz, E. A., and Haughn, G. W. (1991). *leafy,* a homeotic gene that regulates inflorescence development in *Arabidopsis. Plant Cell* **3,** 771–781.

Sinnott, E. M. (1960). *Plant Morphogenesis.* McGraw-Hill, New York.

# 3

# The Size and Complexity of Plant Genomes

In chapter 2 we presented some of the evidence that developmental information is contained in the nucleus and encoded in the nuclear DNA. The fundamental dogma of molecular biology states that the genetic information encoded into DNA is **transcribed** into complementary RNA molecules and these in turn are **translated** into proteins. In this chapter we examine the structure of DNA and see how the linear sequence of bases in the DNA molecule encodes proteins. In addition, we learn how much DNA there is in the nucleus of a higher plant and what fraction of this DNA actually represents functional genes. Plant nuclear genomes vary greatly in size and all of them contain a significant fraction of DNA that contains no genetic information and has no known function. In addition, plant genomes contain a variety of genetic elements that are mobile and can interfere with the expression of functional genes. Armed with this information, we can begin to ask the questions: How many of these genes are involved in development? How are the genes concerned with development organized? In future chapters we expand on this information to consider the mechanism by which genetic information encoded in the nuclear genome is transcribed and translated to form cellular proteins.

# The Structure of DNA and RNA

Genetic information is encoded in the sequence of bases in DNA. DNA is a polymeric molecule that is constructed by the sequential assembly of individual subunits, known as **nucleotides.** Each nucleotide is itself a complex of three other molecules: a deoxyribose sugar, a nucleic acid base, and phosphoric acid. Only four bases are found in DNA: adenine, guanine, cytosine, and thymine. RNA also is composed of nucleotides, although the RNA nucleotides contain ribose sugar instead of deoxyribose, and RNA bases include uracil, but not thymine (Fig. 3.1). The nucleotides are chemically bonded to each other through ester linkages between phosphoric acid and the hydroxyl groups on the deoxyribose sugars of adjacent subunits (Fig. 3.2A). These phosphodiester bonds link the 5'-deoxyribose carbon of one nucleotide with the 3'-deoxyribose carbon of the next nucleotide. DNA is a linear, double-helical molecule made up of two deoxyribose sugar phosphate backbones, from which the nucleic acid bases project (Fig. 3.2B). The two strands of the double helix are held together by hydrogen bonds between complementary bases, according to certain simple base pairing rules, as illustrated in Fig. 3.2C. Adenine pairs with thymine by the formation of two hydrogen bonds, and guanine pairs with cytosine by the formation of three hydrogen bonds.

Because the phosphodiester bonds linking the nucleotides extend from the 5' carbon of one nucleotide to the 3' carbon of the next nucleotide, each strand of the double helix has an inherent polarity. One end will have a free 5'-hydroxyl or phosphate group, whereas the opposite end will have a free 3'-hydroxyl group. Furthermore, the two strands of the DNA double helix are antiparallel. At a given end of the double helix, one strand has a free 5' end and the opposite strand has a free 3' end.

Theoretically, the two strands of a DNA molecule could form either a left-handed or a right-handed double helix. In fact, DNA most often assumes the right-handed configuration in which the helix makes one complete turn every 3.4 nm, with 10 bases per turn (Fig. 3.2B). DNA that has a high proportion of guanine and cytosine bases can, however, assume a left-handed configuration, known as Z-DNA (Fig. 3.3A). Regions of Z-DNA configuration could represent important signals for gene transcription. In addition, the binding of proteins to DNA can alter its configuration (Fig. 3.3B).

**A**

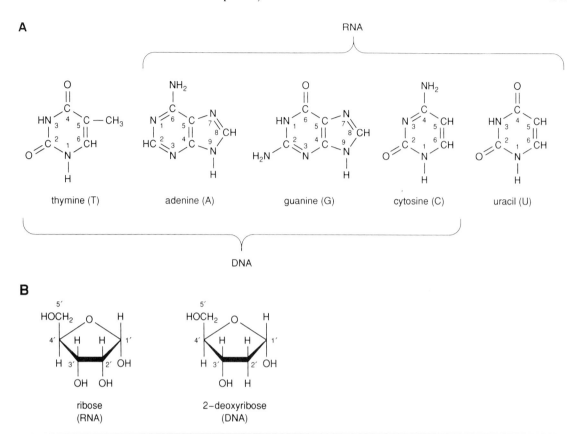

**Figure 3.1**  **Chemical components of nucleic acids**
Both RNA and DNA are composed of nucleotides. Each nucleotide consists of a base, a
pentose sugar, and phosphoric acid. (A) Chemical structures of nucleic acid bases.
RNA contains uracil, but not thymine, whereas DNA contains thymine and not uracil.
(B) Chemical structures of the sugars found in nucleic acids. The pentose sugar of
RNA is ribose, whereas that of DNA is deoxyribose. The carbon atoms in the pentose
sugars are numbered with primes.

 The two strands of the double helix contain completely different, although
complementary, sequences of bases. As a result, when the DNA is replicated,
either or both strands of the double helix can act as a template for the assembly
of the other strand (Fig. 3.4). The two strands are separated and nucleotides are
lined up along each of the strands according to the base pairing rules: adenine
pairs with thymine, guanine pairs with thymine. Similarly, when a gene is transcribed, one strand of the DNA acts as a template for the formation of a sequence

of complementary RNA bases, again according to these same simple base pairing rules. Only one strand of the DNA contains the information necessary to construct a particular protein. The complementary strand would encode a completely different protein, if it also represented a structural gene.

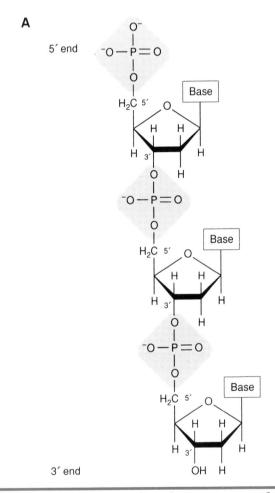

**Figure 3.2   Chemical bonding between nucleotides in DNA**
(A) Nucleotides are joined through phosphodiester bonds linking the 5′ carbon of one nucleotide with the 3′ carbon of the next nucleotide. Note that one end of the structure has a free 3′-hydroxyl group while the other has a free 5′ phosphate. (B) Skeletal model of a DNA double helix. (C) The base pairing that holds the two strands of the double helix together. Adapted from Darnell *et al.* (1990).

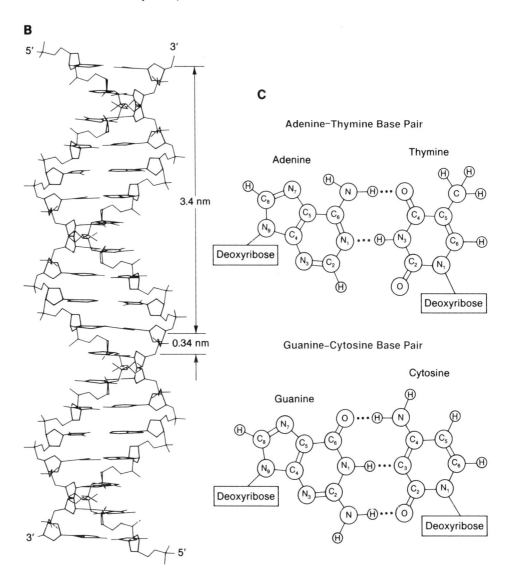

**B**

5′  3′

3.4 nm

0.34 nm

3′  5′

**C**

Adenine–Thymine Base Pair

Adenine                                        Thymine

Deoxyribose

Deoxyribose

Guanine–Cytosine Base Pair

Cytosine

Guanine

Deoxyribose

Deoxyribose

## The Structure of Proteins

Proteins also are polymers, although their structure is considerably more complex than that of nucleic acids. They are composed of 20 different amino acid building blocks. All except one of these amino acids contains an amino group and a carboxyl group attached to the same $\alpha$-carbon atom (Fig. 3.5A). Proline, which

## Table 3.2    Amounts (pg) of DNA in eukaryotes and prokaryotes

| | DNA per 2c nucleus | Chromosome number, 2n |
|---|---|---|
| | Eukaryotes | |
| **Animals** | | |
| *Amphiuma* | 168.0 | 24 |
| *Protopterus* (lungfish) | 100.0 | 38 |
| *Salamandra salamandra* | 85.3 | 24 |
| *Triturus viridescens* | 72.0 | 22 |
| *Bombina bombina* | 28.2 | 24 |
| *Rana esculenta* | 16.8 | 26 |
| *Rana temporaria* | 10.9 | 26 |
| *Bos taurus* (ox) | 6.4 | 60 |
| Man | 6.4 | 46 |
| Sheep | 5.7 | 54 |
| Mouse | 5.0 | 40 |
| *Drosophila* | 0.2 | 8 |
| **Flowering plants** | | |
| *Fritillaria davisii* | 196.7 | 24 |
| *Lilium longiflorum* | 72.2 | 24 |
| *Tulipa gesneriana* | 51.5 | 24 |
| *Allium cepa* | 33.5 | 16 |
| *Vicia faba* | 28.0 | 12 |
| *Lathyrus latifolius* | 21.8 | 14 |
| *Ranunculus ficaria* | 19.2 | 16 |
| *Secale cereale* | 18.9 | 14 |
| *Zea mays* | 11.0 | 20 |
| *Crepis capillaris* | 4.2 | 6 |
| *Vicia sativa* | 4.0 | 12 |
| *Antirrhinum majus* | 3.6 | 16 |
| *Linum usitalissimum* | 1.4 | 30 |
| *Aesculus hippocastanum* | 0.3 | 40 |
| *Arabidopsis thaliana* | 0.14 | 10 |

| | DNA per haploid nucleus | Chromosome number, n |
|---|---|---|
| **Fungi** | | |
| *Dictyostelium discoideum* | 0.384 | 7 |
| *Ustilago maydis* | 0.208 | 2 |
| *Aspergillus nidulans* | 0.048 | 8 |
| *Saccharomyces cerevisiae* | 0.026 | 15 |

| | | |
|---|---|---|
| | Prokaryotes | |
| **Bacteria** | | |
| *Salmonella typhimurium* | DNA per cell | |
| *Escherichia coli* | 0.0143 | |
| *Diplococcus pneumoniae* | 0.0040 | |
| *Mycoplasma hominis* | 0.0022 | |
| | 0.0009 | |
| **Viruses** | | |
| T2 | 0.000 220 | |
| λ | 0.000 055 | |
| P4 | 0.000 016 | |
| φX174 | 0.000 005 | |

Histone proteins have a high affinity for each other and they will aggregate in a specific way to form an octamer. Because of their basic character, the acidic DNA has a high affinity for these histone octamers. The DNA double helix is wound around a globular core of histone proteins to form a spherical particle known as a **nucleosome** (Fig. 3.11A). Each nucleosome contains an octamer of eight histone proteins around which approximately 200 bp of DNA are coiled. When completely relaxed, the chromatin fiber has the appearance of a beaded string, where each bead is a nucleosome (Fig. 3.11B); however, the chromatin fiber usually is not in a relaxed state, but rather is tightly coiled, even in interphase. When cells divide mitotically, the chromatin fibers condense even further to take on the configuration we call chromosomes. In the condensed state, the chromatin fiber is coiled on itself, as shown in Fig. 3.11C. The chromatin fibers appear to occupy a specific location within the nucleus. The inner surface of the nuclear envelope contains special proteins known as **lamins** which form a fibrous network, to which chromatin fibers are attached.

### Number of Genes Required to Encode the Functions of an Angiosperm

The *Arabidopsis thaliana* genome contains $7 \times 10^7$ bp of nuclear DNA. If all of this DNA had a coding function, how many genes would the *A. thaliana* genome contain? As a preliminary estimate, we will assume that an average protein has a molecular weight of about 50,000 daltons (Da) and an amino acid has a molecular weight of about 100 Da. This means that the average protein would contain about 500 amino acids. As the genetic code is read in groups of three bases, the average protein would require a minimum of 1500 bp of DNA to encode all the amino acids in its primary sequence. Genes must be larger than this because they consist of more than just the coding sequence. Almost all expressed genes contain regulatory elements at their 5′ ends that control their transcription. The sequence containing these regulatory elements often is 500 bp or longer. In addition, genes contain another 200 to 300 bp of sequence beyond the 3′ end of the coding sequence that is translated into RNA when the gene is expressed. Finally, the coding sequence of nearly all expressed genes is divided into segments, called **exons,** which are interrupted by sequences that have no coding function. These noncoding regions are called **introns,** and there may be from one to many introns in each gene. The human dystrophin gene, which is the largest gene characterized so far, contains $2.3 \times 10^6$ bp and has at least 75 introns; however, most plant genes, and even most animal genes, are much smaller than the dystrophin gene and often contain only three or four introns. For the purpose of our calculations here we will assume that an average gene contains three introns and that the introns average 500 bp each. This would mean that each gene on average, would

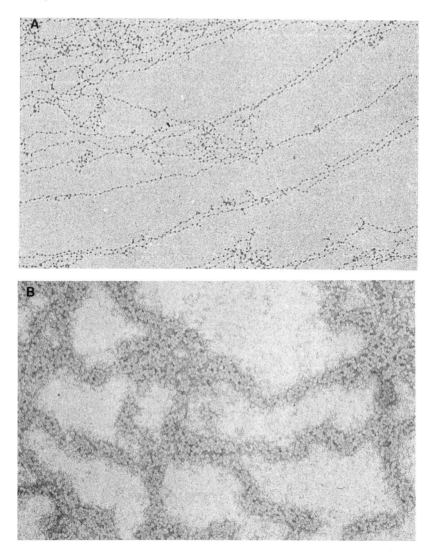

**Figure 3.11 Chromatin fibers**

DNA in a nucleus is complexed with proteins to form the chromatin fiber shown in these electron micrographs. (A) In a decondensed state, the chromatin fiber has the appearance of a beaded string in which the beads represent octamers of histone proteins around which the DNA strand is coiled. (B) However, in the nucleus, this beaded string is tightly coiled on itself to form the 30-nm fiber. (C) Illustrates the basic components of the nucleosome "bead" and demonstrates the way these components interact. (A,B) Courtesy of Barbara Hamkalo and (C) redrawn with permission from Alberts *et al.* (1989).

**c**

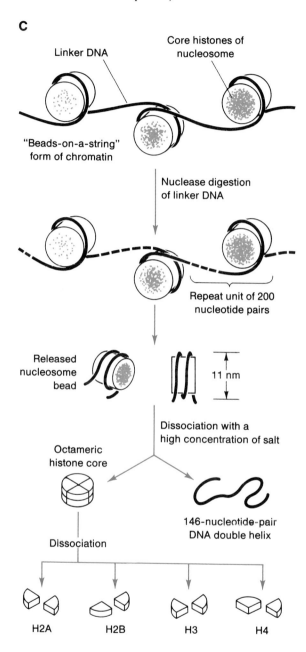

Linker DNA

Core histones of
nucleosome

"Beads-on-a-string"
form of chromatin

Nuclease digestion
of linker DNA

Repeat unit of 200
nucleotide pairs

Released
nucleosome
bead

11 nm

Dissociation with a
high concentration of salt

Octameric
histone core

146-nucleotide-pair
DNA double helix

Dissociation

H2A     H2B     H3     H4

be a 3800-bp-long DNA sequence. If all ($7 \times 10^7$ bp) of the *Arabidopsis* nuclear DNA-encoded proteins, it could contain nearly 18,500 genes. Other studies (see later) tell us that about 10% of the *Arabidopsis* genome consists of highly repetitive DNA that is unlikely to represent genes. If the remaining 90% of the genome had a coding function, it could contain 16,500 genes.

*Arabidopsis* has an unusually small genome and it could be expected to contain many fewer genes than other flowering plants. The maize genome is nearly 100 times larger, for example. Does it contain 100 times more genes? The answer is clearly no. There are more genes in the maize genome than in the *Arabidopsis* genome, but this difference is unlikely to be more than a factor of 2 to 3. There are other ways to estimate the number of expressed genes. One way is to determine the number of proteins found in an organism because each protein is encoded by a gene. In practice this is very difficult, but the technique of two-dimensional (2D) polyacrylamide gel electrophoresis can be used to estimate the number of proteins found in different plant organs. Under the best conditions, up to 2000 proteins can be separated as individual spots using this technique. Most organs contain a few proteins that are very abundant, each of which is present in 50,000 copies or more per cell. Other proteins that can be detected in 2D gels are moderately abundant, but rare proteins are not easily detected. There are many more different kinds of rare proteins than there are abundant or moderately abundant proteins. In fact, the sensitivity of 2D gel electrophoresis is such that only about 10% of the total cellular proteins are detected. This suggests that an organ such as a leaf of maize or tomato might contain 20,000 different proteins, which of course would be encoded by 20,000 different genes. The 2D gel profiles of different plant organs are rather similar, with only approximately 5% of the proteins being different. That is, many of the same genes are expressed in leaves, stems, and roots. Because any given plant has 8 to 12 different organ systems, we can add another 8000 to 12,000 organ-specific genes to our estimate of 20,000 generally expressed genes for a total of 28,000 to 32,000 genes. This can be considered a ballpark approximation of the number of genes in most higher plants.

Although this number is larger than our estimate for the number of genes in the *Arabidopsis* genome, it is far from the $1.4 \times 10^6$ genes that could be present in the maize genome if all $5.5 \times 10^9$ bp of nuclear DNA (haploid amount) had a coding function (assuming an average size of 3000 bp/gene). This raises two questions. (1) How do we account for the large discrepancy between the potential coding capacity of many plant genomes and our best guess as to the numbers of genes that might be present in these genomes? (2) Why would a plant such as maize have nearly twice as many genes as *Arabidopsis,* assuming that our estimates for the actual numbers of genes in these two genomes are correct? Complete answers to these questions are not yet available, but partial answers are. Our answer to the first question is that a substantial fraction of most plant genomes has no coding function. It consists of repetitive-sequence DNA that does not encode proteins, and plants differ in the fraction of their genomes that is made up of repetitive-sequence DNA. This concept is developed at some length later. As far as the second question is concerned, plants differ substantially in the numbers of genes in their genomes because there is a varying degree of redundancy in many important biosynthetic pathways and in the genes encoding many enzymes and other proteins.

## Important Proteins Encoded by Multigene Families

Many proteins are encoded by small multigene families in higher plants and other eukaryotic organisms. For example, glycinin is one of the major storage proteins found in the seeds of soybean. Glycinin serves as a food reserve used by the germinating seedling. In the soybean genome five genes encode glycinin. These five genes can be shown to represent two gene subfamilies. The members of each of these subfamilies are 80 to 90% identical in nucleotide sequence, but when members of the two different families are compared, they are only about 60% identical. The seed storage proteins of *Arabidopsis* are known as 2 S albumins. Four genes encode the 2 S albumin proteins in the *Arabidopsis* genome. At the other extreme, in the maize genome nearly 100 genes encode the seed storage proteins of this plant, which are known as *zeins*. In general, organisms with large genomes tend to have larger multigene families than plants with small genomes. This is not always the case, however. The *Arabidopsis* β-tubulin multigene family has nine members and other angiosperms appear to contain a similar number of β-tubulin genes. It is possible that plants with small genomes have evolved through the elimination of unnecessary DNA. If this is true, then during evolution of the *Arabidopsis* genome, multigene families whose members perform some essential function may have been retained, whereas other multigene families were reduced in number.

## Correlation of Genome Size with Organismic Complexity

In general, complex organisms have larger and more complex genomes than simple, small organisms. For example, many viruses have genomes that contain around $10^4$ to $10^5$ bp of DNA, whereas an average bacterium has about $10^6$ bp of DNA and humans have $2.8 \times 10^9$ bp of nuclear DNA, or about 1000 times more than that of the bacterium. The relationship between DNA content and organism complexity appears to be valid for these simple organisms. This supports the reasonable assumption that, as organisms become more complex, they acquire more nuclear DNA. If, however, we consider the variation in amounts of nuclear DNA for any group of organisms, this relationship quickly breaks down. Many higher plants have as much or more nuclear DNA than humans and the range of variation in DNA content within any group of organisms of similar complexity, such as the angiosperms, varies over several orders of magnitude (see Table 3.2). For example, different species of algae have DNA contents that range from $10^8$ nucleotide pairs, to $10^{11}$ nucleotide pairs, or about 100 times larger than the nuclear DNA content of humans. It is difficult to see why some algae would need 100 times more DNA than many angiosperms, as morphologically, developmentally, and biochemically they are much simpler than the flowering plants. Even more puzzling is why closely related species have vastly different DNA contents. For example, the broad bean, *Vicia faba*, has nearly 100 times more DNA/haploid nucleus than does *Lotus tenuis*. Yet both are members of the same family, the Leguminosae, with many of the same structural characteristics. Both of these species have six chromosomes, although the *Vicia faba* chromosomes are very much larger (Fig. 3.12).

It may be possible that *Vicia faba* needs four times more DNA than humans because it is biochemically more complex, although this seems very unlikely. It is also very difficult to understand why it would have much more DNA than its close relative, *Lotus tenuis*. The fact that some simple organisms have much more nuclear DNA than more complex organisms is a paradox. In fact, it is known as the **c-value paradox.** Why do some organisms have so much DNA while other related organisms get by with so much less? The answer to this question is complex, but in brief there are two main reasons, both of which are developed fully in the following paragraphs. In brief, the first reason is that many plant species are polyploid. Polyploidy is an important evolutionary mechanism in angiosperms, so genome sizes of related species can differ substantially if they have diverged through an increase in ploidy. Second, a fraction of the genomic DNA of nearly all eukaryotes is highly repetitive in its base sequence. Much of this repetitive-sequence DNA has no coding function and the fraction of the genome that is repetitive varies from only a few percent to more than 90%.

## Polyploidy

One reason why some plant species have much more DNA than other closely related species is that they are polyploid while the related species are diploid. Many plant species have become polyploid during evolution, whereas crop plants often have been made polyploid by plant breeders (Table 3.3). The term **ploidy** refers to the number of copies of chromosomes found in each nucleus in a given organism. A haploid cell has a single copy of each chromosome; a diploid cell has two copies of each chromosome. An organism that is diploid will have two sets of chromosomes in all the cells that make up its body, whereas its reproductive cells (sperm and egg) will have a single set of chromosomes. In a polyploid species, the basic chromosome number is increased by some factor beyond the

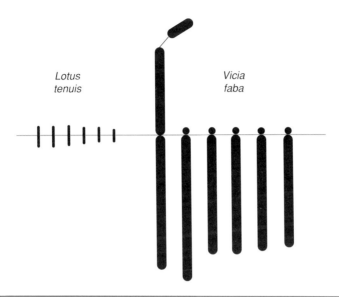

**Figure 3.12** **Comparison of the chromosomes of two members of the same family (Leguminosae),** *Lotus tenuis* **and** *Vicia faba*, **which have the same chromosome number**
Redrawn with permission from Stebbins (1971).

usual diploid level. A tetraploid species would be one in which, for some reason, the body or somatic cells contained four copies of each chromosome and the reproductive cells contained two copies of each chromosome.

The domestic potato has 24 chromosomes in its reproductive cells and 48 chromosomes in dividing, meristematic cells. There is good evidence that this plant is a tetraploid, and that the true haploid chromosome number is 12: wild relatives of the potato, which can be found today growing in Peru, all have a haploid chromosome number of 12. The potato was domesticated by the Inca civilization in this region long before the Spanish conquests. In the process of its domestication, these early plant breeders selected plants that produced larger tubers, which happened to be polyploid. This is not unusual. If we experimentally induce polyploidy, for example, by treating a diploid species with the drug colchicine, the tetraploid offspring usually are larger. Potato is considered to be an **autotetraploid,** and it is an example of an **autopolyploid** species, because the species became polyploid as a result of a doubling of its own chromosomes.

Polyploidy also arises as a result of sexual crosses between two species. For example, the commercial tobacco plant, *Nicotiana tabacum,* arose as a result

## Table 3.3 Polyploidy in flowering plants

| Species[a] | Common name | Constitution | Origin |
|---|---|---|---|
| **Allopolyploids** | | | |
| *Galeopsis terahit* (N) | Hemp nettle | $2n = 4x = 32$ | *G. pubescens* × *G. speciosa* $(2x = 16)$ $(2x = 16)$ |
| *Raphano–Brassica* (C) | | $2n = 4x = 36$ | *R. sativus* × *B. oleracea* $(2x = 18)$ $(2x = 18)$ |
| *Brassica napus* (N) | Rape | $2n = 4x = 38$ | *B. oleracea* × *B. campestris* $(2x = 18)$ $(2x = 20)$ |
| *Triticale* (C) | | $2n = 8x = 56$ | *Triticum aestivum* × *Secale cereale* $(6x = 42)$ $(2x = 14)$ |
| *Triticum aestivum* (C) | Wheat | $2n = 6x = 42$ | *T. monococcum* × *T. speltoides* × *Aegilops squarrosa* $(2x = 14)$ $(2x = 14)$ $(2x = 14)$ |
| *Primula kewensis* (C) | | $2n = 4x = 36$ | *P. floribunda* × *P. verticillata* $(2x = 18)$ $(2x = 18)$ |
| *Nicotiana tobacum* (C) | Tobacco | $2n = 4x = 48$ | *N. sylvestris* × *N. tomentosiformis* $(2x = 24)$ $(2x = 24)$ |

| | | | **Method of propagation** |
|---|---|---|---|
| **Autopolyploids** | | | |
| *Beta vulgaris* (C) | Sugar beet | $2n = 3x = 27$ | Seed, by crossing diploid × tetraploid |
| *Malus pumila* (C) | Apple | $2n = 3x = 51$ | Grafting |
| *Secale cereale* (C) | Rye | $2n = 4x = 28$ | Seed |
| *Allium tuberosum* (N) | | $2n = 4x = 32$ | Bulbs |
| *Ranunculus ficaria* (N) | Celandine | $2n = 2x, 4x, 5x, 6x$ $= 24, 32, 40, 48$ | Bulbils |
| *Solanum tuberosum* (C) | Potato | $2n = 4x = 48$ | Tubers |
| *Hyacinthus orientalis* (C) | Hyacinth | $2n = 3x, 4x = 24, 32$ | Bulbs |
| *Fritillaria camschatensis* (N) | | $2n = 3x = 36$ | Bulbs |

[a] N, natural; C, cultivated.

of a sexual cross between two different tobacco species, *N. tomentosiformis* × *N. sylvestris,* both of which have a haploid chromosome number of 12.

$$N.\ tomentosiformis\ (1N\ =\ 12)\ \times\ N.\ sylvestris\ (1N\ =\ 12)$$
$$F1\ hybrid:\ 2N\ =\ 24$$

This hybrid will contain one set of chromosomes from each parent. It is sterile because these chromosomes are not homologous and cannot form pairs at meiosis. If, however, the chromosomes are first doubled so that the somatic cells contain two sets of chromosomes from each parent, then meiosis will be normal and the F1 hybrid will be fertile. In addition, of course, the new fertile plant will be a tetraploid. This type of polyploidy is called **allopolyploidy** and *N. tabacum* is an example of an **allotetraploid** species. This mechanism is illustrated in Fig. 3.13.

Polyploid species, such as the commercial varieties of tobacco and potato, are very common among crop plants. Other examples include domestic wheat, which is an allohexaploid, and apple and sugar beet, which are both triploids. Polyploid species are known to occur in the wild as well. Some ferns have more than 1000 chromosomes and are almost certainly polyploid species.

Both allopolyploidy and autopolyploidy lead to an increase in genome size without an increase in organismic complexity. Although polyploidy helps explain why some similar species have much higher DNA contents than other species, it cannot explain why two closely related species with the same chromosome number have vastly different DNA contents. In the example cited earlier, *Vicia faba* and *Lotus tenuis* both have six chromosomes but differ by a factor of 100 in their DNA content. Even more puzzling is the case of *V. faba* and *V. sativa,* members of the same genus, each with six chromosomes; yet *V. faba* has six times more DNA than *V. sativa.* The explanation for these differences can be found in the composition of the genomic DNA. Related species with very different amounts of genomic DNA usually differ most in the amount of repetitive-sequence DNA in their genomes.

## Repetitive-Sequence DNA

The nuclear genomes of all eukaryotic organisms are highly complex. It is possible to analyze this complexity by studying the kinetics of DNA renaturation after the genomic DNA is denatured. Double-stranded DNA can be denatured by heating it above its melting point or by treating it with sodium hydroxide. In either case, the hydrogen bonds holding the double helix together are ruptured and the two strands separate. The denatured DNA will renature if the reaction is allowed to cool or the mixture is neutralized (Fig. 3.14). The extent of this renaturation can be determined by passing the solution through a column of the

*Brassica oleracea*
($2n = 2x = 18$)

*Raphanus sativus*
($2n = 2x = 18$)

Sterile F1 hybrid

Chromosome
doubling

Fertile allotetraploid
($2n = 4x = 36$)

**Figure 3.13 Chromosome pairing at meiosis in *Brassica oleracea* and *Raphanus sativus*, their F1 diploid hybrid, and the allotetraploid**

Chromosomes pair with their homologs at metaphase of the first meiotic division; however, the chromosomes of *B. oleracea* and *R. sativus* lack sufficient homology to permit pairing during meiosis in their F1 hybrid. As a result, meiosis fails and the hybrid is sterile. If, however, the chromosomes are doubled, each has a homolog with which it can pair at meiosis and this allotetraploid is fertile. Redrawn with permission from Rees and Jones (1977).

mineral hydroxyapatite. Double-stranded DNA binds to hydroxyapatite, whereas single-stranded DNA does not. As a result, the extent of DNA renaturation can readily be determined at any given time (Fig. 3.15).

The rate and extent of renaturation of DNA strands depend on several

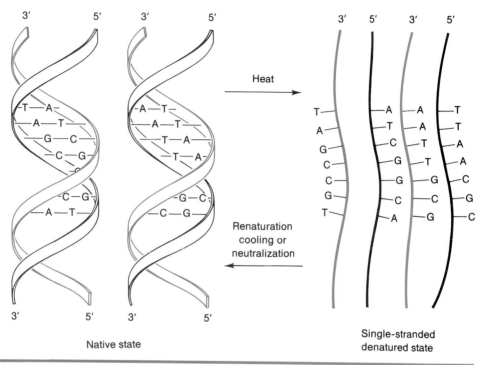

**Figure 3.14 Denaturation and renaturation of DNA**
Heating a solution of DNA above its melting point ($T_m$) ruptures the hydrogen bonds holding the two strands of the double helix together. If the solution is allowed to cool, the strands will come together, re-forming the hydrogen bonds to renature the double helix.

variables, the most important of which are

- the concentration of a given DNA sequence in the solution
- the time allotted for the reaction
- the size of the DNA fragment
- the temperature of the reaction
- the salt concentration

For experimental purposes, many of these factors can be held constant. The DNA can be sheared so that all of the molecules are about the same size. The temperature of renaturation can be held constant at 20 degrees below the melting temperature ($T_m$), and the reaction can be carried out in a standard salt solution. Under these conditions, the extent and rate of the reaction are a function of the concentration of any particular DNA sequence and the time allotted for the reaction. Britten and Kohne suggested that we use the term $C_0t$ to characterize a particular kind of DNA.

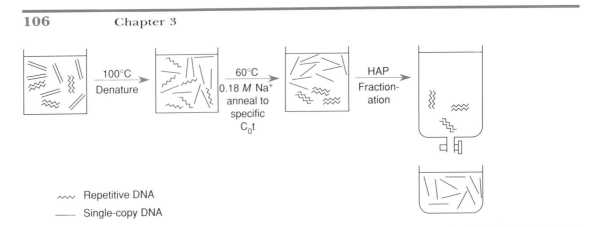

~~~ Repetitive DNA

——— Single-copy DNA

**Figure 3.15 Analysis of the extent of DNA renaturation using hydroxyapatite (HAP) chromatography**
Redrawn with permission from Walbot and Goldberg (1979).

$C_0t$ = concentration of DNA (in moles of nucleotides/liter) × time (in seconds)

The $C_0t$ value for any particular DNA is a measure of its base sequence complexity. If the denatured DNA is a simple repeating structure so that the concentration of any particular base sequence is very high, the molecules will renature very quickly after denaturation, giving a low $C_0t$ value. On the other hand, if the genome consists of one long sequence of DNA that is unique throughout so that no sequence of bases is repeated within the molecule, then it will renature with a much higher $C_0t$ value. The bacteriophage T4 and the bacterium *Escherichia coli* each have genomes that consist of a single DNA molecule with a unique base sequence. If these genomes are sheared into fragments, each consisting of approximately 500 bp, and then denatured, after denaturation, each fragment will have only one strand with which it can renature. As a result, the time required for any given sequence to find its complementary strand and renature is a measure of the amount of DNA with unique sequence in the genome. *E. coli* genomic DNA has a higher $C_0t$ value than T4 because it is a much larger genome.

The $C_0t$ curves for the nuclear DNA of eukaryotes are much more complex than those for simple organisms. As a eukaryote contains much more DNA than a simple prokaryote, such as *E. coli*, you would expect its DNA to reassociate at a much higher $C_0t$ value than that of *E. coli* if its genomes consisted entirely of unique-base-sequence DNA. This is in part true. A fraction of the DNA of most eukaryotes reassociates slowly at a high $C_0t$ value; however, most eukaryotic DNA is more complex and behaves as though it consisted of at least two or frequently three components. A hypothetical analysis of DNA sequence organization in a eukaryote is shown in Figure 3.16.

The genomic DNA of an angiosperm, such as *Nicotiana tabacum* (tobacco), not only contains a fraction that reassociates much more slowly than those of *E. coli*, as you would expect, but also contains a fraction that reassociates much more

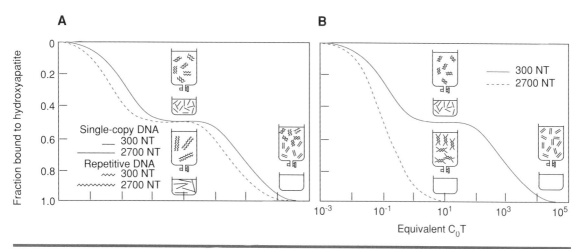

**Figure 3.16 Hypothetical experiment to evaluate DNA sequence organization**

Analyze the DNA sequence organization of a hypothetical haploid plant with a genome containing 1 pg of DNA in which 50% of the genome is single-copy DNA and the remainder is repetitive-sequence DNA. The repetitive DNA consists of a single sequence and there are 10,000 copies of this sequence in the genome; however, it is unknown how these sequences are arranged within the genome. It is possible that these repetitive sequences are interspersed among the single-copy sequences (short-period interspersion). Alternatively, the repetitive DNA sequences may be clustered in large blocks on chromosomes, such as at the centromeres and telomeres of chromosomes, while other large stretches of the genome consist of many single-copy sequences (long-period interspersion or no interspersion). To distinguish between these two possibilities experimentally, the DNA is sheared into fragments that are either 2700 nucleotide pairs (NT) or 300 NT long. In either case, the DNA is denatured and then allowed to renature. The rate of reassociation is determined by periodically analyzing the solutions for the percentage of the DNA that remains single stranded by means of hydroxyapatite (HAP) chromatography. Only single-stranded DNA is retained by the HAP under these conditions. If the genome was organized with a long-period interspersion pattern or into large blocks in which the repetitive DNA and single-copy DNA were separated by at least 50,000 bp, the reassociation kinetics would be similar to those shown in (A). Alternatively, if the genome had a short period interspersion pattern, the reassociation kinetics would be as is shown in (B). Redrawn with permission from Walbot and Goldberg (1979).

rapidly. The fact that much of the *N. tabacum* DNA reassociated rapidly indicates that this fraction consisted of DNA with repeated base sequences. This means that certain base sequences are repeated many times in the genome. Approximately 60% of the tobacco genomic DNA consists of sequences that are repeated, and a portion of this repeat-sequence DNA is present in a very large number of copies. Somewhat under 10% of the DNA consists of a simple DNA sequence that is repeated more than 12,000 times, and another 60% of the DNA consists of sequences that are repeated about 250 times in the genome. The remaining 30% of the genome consists of unique-sequence DNA.

From our previous discussions, you will remember that *N. tabacum* is an allotetraploid species which actually has two genomes, one genome derived from *N. sylvestris* and one genome derived from *N. tomentosiformis*. Analysis of the $C_0t$ curves also allows us to determine the true size of the genome, as well as the fractions of the genome consisting of unique-sequence DNA and repetitive-sequence DNA.

### Fraction of DNA Containing Functional Genes

It is possible to isolate the messenger RNAs from polyribosomes where they are being translated into protein. Furthermore, one can make a single-stranded DNA copy of this RNA with a viral enzyme known as reverse transcriptase. Reverse transcriptase uses an RNA template to construct a single-stranded DNA molecule with a base sequence complementary to the RNA. These single-stranded DNAs are known as complementary DNAs (cDNAs). After the other strand of the cDNAs is synthesized, they can be cloned, and specific cDNAs representing a particular gene can be identified, subcloned, sequenced, and otherwise character-ized. These cDNAs are exceedingly powerful tools for the analysis of genome organization and gene expression. Any particular cDNA represents an active gene that was expressed in the tissue from which the mRNA was extracted. If the cDNA is synthesized with $^{32}$P-labeled nucleotides, it will be radioactive and hence readily detected. The radioactive cDNA can form a hybrid with DNA of the gene from which it was initially transcribed, and these hybrids can be detected by autoradiography or a variety of other techniques.

By the use of these labeled cDNA probes, it can be shown that most tran-scribed genes are found in the unique-sequence DNA. This is demonstrated by hybridizing cDNAs prepared from tobacco leaf polysomal messenger RNA with denatured tobacco genomic DNA representing the different $C_0t$ fractions. Hy-brids are formed almost exclusively between the cDNAs and the DNA of the unique-sequence fraction. That is, the unique-sequence fraction of the DNA contained most of the genes expressed in tobacco leaves. The tobacco genome contains $1.65 \times 10^9$ bp of DNA ($1c$ DNA content), of which approximately 40% is unique-sequence DNA. A total of $3.33 \times 10^7$ bp of nuclear DNA are transcribed into mRNA in tobacco leaves. This represents approximately 27,000 average-sized genes, but it accounts for only about 5% of the unique-sequence DNA or approximately 2% of the total tobacco genome. What about the remaining DNA? What purpose does it serve? What function does it perform?

### Repetitive DNA

Repetitive DNA consists of a number of families of similar, but not identical, base sequences. That is, the individual members of each group of sequences would resemble each other as do the members of human family, but they would all have some unique characteristics. That is, they would not all have exactly the same base sequence. How many families of repetitive DNA are there? This value varies greatly with the species. On the basis of DNA reassociation kinetics it has been estimated that there are approximately 1000 different repetitive DNA families in the genome of the cotton plant, 4000 in the tobacco genome, and 40,000 in the soybean genome. Each of these families contains 20 to 300 members. The size of the repetitive sequence averages 200 to 400 bp, but there is also a significant fraction of the moderately repetitive-sequence DNA that is much larger, and consists of a 1000 bp or more. A significant fraction of these moderately repetitive DNA sequences have no coding function and are not transcribed into RNA; however, some of these families of repetitive-sequence DNA have very significant effects on the structure and integrity of the genetic material. This is particularly true of transposons, retrotransposons, and viruses, all of which are present in multiple copies in the genome. In addition, some expressed genes, such as those encoding the histone proteins, are present in a large number of copies in the genome so that they would behave as repetitive-sequence DNA during $C_0t$ analysis of genomic DNA. Most importantly, the genes for ribosomal RNA are highly reiterated in most eukaryotes. There may be as many as 20,000 copies of the ribosomal RNA genes in some plants and most plants have at least a few thousand copies of ribosomal RNA genes.

## Location of Repetitive DNA in Plant Genomes

There are two main classes of repetitive-sequence DNA in terms of their distribution within the genome: tandemly repeated sequences and interspersed sequences. Interspersed repetitive-sequence DNA is scattered throughout the genome and each copy of this type of repeat sequence may be many thousands of bases away from any related sequence. In contrast, tandem linked repeats are physically closely associated and many hundreds or even thousands of copies of a particular repeat may be immediately adjacent to each other, or each repeat may be separated by a short linker sequence. Tandemly linked repeat-sequence DNA behaves as satellite DNA when genomic DNA is analyzed by ultracentrifugation, if it is present in a minimum number of copies. When a solution of cesium salts is subjected to high centrifugal forces during ultracentrifugation, a linear gradient of cesium concentration is formed. If the solution contains DNA, the DNA will seek its buoyant density within the gradient, forming a band. DNA molecules rich in guanine and cytosine (G + C) have a greater density than DNA molecules rich in adenine and thymine (A + T). Most higher-plant DNAs have an average G + C content of 34 to 46% and a buoyant density of about 1.69. Analysis of higher-plant nuclear genomes by ultracentrifugation typically shows that, although the bulk of the DNA migrates as a single band, there is often one or more smaller bands that are either less or more dense than the bulk of the DNA. These have been given the name **satellite DNAs.** If they have a higher G + C content than the bulk of the DNA, they will be more dense than the main-band DNA. If they are A + T-rich, they are less dense than the main-band

DNA. Satellite DNAs can be isolated and separated from the bulk of the nuclear DNA by the technique of preparative ultracentrifugation. The reassociation kinetics of denatured satellite DNAs demonstrates that they are highly repetitive, and direct sequencing has established the nature of the repeated sequence in many cases. One of the most extensively studied satellite DNAs is an A + T-rich satellite DNA in the mouse genome that makes up about 10% of the total nuclear DNA of this organism. The basic repeating unit of this satellite DNA is 20 bp long and has the following sequence:

```
A A A T T G C A T A G A T T T T G A A T
T T T A A C G T A T C T A A A A C T T A
```

The mouse genome contains approximately $15 \times 10^6$ copies of this sequence. They are, however, not exact copies. Point mutations have occurred in many of the copies so that they are slightly different from each other.

Many plant genomes have satellite DNAs and most of these bear little relationship to each other. Even related species may have vastly different amounts of a particular satellite DNA, and the sequences of their satellite DNAs may be very different. Melon satellite DNA, which is the best-studied plant satellite, consists of a basic 400-bp sequence that is repeated about $10^6$ times in the melon genome. Satellite DNAs have also been characterized from several species in the monocot genus *Scilla* (bluebells). The satellite represents 19% of the genome of *S. siberica,* whereas it is 10.5% in *S. ingridae* and 6.5% in *S. amoena.* In all three species the satellites are 78% G + C and the repeating unit is 34 bp long, but the repeats frequently are interrupted by the insertion of variable numbers of the tetranucleotide GTCC (Fig. 3.17).

### Chromosomal Localization of Satellite DNAs

As satellite DNAs are tandemly linked, obviously they must be clustered in specific regions of chromosomes. To identify these regions, a procedure known as *in situ* hybridization has been used. Satellite DNA is not transcribed in living cells. In the laboratory, however, one can synthesize RNA transcripts that have a nucleotide sequence complementary to a given isolated satellite DNA with a bacterial RNA polymerase. If the complementary RNA (cRNA) is made with radioactive RNA nucleotides, the resultant radioactive probe can be used to determine the cytological localization of the satellite DNA from which it was transcribed. Cells of the organism from which the satellite was isolated are squashed onto a glass slide to spread the chromosomes out so that they may be observed. The DNA is denatured with a dilute alkali solution. Then a solution containing the radioactive cRNA, complementary to satellite DNA, is placed over the squashed cells and the cRNA is allowed to hybridize with the chromosomal DNA. After the cRNA has hybrid-

I    5' - CCCÅTGCACC GAACCGCCCG CGGCTCGTCC GTGGG

II   5' - CCCACGCACC GAACCGCCTG CGCGGTTCGT CCGTCCGTCC GGCCCGTGGG

III  5' - CCCACGCACC GAACCGCCCG CGGCCCGTCC GTCCGTCCGT CCGTCCGTCC GCCCGTGGG

IV   5' - CCCATGCACC GAACCGCCCG CGGTTCGTCC GTGGCCATGC ACCGAACCGC CCGCGGCTCG TCCGTGGG

PS   5' - CCCA<sup>C</sup>GCACC GAACCGCCCG CGGCTCGTCC GTGGG

**Figure 3.17 Nucleotide sequences of some restriction fragments of the tandemly repeated *Scilla* satellite DNA**

Sequences of five *Hae*III fragments are shown (I–IV), along with the prototype sequence (PS), from which the repeats are thought to be derived. The tetranucleotide GTCC (underlined) was inserted in many of the repeats. The arrow indicates the axis of symmetry of repeat IV. From Deumling (1981).

ized with the satellite DNA in the chromosomes, its localization can be revealed by autoradiography. Most often the *in situ* hybridization technique demonstrates that the satellite is localized near the centromeres of chromosomes or at their ends (known as telomeres). In *S. siberica* the satellite is associated with the telomeres of some chromosomes and the centromeres of others; other chromosomes of this species contain little or no satellite DNA (Fig. 3.18A). The closely related species *S. ingridae* has a different pattern of distribution of the satellite, with most of this repetitive DNA localized in bands along the chromosome arms and not near the telomeres or centromeres (Fig. 3.18B). The function of this repetitive, tandemly repeated DNA is not known, if indeed it has a function. It has been suggested that satellite DNA represents "selfish" DNA whose only function is to ensure its own replication.

**Heterochromatin is a portion of the nuclear DNA that remains condensed in interphase** The chromatin fibers condense during mitosis during the formation of the chromosomes. Following mitosis, most of the nuclear chromatin decondenses as the cells resume interphase and DNA transcription; however, a fraction of the nuclear DNA condensed during mitosis remains condensed in interphase. This DNA is called **heterochromatin.** It is called heterochromatin because it stains more intensely than the rest of the interphase nucleus when cytological preparations are treated with dyes that bind to DNA. Heterochromatic regions stand out as intensely stained portions of the nucleus on a background that is more lightly stained. Heterochromatin has a number of unique properties. First, it is replicated later than the rest of the nuclear DNA during the DNA replication cycle. In addition, no genes can be mapped to regions of chromosomes that are heterochromatic. *In situ* hybridization with labeled satellite DNA often shows that the chromosomal regions in which the satellite is localized are hetero-

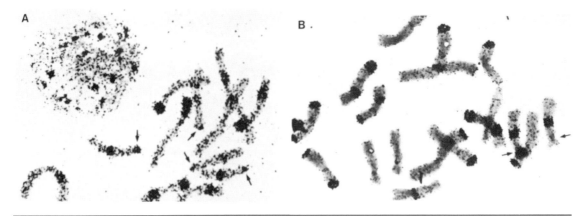

**Figure 3.18 Use of *in situ* hybridization to localize *Scilla* satellite DNA in the interphase nuclei and chromosomes of different *Scilla* species**

An RNA copy of satellite DNA isolated from *Scilla siberica* or *Scilla mischtschenkoana* was synthesized *in vitro* with an RNA polymerase and ³H-labeled nucleotides. The respective radioactively labeled satellite cRNA was hybridized to cytological preparations of root tip cells of *S. siberica* (A) and *S. mischtschenkoana* (B). In both cases there is heavy labeling of the telomeres and other regions representing heterochromatin. The arrows indicate the location of the nucleolar organizer region containing the genes encoding the large ribosomal RNAs, but also containing some constitutive heterochromatin. An interphase nucleus also is present in the lower left of (B) in which the labeled probe has hybridized to regions representing heterochromatin. Reprinted with permission from Deumling and Greilhuber (1982).

chromatic in interphase nuclei (Fig. 3.18A). This is consistent with the fact that most satellite DNAs whose sequence we know do not contain genes. The base sequence of these satellite DNAs is such that, if an RNA was transcribed from it, it would contain a very high frequency of stop codons. This would prevent it from being translated into a protein, should it ever be transcribed.

**Repetitive-sequence DNA could be involved in speciation**     Repetitive-sequence DNA may be involved in the evolutionary process by which species become differentiated from each other. This is suggested by the observation that closely related species that differ in DNA content often have widely different amounts of repetitive-sequence DNA, as we have seen in the case of the three species of *Scilla*. Nagl has shown that where two closely related species have widely different amounts of DNA, the species with the larger amount of DNA has proportionally more heterochromatin and thus more repetitive-sequence DNA. Two *Allium* (onion) species, *A. cepa* and *A. fistulosum,* have the same chromosome number ($2n = 16$) but very different DNA contents.

- *A. cepa:* 34 pg = 2*n*
- *A. fistulosum:* 26 pg of DNA = 2*n*

*A. cepa* has more repetitive-sequence DNA than *A. fistulosum*, most of which consists of tandemly repeated sequences. Although these are separate species, they can be sexually crossed and the hybrid plants resulting from the cross are fertile. As a result, it is possible to examine the meiotic chromosomes during the stage of meiosis when the homologs have paired. Because the interspecific hybrid will have one set of chromosomes from each parental species, each pair will contain one *A. cepa* and one *A. fistulosum* chromosome. *A. cepa* chromosomes are longer than *A. fistulosum* chromosomes, so the regions of homologous DNA pair, while the divergent, heterochromatic DNA regions of the *A. cepa* chromosomes loop out from the synaptonemal complex.

# Transposons

**Transposons** are genetic elements that can change position, or transpose, within the genome. Most genomes contain several kinds of transposable elements. Together, these transposable elements can account for up to 10% of the total genomic DNA. Transposable elements range in size from a few hundred to thousands of base pairs. Usually they are present in many copies per cell and these copies are interspersed throughout the genome. As a result, they represent one type of repetitive-sequence DNA. The number of copies varies greatly from only a few per genome to many hundreds, depending on the element. Transposons typically have the structure shown in Fig. 3.19. They can be recognized by relatively short DNA sequences at either end of the element that are the same, but are in an inverted order. Transposition is brought about by an enzyme known as a **transposase** which in many cases is encoded by the transposon. Transposons that encode a transposase that is capable of initiating its own movement are said to be **autonomous.** The maize transposon *Ac* is autonomous, whereas *Ds* is not; however, the transposase encoded by *Ac* can cause a *Ds* element to move, as well as bring about its own movement. As transposons do not move very often, their presence in a given genomic region can be inferred by the inverted repeats at either end of the element. Transposition is initiated when the gene encoding the transposase is expressed and its mRNA is translated in the cytoplasm. The transposase then enters the nucleus and binds to the inverted repeat sequences at either end of the transposon (see Fig. 3.19). There are two types of transposition events. In one case, the transposon moves without replicating. The transposase simply breaks the DNA strand at the ends of the repeats and then inserts the transposon at another location within the genomic DNA (Fig. 3.20). In other cases, the transposon moves by replicating itself at a new site within the genome (Fig. 3.21).

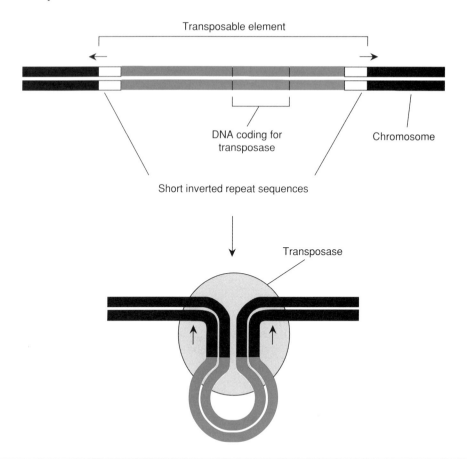

**Figure 3.19 Structure of a transposable element**
Transposable elements typically are several thousand base pairs long and have short inverted repeat sequences at either end. The position and orientation of these inverted repeat sequences are shown by arrows. Transposable elements often contain a gene that encodes an enzyme known as a transposase that is responsible for their transposition. The transposase recognizes and binds to the inverted repeat sequence. Even transposons that lack a transposase gene contain the inverted repeats that may be recognized by a transposase encoded by another transposon. Redrawn with permission from Alberts *et al.* (1989).

## Transposons and Mutations

Transposons cause mutations when they enter a gene and also when they transpose from the gene. The insertion of a transposon into a gene can prevent its expression, although this does not invariably occur. Whether it causes a mutation depends on where within the gene it inserts and what happens when it transposes from the gene. Insertion within the region of the gene encoding the protein

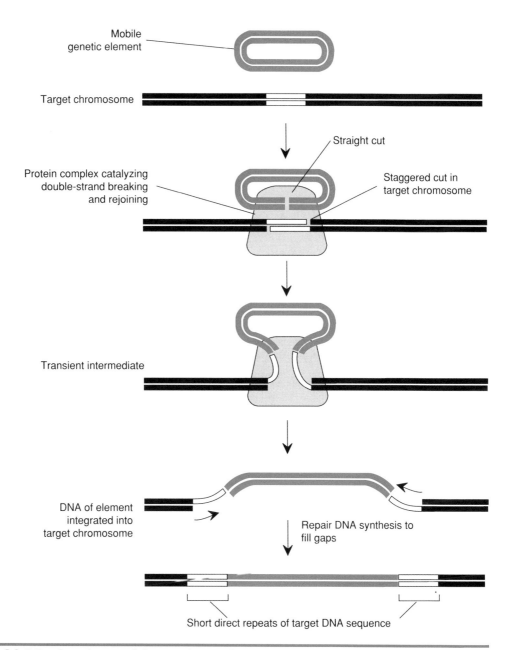

Mobile
genetic element

Target chromosome

Straight cut

Protein complex catalyzing
double-strand breaking
and rejoining

Staggered cut in
target chromosome

Transient intermediate

DNA of element
integrated into
target chromosome

Repair DNA synthesis to
fill gaps

Short direct repeats of target DNA sequence

**Figure 3.20 Mechanism of insertion of a mobile genetic element
such as a transposon into chromosomal DNA,
without duplication**

The transposase, which has excised the transposon from its original site by making
straight cuts at the ends of the inverted repeat sequences, remains bound to the
element. The transposase then binds to DNA in a new chromosomal location and
makes a staggered cut in the target DNA. The cut ends of the transposon are then
joined to the single-stranded ends of the staggered cut in the target DNA. DNA
synthesis then fills in the gaps, creating target site duplications. Redrawn with
permission from Alberts *et al.* (1989).

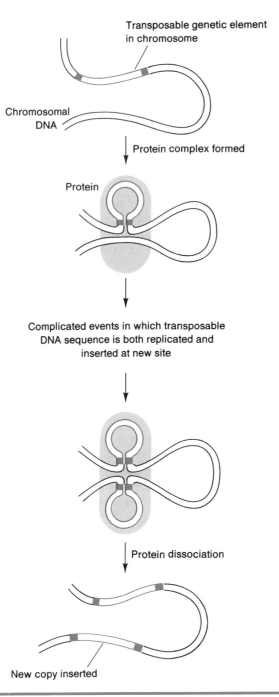

Transposable genetic element
in chromosome

Chromosomal
DNA

↓ Protein complex formed

Protein

Complicated events in which transposable
DNA sequence is both replicated and
inserted at new site

↓

↓ Protein dissociation

New copy inserted

**Figure 3.21 Transposon movement by replication and insertion of the replicate at a new location**

The inverted repeats at either end of the transposon are indicated by the green squares. The transposase binds to the inverted repeats and cuts the DNA at their ends. The transposon then is replicated and one copy is inserted back into the chromosome at the original site while the other copy is inserted at a new location. Redrawn with permission from Alberts *et al.* (1989).

usually blocks the normal expression of the gene, whereas insertion in the regulatory regions of the gene may change the tissues or organs in which the gene is expressed, or may affect the amount of expression (Fig. 3.22). Also, the mechanism of transposon insertion results in the duplication of a short sequence of the gene at either end of the inverted repeat. These duplicated sequences may remain after the element transposes out of the gene (Fig. 3.23). If the element was in a coding portion of the gene, the presence of these duplicated sequences can disrupt the reading frame or insert additional codons into the gene, causing mutations. Insertion of the transposon into a gene, however, often prevents it from being expressed, whereas transposition from the gene may restore the ability of the gene to be expressed.

## Transposons of Maize

The maize genome contains at least six families of transposons. We know the most about the *Ac–Ds* (Activator–Dissociation) and *Spm* (Suppressor–Mutator) families. Both of these families of transposons contain a unique autonomous (transposition-competent) element and a large number of related nonautonomous (transposition-incompetent) elements that are capable of movement only through the action of the transposase encoded by the related competent element. In the *Ac–Ds* transposon family, *Ac* encodes a transposase, whereas the *Ds* elements can move only in the presence of an active *Ac* element. *Ds* originally was identified and characterized by McClintock, who showed that mutations caused by *Ds* were stable if the genome lacked *Ac*. *Ac* is 4.6 kb long and is flanked by an 11-bp inverted repeat (Fig. 3.24A). It contains a single gene that encodes the transposase. Several *Ds* elements have been cloned and sequenced (Fig. 3.24B). Comparison of the sequences of these *Ds* elements with that of *Ac* demonstrates that the *Ds* elements were derived from *Ac* by some unknown mechanism which has preserved the framework of *Ac*, but not its functional machinery. All sequenced *Ds* elements contain regions that are very similar to *Ac*, including the *Ac* 11-bp inverted repeat, but frequently all or part of the region of the element encoding the transposase has been deleted (see Fig. 3.24B). Some *Ds* elements contain rearrangements and/or duplications of the internal *Ac* sequence. The *Ds* element known as *double Ds* contains a second *Ds* element inserted within it. *Double Ds* breaks chromosomes in the presence of an active *Ac*, and it was this property that led McClintock to call this group of elements *Dissociation*. *Ac* and most *Ds* elements are nonreplicative during transposition. They simply leave one site and enter another; however, the *double Ds* element is replicative during its transposition in the presence of an active *Ac*.

The *Spm* family of transposable elements consists of *Spm*, which is autonomous and is present in only one or a few copies per genome, as well as a large number of nonautonomous elements, which are designated *dSpm*. *Spm* is an 8.3-kb element flanked by a perfect 13-bp inverted repeat (Fig. 3.25). The *Spm* transposase gene contains 11 exons and 10 introns. There are two open reading

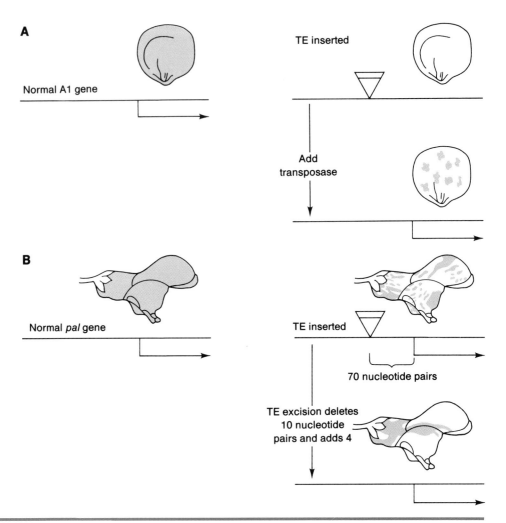

**Figure 3.22 Mutation of genes after entry or exit of transposons**
(A) The A1 gene encodes an enzyme that is necessary for anthocyanin synthesis in the aleurone layer of the developing maize seed. A1 gene expression during seed development leads to the synthesis of anthocyanins in the aleurone layer, resulting in purple kernels. Insertion of a *Ds* transposable element (TE) into the A1 gene blocks its expression, so anthocyanins are not synthesized and the kernels are yellow. In the presence of an active transposon such as *Ac,* transposase causes the element to move from the A1 gene at variable times in aleurone development, restoring the activity of the A1 gene and causing anthocyanin synthesis in patches of cells. (B) *pal* gene expression is necessary for flower color in snapdragon, as the PAL enzyme is required for anthocyanin synthesis. Insertion of the transposon *Tam3* into the regulatory region of the *pal* gene blocks its expression in the flower so the flower is white, except in patches where the element has transposed out of the gene during petal development. In one case the element transposed from the gene early in embryogenesis so none of the cells in the plant had a *Tam3* transposon in their *pal* gene. When the element transposed from *pal,* however, it deleted ten bases from the gene and left four from the ends of the element, altering the pattern of expression of the gene in the flowers. Redrawn with permission from Alberts *et al.* (1989).

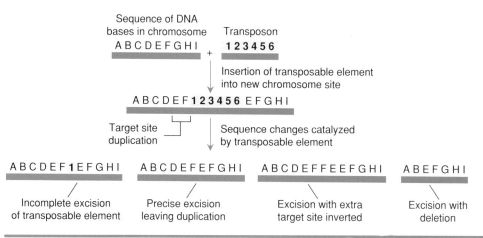

**Figure 3.23 Transposons induce mutations in the target gene DNA sequence**
The insertion of a transposon into any DNA sequence causes the duplication of a short region of the target DNA ranging from 3 to 12 bp. Transposition of the element out of the target DNA sequence may not restore the original sequence and a variety of target DNA sequence mutations can occur. Redrawn with permission from Alberts *et al.* (1989).

frames within the first intron. An **open reading frame** is a DNA sequence consisting of series of codons specifying amino acids, without interruption by stop codons. Potentially an open reading frame is a gene encoding a protein, but many times we do not know enough about it to say either that it is transcribed or what the function of its product might be. Thus, either or both of the open reading frames within the first intron of the *Spm* transposase gene may themselves be genes with unknown functions (see Fig. 3.25). All of the *dSpm* genes that have been cloned are deletions of *Spm* (see Fig. 3.25). One group of *Spm* deletions, known as *Spm-w,* have weak autonomous transposition. That is, they are able to move, but with a frequency much lower than that of *Spm* itself. These *Spm-w* elements have lost one or both of the open reading frames from the first intron of the transposase gene, suggesting that these may be genes whose products assist transposition in some unknown manner.

## Viruses

There are some similarities between transposons and certain viruses. Both are genetic elements that can move about and both contain information that facilitates their replication within a host cell. Viruses differ from transposons chiefly in that at some stage in their life cycle, viruses usually are enclosed in a protein coat known as the **capsid.** Viral genomes can consist of either RNA or DNA. By far the largest number of viruses affecting plants have single-stranded RNA genomes,

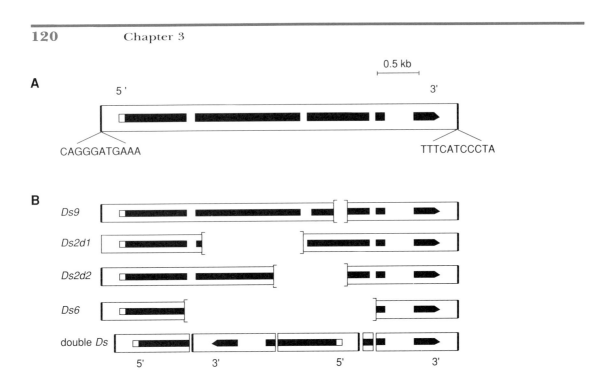

**Figure 3.24 Structures of the maize transposon *Ac* and several of its *Ds* derivatives**
(A) The sequence of the *Ac* element is represented as an open box with the inverted repeat at either end indicated as a bar. The sequences of the inverted repeats are shown expanded below the bar. The smaller rectangles within the larger box represent the exons of the gene encoding the *Ac* transposase. (B) Structures of several *Ds* elements that have been cloned and sequenced are diagrammed by the convention used for *Ac*. The gaps shown in these elements present the relative sizes of the internal deletions found in each of these. The double *Ds* elements contains a copy of the *Ds6* element inserted within it in an inverted order. Redrawn with permission from Federoff (1989).

and in some cases they have multipart genomes that are composed of more than one strand of RNA. In most cases, these genomes consist of single-stranded RNA with the same nucleotide sequence as the viral mRNA (plus-strand RNA viral genomes). Viruses with plus-strand RNA genomes may first be transcribed into minus-strand subgenomic RNA molecules which have a nucleotide sequence complementary to mRNA. These subgenomic molecules are then transcribed into the mRNAs, which are translated into viral proteins. **Tobacco mosaic virus** (TMV) is an example of a common, well-studied plant virus with a genome consisting of a single molecule of plus-strand RNA. **Reoviruses** have double-stranded RNA genomes, and the plant reovirus known as **wound tumor virus** consists of several molecules of RNA. **Cauliflower mosaic virus** has a double-stranded DNA genome, whereas the **geminiviruses** have single-stranded circular

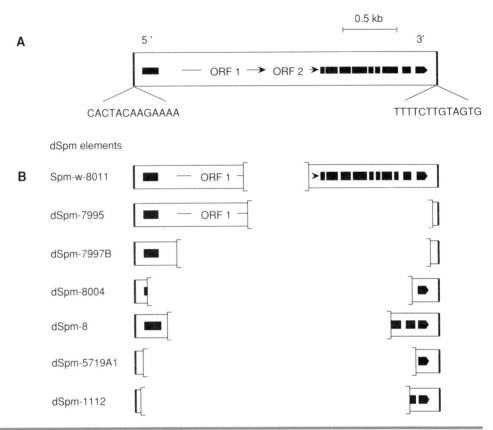

**Figure 3.25** **Structures of the maize transposon *Spm* and several of its nonautonomous derivatives**

(A) The transposon *Spm* is represented as an open box. The terminal inverted repeats at either end of the element are indicated as solid bars. The sequence of the inverted repeat is shown expanded below the bar. The smaller rectangles within the larger box represent the exons of the gene encoding the *Spm* transposase. The first and second exons are interrupted by a large intron that contains two open reading frames (ORF1 and OFR2), potentially active genes. (B) Structures of several of the elements derived from *Spm* are diagrammed by the convention used for *Spm*. The gaps in these open bars present the relative sizes and positions of the internal deletions found in each of the elements. In most cases, the deletions have eliminated a major portion of the coding sequence for the transposase gene. These elements, labeled *dSpm*, are completely nonautonomous. The element *Spm-w* has lost only a portion of the first intron of the transposase gene and retains some autonomous activity. Redrawn with permission from Federoff (1989).

DNA genomes. Viral genomes encode a **replicase,** which is used to replicate their genomes inside the host cell. Their genomes also contain genes for the coat protein(s).

The relationship between viruses and transposons may go beyond analogy. Some transposons appear to be derived from viruses, or the viruses were derived from transposons. Certainly the relationship between transposons and viruses is very direct in the case of a class of transposons known as **retrotransposons** and the **retroviruses.** To understand better how retrotransposons function and how these mobile genetic elements might have arisen, we consider the structure and replication of retroviruses.

## Retroviruses

Retroviruses have single-stranded plus-sense RNA genomes, similar to the largest group of plant viruses; however, the known retroviruses infect only animals. They differ from other viruses with plus-sense RNA genomes in that they replicate via a DNA intermediate. The life cycle of retroviruses includes a stage in which they insert a DNA copy of their genome into the host's genome (Fig. 3.26). Infection with some kinds of retroviruses, such as herpes simplex, is fairly benign. The host cells survive retroviral infection, although they continue to produce virus particles. Other retroviruses, however, cause some of the most devastating human diseases, including acquired immune deficiency syndrome (AIDS), which is caused by human immunodeficiency virus; hepatitis B; and several types of tumors. The retrovirus' genome is encased in a protein coat, which in turn is surrounded by a membranous envelope. The protein coat, or capsid, is composed of several different proteins. In addition to its genome, a retrovirus contains several copies of the enzymes reverse transcriptase and integrase. On entering a cell the virally encoded reverse transcriptase makes a single-stranded complementary DNA copy of its RNA genome. The first DNA strand remains associated with the RNA template to form an RNA–DNA heteroduplex. The reverse transcriptase then uses the RNA–DNA hybrid molecule as a template to synthesize the second strand of the DNA double helix. The viral integrase then integrates this double-stranded DNA copy of the viral genome into the host cell genomic DNA, where it is known as the **proviral DNA.** The proviral DNA has a characteristic structure that includes **long terminal repeats** (LTRs) at either end of the element and three genes, *gag, pol,* and *env* (Fig. 3.27). The sequences of the LTRs at either end of the element are exactly the same, but the left and right repeats have different functions. The left LTR functions as the signal for the transcription of the proviral DNA; the right LTR contains signals for the processing of the mRNA transcript. Both LTRs also contain signals for proviral DNA integration into the host genomic DNA. The *gag* gene encodes four proteins that make up the capsid, the *pol* gene encodes both the reverse transcriptase and the integrase, and the *env* gene encodes a glycoprotein that becomes part of the envelope (see Fig. 3.27). The proviral DNA is replicated along with the rest of

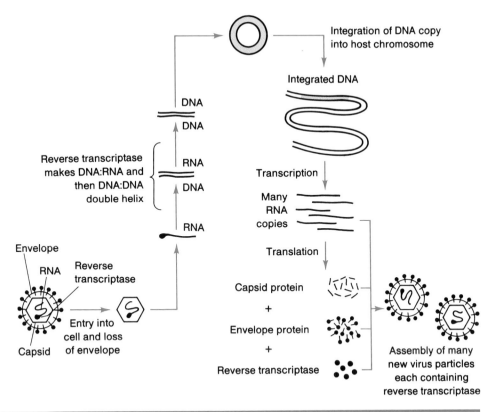

**Figure 3.26 Life cycle of a retrovirus**
Retroviruses have single-stranded RNA genomes and the virus particles are surrounded by an envelope and a proteinaceous capsid. Retroviruses also contain the enzyme reverse transcriptase. When a retrovirus enters a cell, its RNA genome is transcribed into DNA by the reverse transcriptase and these DNA copies are integrated into the host genomic DNA. The viral DNA integrated into the host cell chromosome is known as proviral DNA. Transcription of the proviral DNA results in synthesis of both the viral RNA genome and the mRNAs used to synthesize viral proteins, including the reverse transcriptase. Redrawn, with permission, from Alberts *et al.* (1989).

the nuclear DNA by the host cell's DNA synthetic machinery when the cell goes through its cell division cycle. Host cell RNA polymerase also transcribes the proviral DNA into RNA, which is translated by the host cell's protein synthetic machinery into a number of viral proteins, including those of the capsid, integrase, and reverse transcriptase. These proteins and the viral RNA self-assemble to form more virus particles which are then secreted from the host cell. As they leave the host cell, they become surrounded by some of the host's plasma membrane, which remains on the virus particle surface and helps it gain entry into other host cells (see Fig. 3.26).

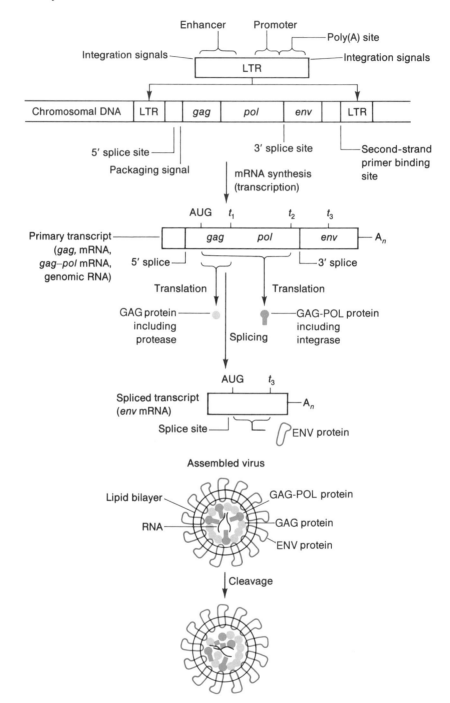

**Figure 3.28 Wound tumor virus genome**

The wound tumor virus genome consists of 12 molecules of double-stranded RNA. Each segment has the terminal 6- or 4-bp sequence shown at the top. This conserved sequence is followed by a 6- to 14-bp inverted repeat (IR) which flanks a coding region that ranges from 534 to 2412 nucleotides (NT) long. Redrawn with permission from Nuss and Dall (1990).

### Vertebrate Tumor Retroviruses

Some vertebrate retroviruses cause tumors because, in the course of inserting into the genome, they pick up normal cellular genes that regulate cell growth. These may become **oncogenes,** or cancer-causing genes, if they are modified so that their activity can no longer be controlled by the infected cell. The provirus of the oncogenic retroviruses contains oncogenes. The expression of the oncogene stimulates the host cell to grow and divide, initiating the formation of a tumor. One group of plant viruses causes tumors, namely, the **reoviruses;** however, plant reoviruses contain double-stranded RNA genomes, rather than the single-stranded RNA genomes of the retroviruses. Furthermore, the reovirus genomes do not encode a reverse transcriptase, nor are they incorporated as proviral DNA copies into the host plant genome. The mechanism by which reoviruses cause tumors is not known. Wound tumor virus is a reovirus whose genome consists of 12 molecules of double-stranded RNA ranging in size from 851 to 2565 bp. Each genomic segment is flanked by inverted repeats (Fig. 3.28), and some of the segments encode the structural proteins of the virus, including the capsid, but none of the proteins encoded by any of the segments is similar in amino acid sequence to reverse transcriptase and they do not reproduce via a DNA

**Figure 3.27 Structure of the proviral DNA of a retrovirus integrated into a chromosome of a vertebrate and its transcription**

The long terminal repeats (LTRs) serve both as integration signals and as signals for proviral transcription. The RNA transcribed from the proviral DNA has several functions. It serves as the mRNA for the synthesis of the reverse transcriptase and integrase proteins from the *pol* gene, the envelope protein from the *env* gene, and the coat protein from the *gag* gene. The mRNA also is the viral genome. Only about 50% of the mRNA is used as the genome for the new virus particles. The remainder is spliced into smaller fragments which are used as mRNAs that will direct the synthesis of viral proteins. These components self-assemble to form the virus particles, after which the gag and gag-pol proteins are modified by a viral protease. Redrawn with permission from Alberts *et al.* (1989).

intermediate. Thus, although reoviruses produce symptoms in plants that are superficially similar to those of vertebrate tumor viruses (they cause plant tumors), there is no direct relationship between them.

### The Reverse Transcriptase Gene in Cauliflower Mosaic Virus

There do not appear to be retroviruses that cause plant diseases, but there are some striking similarities between cauliflower mosaic virus (CaMV) and the retroviruses. CaMV causes lesions on the leaves of cauliflower and related plant species. It does not cause tumors. It does, however, reproduce via an RNA intermediate which is transcribed into DNA by a virus-encoded reverse transcriptase. The genome of CaMV is an approximately 8000-bp double-stranded circular DNA molecule, with two discontinuities. It loses its protein coat after entering the cytoplasm of infected cells and the viral DNA enters the nucleus. On entering the nucleus, the gaps in the two strands of the DNA are repaired so that the genome becomes a complete double-helical molecule, and histones associate with it so that it has a chromatin-like structure. It is transcribed into two RNA molecules by the host cell RNA polymerase. The smaller one is designated 19 S RNA and the larger transcript, known as 35 S RNA, is a complementary copy of one strand of the entire genomic DNA. The 19 S RNA and some of the 35 S RNA are translated in the cytoplasm into viral proteins, one of which is a reverse transcriptase. A portion of the 35 S RNA is then transcribed by the viral reverse transcriptase to produce the genomic DNA (Fig. 3.29). There are striking similarities in the organization of the *gag* and *pol* genes in CaMV and the retroviruses, as well as in the amino acid sequences of their reverse transcriptases, suggesting that CaMV may be related to the retroviruses.

### Retrotransposons

Retrotransposons, often simply called *retroposons*, are another class of mobile genetic elements residing in the genomes of all classes of organisms. They differ from the transposons discussed earlier in that they transpose via an RNA intermediate. Retrotransposons are transposons that behave much like retroviruses and have a retrovirus-like structure. They contain a *pol* gene, which encodes a reverse transcriptase with considerable homology to the retroviral reverse transcriptase. Retrotransposons also are flanked by long terminal repeats which carry the signals for the transcription of the retrotransposon by host cell RNA polymerase. Transcription of the retrotransposon is followed by translation of some of the mRNA into proteins, as was the case for the retroviruses; however, the retrotransposons do not encode the structural proteins of the viruses. The reverse transcriptase makes a cDNA copy of the retrotransposon RNA which is then integrated back into the host genome at a new location. The retrotransposon RNA is never packaged into virus particles. In many respects these mobile genetic elements can be viewed as defective retroviruses.

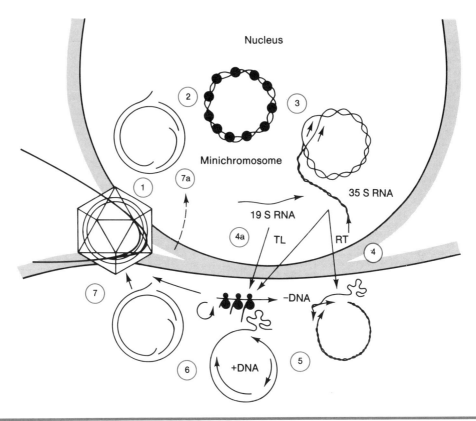

**Figure 3.29 Life cycle of cauliflower mosaic virus (CaMV)**
The virus particles release their DNA after they enter the host cell's cytoplasm (1). The double-stranded DNA genome contains breaks. These are repaired after the DNA enters the plant cell nucleus, where it interacts with histones and takes on the structure of chromatin fibers (2). The viral minichromosome is transcribed to produce 19 S and 35 S RNA transcripts (3). The 19 S and some of the 35 S RNA transcripts enter the cytoplasm, where they are translated (TL) into viral proteins (4a). The remainder of the 35 S RNA also enters the cytoplasm, but serves as a substrate for reverse transcriptases (4). Reverse transcriptase (RT) first generates the minus strand complementary to the mRNA (5). This complex is then used to generate the plus DNA strand (6). The two strands of the viral genome are packaged into virus particles (7). Redrawn with permission from Pfeiffer *et al.* (1987).

*Arabidopsis* contains a family of retrotransposons, *Ta*1–10, which together constitute 0.1% of the genome (Fig. 3.30). Although these elements have the sequences necessary for transposition, it has not actually been demonstrated that they transpose. The tobacco genome contains the retrotransposon *Tnt*1, which was isolated after it transposed into a nitrate reductase gene, inactivating it. *Tnt*1 is 5334 bp long. The repeats at either end of the element are 610 bp long, and *Tnt*1 contains one 3984-bp open reading frame that encodes a polyprotein with

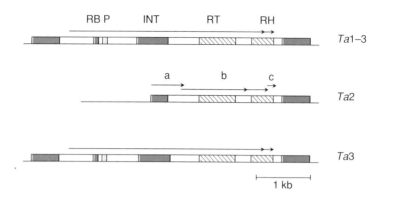

**Figure 3.30 Structures of three *Arabidopsis* retrotransposons: Ta1-3, Ta2, and Ta3**
The long terminal repeats (LTRs) are indicated by the green boxes. The arrows over the boxes indicate open reading frames, whereas the shaded regions of the boxes show portions of the genes where there is strong conservation of amino acid sequence between the different retrotransposons and retroviruses. RB, RNA binding domain; P, protease; INT, integrase; RT, reverse transcriptase; RH, RNase H. Redrawn with permission from Konieczny *et al.* (1991).

the functions necessary for transposition, including a reverse transcriptase and an integrase. *Tnt*1 is present in the genomes of many other members of Solanaceae, the family of angiosperms to which tobacco belongs. Plants belonging to the lily family contain a retrotransposon known as *del*. There are at least 13,000 copies of *del* in the genome of *Lilium longiflorum*. Both *Tnt*1 and the *Arabidopsis Ta* elements are similar in sequence to a *Drosophila* retrotransposon known as *copia,* suggesting that they may have been transmitted laterally from insects to plants, probably from a group of insects known as leaf hoppers. Plant reoviruses also are thought to have arisen from insect viruses.

### Nonviral Retrotransposons

Nonviral retrotransposons are partial or complete DNA copies of cellular RNA molecules that have been inserted at random into the genome. The nonviral retrotransposons exhibit no consistent unifying structural similarities. They are related because they are all derived from cellular RNA molecules, as opposed to viruses, and they were generated via a mechanism involving reverse transcriptase. All classes of cellular RNA, except ribosomal RNAs, have given rise to nonviral retrotransposons. Those derived from mRNA represent pseudogenes. They differ from functional genes in that they lack the introns normally present in expressed genes and because they contain a 3′ poly(A) tail (Fig. 3.31). Retropseudogenes contain the coding sequence of the mRNA from which they were produced, although they may be truncated at either their 3′ or 5′ end. As the

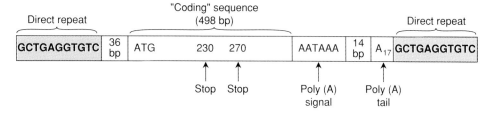

**Figure 3.31** **Structure of a nonviral retrotransposon found in the human genome**
The element diagrammed is a processed β-tubulin pseudogene that appears to have arisen through reverse transcription of the β-tubulin mRNA, with the subsequent insertion of the cDNA into the genomic DNA of germ line cells. This is a pseudogene because it lacks the regulatory sequences necessary for the transcription of the gene by RNA polymerase that reside outside the sequence transcribed into mRNA. Also, it lacks the introns found in all expressed β-tubulin genes, and two mutations, at positions 230 and 270, have created stop codons within the coding sequence. It is also flanked by direct repeats of the target DNA, as is the case with other mobile genetic elements. Redrawn with permission from Wilde *et al.* (1982).

regulatory regions of the gene that are necessary for its transcription are not translated into the mRNA, the retropseudogenes also lack the signals necessary for their transcription. Most nonviral retrotransposons transpose passively. They do not encode a reverse transcriptase or any other protein responsible for their transposition. The differences between viral and nonviral retrotransposons are summarized in Table 3.4.

**Table 3.4** **Characteristics of viral and nonviral families of retrotransposons**

| Viral superfamily | Nonviral superfamily |
| --- | --- |
| Dispersed in genome | Dispersed in genome |
| Bound by long terminal repeats | May not have terminal direct or inverted repeats |
| Active transposition (element encodes reverse transcriptase and/or integrase) | Passive transposition |
| Generate 4- to 6-bp target site duplications characteristic of the retrotransposon | Generates 7- to 21-bp target site duplication |
| No 3′-terminal poly(A) tract | Often have a 3′-terminal poly(A) tract |
| May contain introns that are removed after transcription | Do not contain introns even if the parental gene from which they are derived contains introns |

Reprinted with permission from Britten *et al.* (1989).

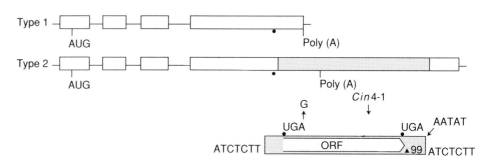

**Figure 3.32 Insertion of a nonviral retrotransposon, *Cin*4, into the A1 gene of maize**

The structure of the 1.1 kb *Cin*4 element, *Cin*4-1, is shown at the bottom. *Cin*4-1 differs from other *Cin*4 elements in that a codon near the start of translation (UGG) designating the amino acid tryptophan has been changed to the stop codon UGA by a point mutation. Other *Cin*4 elements contain an open reading frame (ORF) encoding a protein with homology to the reverse transcriptase of retroviruses and cauliflower mosaic virus. Its identity as a nonviral retrotransposon is inferred by the presence of a direct repeat of a short portion of the A1 gene sequence at either end of the element. The structure of the A1 gene is shown in the upper diagram. Open boxes indicate the four exons of this gene. The shaded region in the middle diagram shows the position of the *Cin*4 element within the fourth exon of the A1 gene. Redrawn with permission from Schwarz-Sommer *et al.* (1987).

The *Cin*4 element of maize is an example of a plant nonviral retrotransposon. It was found inserted into the A1 gene (Fig. 3.32). Although the exact mechanism by which the nonviral retrotransposons are generated is not known, it is clear that it involves the transcription of cellular RNA into DNA by reverse transcriptase. A mechanism by which a nonviral retrotransposon is introduced into a genome has been proposed (Fig. 3.33).

## Genomes of Chloroplasts and Mitochondria

Both chloroplasts and mitochondria appear to be the remnants of organisms that established symbiotic relationships with the ancestors of modern eukaryotic organisms. This endosymbiotic theory of the origin of these organelles postulates that chloroplasts are derived from photosynthetic cyanobacteria, whereas mitochondria are the decendants of aerobic bacteria. The modern chloroplast or mitochondrion cannot live outside a eukaryotic cell. Nevertheless, both are selfreplicating organelles that have retained many of the characteristics of their prokaryotic ancestors. Primitive eukaryotes lacking chloroplasts or mitochondria probably could live only in anaerobic environments. Oxygen is toxic to most

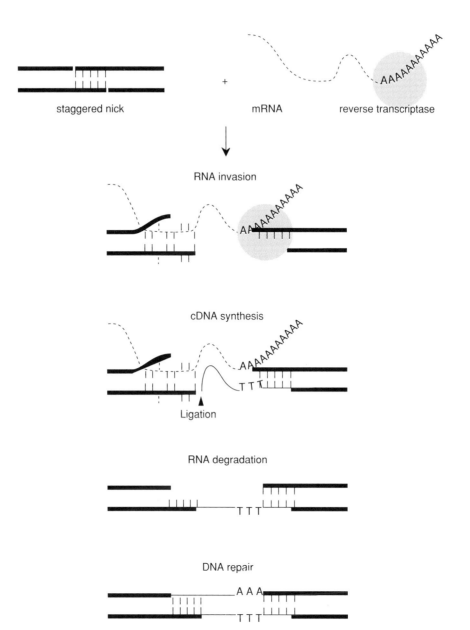

staggered nick       +       mRNA       reverse transcriptase

RNA invasion

cDNA synthesis

Ligation

RNA degradation

DNA repair

**Figure 3.33 Model for the mechanism of *Cin*4 integration**

Thick lines represent chromosomal DNA, dotted lines represent *Cin*4 mRNA, and thin solid lines represent cDNA transcribed from the *Cin*4 mRNA. Reverse transcriptase complexes with the 3′ end of the *Cin*4 mRNA and invades a staggered nick in the chromosomal DNA. the mRNA for the *Cin*4 element is transcribed by reverse transcriptase, with the ligation of the cDNA to the chromosomal DNA. After the RNA is degraded, the second strand of the DNA is generated by DNA repair synthesis. Redrawn with permission from Schwarz-Sommer *et al.* (1987).

anaerobes and they are rapidly killed if exposed to the air. Several hundred species of eukaryotic microorganisms that fit the descriptions of these primitive ancestral eukaryotes can still be found on the earth today in specialized environments. The formation of a symbiotic relationship with the aerobic bacterial precursors of mitochondria enabled these organisms to live in an atmosphere containing oxygen, whereas a symbiotic relationship with a photosynthetic organism would enable the eukaryote to use light energy to drive synthetic chemical reactions. There is considerable evidence to support this theory, including the fact that both mitochondria and chloroplasts contain their own genomes. These genomes are relatively small compared with the nuclear genome. For example, most chloroplast genomes range in size from 120 to 217 kilo-base pairs (kb). Many higher-plant mitochondrial genomes are considerably larger than their chloroplast genomes, ranging in size from 150 to 2500 kb. Even so, the largest mitochondrial genome is less than 0.1% the size of the average plant nuclear genome. These organellar genomes are also smaller than most bacterial genomes, but they are organized and transcribed like bacterial genomes.

### Plastids

The chloroplast is one member of a class of interrelated organelles in higher plants known as **plastids.** These include amyloplasts, chromoplasts, etioplasts, and proplastids, in addition to chloroplasts. Plastids are self-replicating organelles. The proplastid is the precursor of all members of this family of organelles, all of which contain the chloroplast genome. The **amyloplast** is a starch storage organelle which usually lacks photosynthetic pigments and enzymes. Often amyloplasts are abundantly present in roots, tubers, and other storage tissues and organs. **Chromoplasts** are organelles that have specialized in the synthesis and deposition of various yellow, orange, and red carotenoid pigments. Many fruits and flower petals owe their bright colors to these organelles. Like amyloplast, chromoplasts lack photosynthetic enzymes. **Etioplasts** are the plastids that form when leaves and other organs grow in darkness. Etioplasts are not photosynthetic organelles, but rather a stage in the differentiation of chloroplasts.

### Chloroplast Genomes

All plastids contain what has come to be called the "chloroplast genome." It is a circular molecule that ranges in size from 120 to 217 kb. Plant cells typically contain many copies of the chloroplast DNA. The number of copies depends on the type of cell and its state of differentiation. Chloroplasts of mesophyll cells in rapidly growing leaves each contain 20 to 40 identical copies of the chloroplast genome, and there are usually about 200 chloroplasts per mesophyll cell. In contrast, amyloplasts in potato tubers or chromoplasts in tomato fruits each contain only a few copies of the chloroplast genome. Nevertheless, the amyloplasts, chromoplasts, and chloroplasts of the same plant all have the copies of the same chloroplast DNA, so they have identical genetic information.

The complete nucleotide sequence has been determined for the chloroplast genomes of tobacco and the liverwort *Marchantia* (Fig. 3.34). These two organisms are very distantly related, although both are thought to have evolved from a common algal ancestor. They are separated by approximately a hundred million years of evolution. Yet, their chloroplast genomes are similar, indicating that the chloroplast genome has evolved very slowly. One unusual feature of the chloroplast genome is the presence of an inverted repeat. That is, each chloroplast DNA molecule contains two copies of a 20- to 50-kb sequence, and these are arranged in inverted order. An inverted repeat is present in the chloroplast DNA of nearly all land plants, except for members of the legume family. Both *Marchantia* and tobacco chloroplast DNAs have this inverted repeat, so it is likely that this feature evolved early and has been highly conserved. The inverted repeats of tobacco are each 25,339 bp and they are identical in base sequence. In all cases, the inverted repeat contains the chloroplast rRNA genes and some of the chloroplast tRNA genes. Except for the inverted repeat, the chloroplast genome consists of unique-sequence DNA. Outside the inverted repeat each gene is present only once in the genome. As a result, the inverted repeat separates the chloroplast genome into large single-copy (LSC) and small single-copy (SSC) regions (see Fig. 3.34).

The tobacco chloroplast genome contains a total of about 90 different genes. These include all 4 genes encoding chloroplast ribosomal RNAs, 20 genes encoding ribosomal proteins, and 30 genes encoding tRNAs. Many of the proteins that are important for photosynthesis consist of several polypeptides, each the product of a different gene. In several cases the genes encoding some of the polypeptides are in the nuclear genome; in other cases, the genes are in the chloroplast genome. Six of the nine genes encoding subunits of the chloroplast ATP synthase enzyme are in the chloroplast, as are all three of the genes encoding subunits of the chloroplast RNA polymerase (Table 3.5); however, relatively few of the genes encoding other important chloroplast proteins are in the chloroplast DNA. The chloroplast DNA encodes genes for three photosystem I proteins, nine photosystem II proteins, and three photosynthetic electron transport proteins. Fifteen of these genes contain introns.

During the evolution of the symbiotic relationship between the chloroplast precursor and the plant cell, many chloroplast genes were transferred to the nuclear genome. At present, the genes encoding most of the chloroplast enzymes are encoded in the nuclear genome. Presumably, these once must have resided in the chloroplast genome if the chloroplast precursor was an autonomous, free-living organism. There is evidence that this gene transfer is still going on and that the gene transfer occurs not only between the chloroplast and the nucleus, but also between the chloroplast and the mitochondrion. Some chloroplast genes, including a gene encoding the large subunit of RUBISCO, have been found in mitochondrial genomes in a few species.

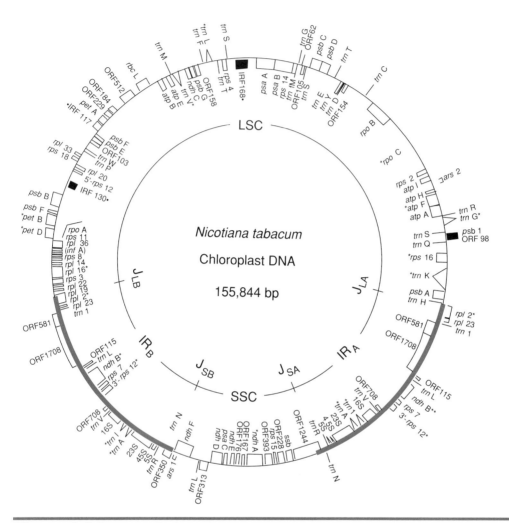

**Figure 3.34 Map of the chloroplast genome of *Nicotiana tabacum***
The chloroplast genomes of most angiosperms are similar in size and in arrangement of genes on their circular chromosome. Usually they contain an inverted repeat, shown here in green and labeled $IR_A$ and $IR_B$. This divides the genome into two single-copy regions, a small single-copy region (SSC) and a large single-copy region (LSC). The positions of known genes are indicated using their accepted designations, with genes transcribed in a clockwise direction shown on the inside of the circle, and genes transcribed in a counterclockwise direction on the outside of the circle. Members of the legume family do not have the inverted repeat. Redrawn with permission from Shinozaki *et al.* (1986).

**Table 3.5**    **Some genes known to be encoded in the chloroplast genome**

| Gene designation | Gene product (protein or RNA encoded by the gene) |
| --- | --- |
| Chloroplast RNA genes | |
| rDNA | Ribosomal RNAs (16 S, 23 S, 4.5 S, 5 S) |
| *trn* | Transfer RNAs (30 species) |
| Photosynthetic proteins | |
| *rbc*L | RUBISCO large subunit |
| *atp*A, B, E | ATP synthetase CF1 $\alpha$, $\beta$, $\varepsilon$ subunits |
| *atp*F, H, I | ATP synthetase $CF_0$ I, III, IV subunits |
| *psa*A, B, C | Photosystem I A1, A2, 9-kDa proteins |
| *psb*A, B, C, D, E | Photosystem II D1, 51 kDa, 44 kDa, D2, Cytb559—9 kDa |
| *psb*F, G, H, I | Photosystem II Cytb559—4kDa, G, 10Pi, I proteins |
| *pet*A, B, D | Electron transport Cytf, Cytb6, IV subunits |
| Respiratory proteins | |
| *ndh*A, B, C, D | NADH dehydrogenase (ND) subunits 1, 2, 3, 4 |
| *ndh*E, F | ND4L, 5 |
| | |
| Chloroplast | |
| Protein synthesis | |
| *rps*2, 3, 4, 7, 8, 11 | 30 S ribosomal proteins (CS) 2, 3, 4, 7, 8, 11 |
| *rps*12, 14, 15, 16, 18, 19 | CS 12, 14, 16, 18, 19 |
| *rpl*2, 14, 16, 20, 22 | 50 S ribosomal proteins (CL) 2, 14, 16, 20, 22, |
| *rpl*23, 33, 36 | CL 23, 33, 36 |
| *inf*A | Initiation factor I |
| Gene transcription | |
| *rpo*A, B, C | RNA polymerase $\alpha$, $\beta$, $\beta'$ subunits |
| *ssb* | ssDNA-binding protein |

## Mitochondrial Genomes

Plant mitochondrial genomes are surprisingly large in comparison with those of other groups of organisms. For example, vertebrate mitochondrial genomes are approximately 16 kb, and even those of fungi are comparatively small (18 to 78 kb). In contrast, plant mitochondrial genomes range from approximately 200 kb in *Brassica* to 2500 kb in watermelon. Furthermore, the size of the mitochondrial genome in most other groups of organisms does not vary much from species to species. Plant mitochondrial genomes are highly variable in size, even within a given family. Frequently the mitochondria of a given plant contain several different-sized DNA molecules. For example, maize mitochondria contain a 570-kb circular molecule and several smaller circular molecules. The smaller molecules are derived from the large molecule through homologous recombination. The large DNA molecule contains five direct repeat sequences and one inverted repeat. The "master" 570-kb molecule breaks and rejoins at these repeats, sometimes regenerating the original 570-kb molecule, but, other times, generating a series of smaller molecules.

Maize, as well as 140 other angiosperms, may exhibit a phenomenon known as **cytoplasmic male sterility,** in which the anthers abort or pollen fails to develop. All other aspects of plant development are normal, including the formation of the ovules, but male reproductive development is blocked. This unusual characteristic is inherited in a uniparental fashion. That is, the trait is carried exclusively by the female parent and it is known to result from genes in the mitochondrial genome. Both mitochondrial and plastid genomes are inherited only through the female parent in most angiosperms. This is because plastids and mitochondria are eliminated from the cytoplasm during the cell division leading to the formation of the sperm cell. More than 80% of the angiosperms eliminate plastids and mitochondria from the sperm, and in only a small minority of angiosperm species is there biparental inheritance of organelle genomes. As a result, a number of plant characteristics, in addition to cytoplasmic male sterility, are maternally inherited, including some types of leaf variegations.

Cytoplasmic male sterility (cms) is a very useful trait for the plant breeder working with self-fertile species because it eliminates the necessity of hand emasculation. In the 1970s most commercially grown maize had the same cms genetic background, known as Texas male-sterile cytoplasm, or cms-T. Unfortunately, the cms-T also carried susceptibility to a disease known as Southern corn leaf blight, caused by the fungus *Bipolaris maydis,* race T, and in 1974 much of the maize crop was destroyed by this disease. *B. maydis* produces a toxin, known as T-toxin, that is responsible for disease symptoms in maize plants carrying the cms-T trait. Susceptibility to T-toxin appears to be caused by a protein encoded by a mitochondrial gene, *T-urf13.* The protein product of the *T-urf13* gene is a 13-kDa membrane protein. The exact function of the 13-kDa protein is not known and it may not have an essential function, as the *T-urf13* gene is present only in the cms-T mitochondrial genome. Most other maize mitochondrial genomes lack the *T-urf13* gene. In those plants carrying the cms-T mitochondrial genome, however, the gene is expressed throughout the plant and the 13-kDa protein is present in the mitochondrial membranes in all of the cells of the plant. Interaction of T-toxin with this receptor protein causes massive loss of ions from the mitochondria, rendering them unable to respire and leading to the death of the tissues. Fortunately for the plant breeder, there are other types of cytoplasmic male sterility that do not carry susceptibility to T-toxin.

# Recombinant DNA Methods in the Analysis of Plant Genomes

Recombinant DNA techniques developed over the course of the last 20 years have greatly increased our understanding of plant genomes, gene structure, and genetic mechanisms. These techniques enable us to clone a DNA sequence, such as a particular gene, which may then be identified, isolated, and characterized. A clone is a population of individuals that are all derived from a single individual (Fig. 3.35). Clonally derived cells, whether they are bacteria or plant cells, are all descendants of a single cell and are genetically identical. DNA molecules are cloned by fusing them with another DNA molecule, known as the **vector,** that can replicate inside bacterial or yeast cells. Not all DNA molecules can replicate when inside cells. Those that can include viruses and plasmids, each of which

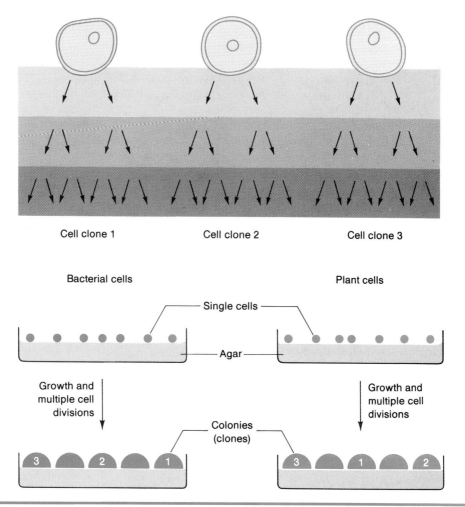

**Figure 3.35 Cloning cells of bacteria or plants**
In either case, the principle is the same. Cells are plated out at a low density so that on average each cell is isolated and separated from neighboring cells. The repeated division of each of these cells gives rise to a population of cells, all of which are identical. Redrawn with permission from Singer and Berg (1991).

contains a relatively short sequence of bases, known as the origin of replication, which is necessary for the initiation of DNA replication by the host cell DNA polymerase. Viruses and plasmids can serve as a carrier, or vector, for cloning DNA molecules. The DNA we want to clone is inserted into the vector, and the vector containing the insert is returned to the host cell where the DNA will replicate.

It is much easier to isolate a particular gene from a relatively small vector than from the larger genomes of a higher plant or even a bacterium. Assume that we want to isolate a particular *Escherichia coli* gene that is $2 \times 10^3$ bp long. A clone of these bacteria that contains $10^6$ cells would also contain a million copies of the gene; however, as the *E. coli* genome contains $4 \times 10^6$ bp of DNA, the gene we are interested in represents only 0.05% of the genomic DNA. In a genome of a plant such as maize with $11 \times 10^9$ bp of genomic DNA ($2c$ level), this gene would obviously represent a much smaller percentage of the total DNA. A clone of the bacterium or the plant cell by itself is not selectively enriched for the gene of interest and it would be difficult to identify the gene among all the rest of the chromosomal DNA. If, however, we inserted our gene into the DNA of a plasmid vector consisting of a circular molecule with $4 \times 10^3$ bp of DNA, finding our gene would be much easier. The plasmid vector containing the gene is placed back into a bacterium and the bacterium is cloned to give us a million copies of the plasmid. We still only have a million copies of the gene, but the gene represents 33% of the plasmid DNA. By isolating the plasmid from a clone of the bacterium we will have obtained approximately a 600-fold enrichment of the gene. If the plasmid contained a gene from a higher plant, the enrichment would be even greater.

### Bacterial Plasmids

Bacterial plasmids are covalently closed circular DNA molecules that contain DNA sequences that permit them to be replicated along with the bacterial chromosome. Frequently they consist of only a few thousand base pairs of DNA and can be readily isolated from the bacteria and separated from chromosomal DNA, so they make good candidates for cloning vectors. Most bacterial plasmids do not contain essential genes, but they may contain genes that cause the bacterium to be resistant to antibiotics. For example, some plasmids carry a gene encoding an enzyme that inactivates the antibiotic chloramphenicol by acetylating it (chloramphenicol acetyltransferase); others contain a $\beta$-lactamase gene that encodes an enzyme that degrades the antibiotic ampicillin. As a result, bacteria that contain a plasmid with an antibiotic resistance gene will have a selective advantage over bacteria lacking the plasmid in an environment containing these fungal antibiotics. Bacteria harboring plasmids with antibiotic resistance genes survive in the presence of the antibiotic, whereas bacteria without the plasmid die. This gives us a simple method for selecting bacterium-containing plasmids to which we have spliced a particular DNA sequence.

In some cases plasmids greatly expand the ecological niches in which the bacterium can grow. Two of the most notable examples of this are the Ti and Ri plasmids of two species of *Agrobacterium* which cause diseases of higher plants known as crown gall (*A. tumefaciens*) and hairy root disease (*A. rhizogenes*). These plasmids enable the bacterium to change the genetic capabilities of plant cells so that the plant cells produce unusual amino acids that only these bacteria can use for their metabolism. The Ti and Ri plasmids contain genes that can be expressed in plant cells that encode enzymes required for the synthesis of a class of amines known as opines. The plasmids also contain other genes that cause these plant-expressible genes to be transferred into plant cells, where they are incorporated into the plant cell nuclear DNA. This *Agrobacterium* plasmid system, which has been developed by plant molecular biologists as a general system to introduce foreign DNA into dicots, is described in more detail later in this chapter and in Chapter 10.

## Nucleases as Tools in Molecular Biology

Nucleases enable us to cut and splice and otherwise modify nucleic acids. Nucleases are enzymes that break phosphodiester bonds in RNA and DNA. Some nucleases use only RNA, whereas others are specific for DNA and still others use either nucleic acid. Also, some nucleases are specific for single-stranded nucleic acids and others cleave only double-stranded nucleic acids. An exonuclease attacks the molecule at its free ends, whereas an endonuclease attacks the nucleic acid in the middle of a long sequence and can cleave circular molecules. For example, **S1 nuclease** is an endonuclease from the fungus *Aspergillus oryzae* with a marked preference for single-stranded DNA or RNA molecules. It cleaves phosphodiester bonds to produce 5′-monophosphate and 3′-hydroxy ends (Fig. 3.36).

Restriction endonucleases are enzymes that cut DNA molecules at a specific DNA sequence. They are found in many bacteria where they are part of a system designed to destroy foreign DNA molecules, such as bacteriophage. A restriction endonuclease recognizes and binds to a specific DNA sequence, often consisting of 4 to 6 bp, and then it breaks the phosphodiester bonds between the bases on each of the two strands of the DNA molecule. The cleavage usually occurs within the sequence recognized by the enzyme, and it may produce either blunt ends or staggered single-stranded ends, depending on where the break occurs, but a given enzyme always cuts the DNA in the same place (Fig. 3.37). A few restriction enzymes bind DNA at one site and cut at another, often at a specified distance from the binding site. The binding site for a restriction enzyme is **palindromic;** that is, the sequence on one strand is the same as the sequence on the other strand read in the opposite direction. The restriction endonucleases break a particular phosphodiester bond within the recognition sequence, generating a 5′-phosphoric acid group at one end of the break and a 3′-hydroxyl group at the other end of the break (Fig. 3.38). Several restriction enzymes make staggered

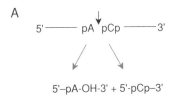

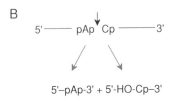

## Figure 3.36 Nucleases break phosphodiester bonds in nucleic acids, yielding either a 3′-hydroxyl end and a 5′-phosphate end or the reverse

(A) Hydrolysis between the phosphate and the 3′-hydroxyl yields 5′phosphomonoester end groups. (B) Hydrolysis between the phosphate and the 5′-hydroxyl yields 3′-phosphomonoester end groups.

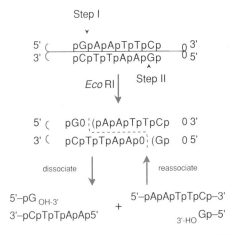

## Figure 3.37 Recognition site and cleavage pattern of the restriction enzyme *Eco*RI

Redrawn with permission from Singer and Berg (1991).

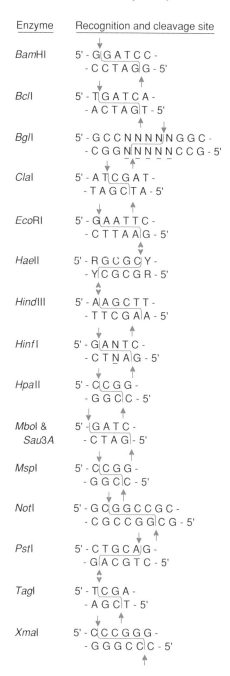

**Figure 3.38 Several restriction endonucleases, showing their recognition sites and cleavage patterns**
From Singer and Berg (1991) with permission.

breaks in DNA, leaving short single-stranded ends. If the ends of two DNA molecules have complementary overhanging ends, they can be recombined. Base pairing will hold the complementary ends together, while an enzyme known as a ligase can restore the broken phosphodiester bonds. This enables us to join a given DNA molecule to a plasmid vector so that it may be cloned (Fig. 3.39).

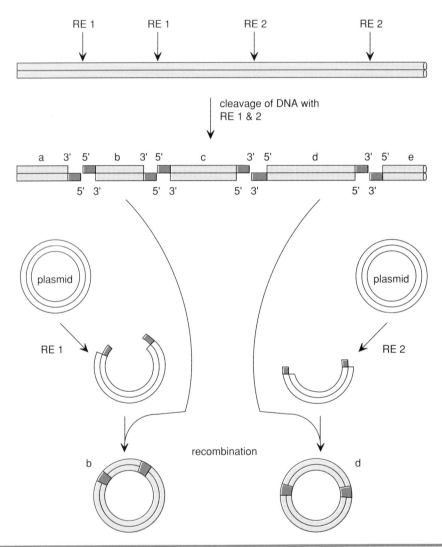

**Figure 3.39 Joining of different DNA molecules**
The staggered cuts made by many restriction endonucleases (RE) make it possible to join different DNA molecules, if they carry complementary single-stranded ends. Redrawn with permission from Singer and Berg (1991).

## Reintroduction of Cloned Genes into Plants by *Agrobacterium*-Mediated Transformation

The soil-dwelling bacterium *Agrobacterium tumefaciens* causes crown gall disease of dicots, which is a type of plant tumor. It does so by inserting a piece of a large plasmid it carries into susceptible plant cells. The plasmid is known as the **Ti plasmid** and the inserted DNA is called **T-DNA** (Fig. 3.40). T-DNA carries genes that may be expressed in plant cells, including genes that initiate the synthesis of two hormones that stimulate plant cell division, auxin and cytokinin. It is the abnormal and unregulated synthesis of these plant hormones that leads to tumor formation, so the genes encoding these hormone-synthesizing enzymes can be considered to be phytooncogenes. T-DNA is flanked by 25-bp direct repeats, known as the right and left borders of the T-DNA. The Ti plasmid has been developed as a vehicle for introducing foreign genes into plants. For this purpose it was necessary to disarm the plasmid so that it does not cause tumors. This was

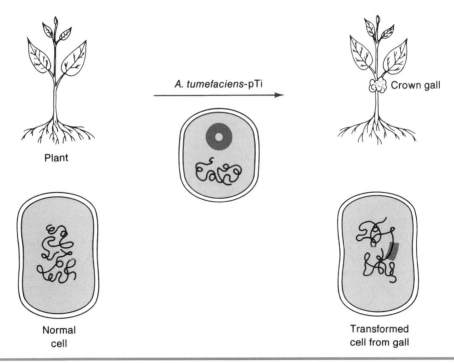

**Figure 3.40 Induction of tumor formation by *Agrobacterium tumefaciens***
The bacterium *A. tumefaciens* induces tumor formation in many angiosperms if it carries a large plasmid known as the Ti plasmid. A portion of the Ti plasmid enters the plant cell and transforms it so that its growth is abnormal, leading to tumor formation. Redrawn with permission from Singer and Berg (1991).

done by deleting those genes in T-DNA that encode enzymes controlling auxin and cytokinin synthesis, the so-called phytooncogenes. In addition, it was necessary to introduce a gene into T-DNA that enables the investigator to select the transformed cells. A gene for antibiotic resistance frequently has been used for this purpose. A desirable cloned gene can then be inserted into the T-DNA of the engineered Ti plasmid and used to infect either cultured cells or leaf disks. The infected cells are placed on a culture medium containing auxin and cytokinin to induce growth, as well as the antibiotic. Only the transformed cells can grow in the presence of the antibiotic. These transformed cells have received the T-DNA containing not only the gene for antibiotic resistance, but also the foreign gene. To obtain a plant containing the foreign gene, it is necessary to regenerate plants from the cultured, transformed cells. Fortunately, methods to accomplish this have been developed for many plants, although not yet all important crop species. At present, both transformation of the cereal grains with *Agrobacterium* and their regeneration from cultured cells are very difficult. Nevertheless, there have been some remarkable successes with this approach. Investigators have introduced a number of foreign genes, including genes for disease and herbicide resistance, into tobacco, cotton, and other plants. In the future, genes for other desirable agronomic characteristics will be transferred to crop plants via this or other techniques. This approach also is very useful for the study of DNA sequences that are involved in the control of gene expression.

### Regeneration of Plants from Transformed Cells

Tissues of a great many plant species can now be cultured as callus tissue. This success is due, in large measure, to the development of superior tissue culture media, particularly the chemically defined, high-nitrogen, high-salt medium of Murashige and Skoog. Although many cultured tissues can be induced to form roots, shoots, flowers, and, in some cases, embryos, by manipulating the levels of auxin and cytokinin in the culture medium, this approach is not commonly used as a commercial method for propagating plants of agronomic or horticultural importance, because chromosome abnormalities accumulate in callus tissue so that plants regenerated from it may have altered characteristics. This can be highly undesirable, as clonal propagation is used not only to induce rapid multiplication of a particular plant, but also to ensure that the offspring all have the same traits as the parent. Nevertheless, this approach has been used extensively in the laboratory to regenerate plants from cells or tissues that have been genetically transformed by *Agrobacterium* or some other method.

In recent years a method, known as **micropropagation** has been developed for the clonal propagation of commercially important plants. Micropropagation is a form of tissue culture that avoids the problems associated with callus tissue, but retains the potential for very rapid multiplication that tissue culture provides. Typically, very young, often embryonic tissue is cultured. The medium used contains levels of two plant hormones, auxin and cytokinin, which were found by trial and error to induce shoot proliferation and growth in that variety, with the formation of little or no callus tissue. The shoots are then dissected out and trasferred to a medium that induces the formation of adventitious roots. When these plantlets are large enough, they are transferred to soil.

## Other Procedures to Introduce DNA into Monocots

Most monocots and some dicots are not easily transformed by *Agrobacterium*. Other mechanisms have been devised to introduce DNA into the cells of these organisms. Two such methods are particle gun transformation (also called biolistics) and electroporation. In **particle gun bombardment,** very small particles of gold or tungsten are used to carry molecules into plant cells. The DNA of interest is precipitated onto the surface of the particles which are then propelled into plant cells by a sudden release of gas. Initially, a charge of gun powder was used to propel the particles, but more recently a high-pressure helium discharge has been used. As the particles are very small, a partial vacuum must be created over the cells before discharge. Only some of the cells in the path of the particles will receive the DNA. Either other cells are not hit or the DNA is degraded in the cytoplasm. As a result, one must introduce a selectable marker into the cells, as well as the DNA of interest. As was the case for *Agrobacterium* transformation, the selectable marker used often is an antibiotic resistance gene. After bombardment, the cells are cultured on medium containing the antibiotic and only the cells that received the antibiotic resistance gene survive.

In **electroporation,** a pulse of high voltage is used to briefly and reversibly make the cell membrane permeable. Both RNA and DNA molecules can enter the cells from the bathing solution during electroporation. Electroporation does not work when cells are surrounded by cell walls. As a result, it is first necessary to convert the cells to protoplasts. A protoplast is a cell lacking its cell wall. Experimental methods have been developed to generate protoplasts from various plant tissues and cultured cells. This is accomplished by a combination of enzymes that degrade the cell wall polymers. Cell wall-degrading enzymes include cellulases, pectinases, and hemicellulases; most are isolated from fungi and are commercially available. As plant cells generate turgor pressure through osmotic water uptake, they will explode when the cell wall is removed if the turgor pressure is not first reduced. This is accomplished by adding to the medium a metabolically inert sugar, such as sorbitol or mannitol.

## Reporter Genes

Reporter genes encode a protein product that can be easily detected when the gene is transcribed. Reporter genes lack their own promoters and they are placed under the control of promoters from other genes when we want to understand how the transcription of that gene is regulated. Typically a reporter gene is selected that is not present in the plant so its transcripts or products can easily be distinguished from the transcripts or products of endogenous genes. The promoter under investigation is fused to the reporter gene so that the reporter can be transcribed and translated to give the protein encoded by the gene.

The bacterial genes encoding chloramphenicol acetyltransferase (*CAT*) and β-glucuronidase (*GUS*) are valuable and widely used reporter genes for the analysis of plant promoters and other DNA sequences regulating gene expression. Chloramphenicol is an antibiotic and *CAT* confers antibiotic resistance on the bacteria that contain it. The *CAT* gene was isolated from a bacterial plasmid. The enzyme encoded by *CAT* inactivates chloramphenicol by introducing two acetyl groups into the molecule. This results in the formation of three derivatives of the antibiotic which may be readily detected in chloramphenicol acetyltransferase assays.

The β-glucuronidase enzyme that is the product of the *GUS* gene hydrolyzes a variety of glucuronides, including a number of commercially available compounds that yield fluorescent or colored products when hydrolyzed. Several characteristics of *GUS* make it a useful reporter gene. First, most plant tissues and organs lack an endogenous glucuronidase enzyme, so if the enzyme is detected in a tissue it must have come from the expression of the reporter gene. Second, the *GUS* protein it encodes is very stable and its activity is not affected by most components of a plant cell homogenate, although it is necessary to protect the enzyme against plant proteases. Additionally, the presence of the enzyme is easily detected by means of a very sensitive fluorescence assay as well as by histochemical tests.

## Summary

Genetic information is encoded in the sequences of bases in DNA. DNA, a linear, double-helical molecule, is composed of four different nucleotides which are chemically joined to form very large polymers. The two strands of a DNA double helix contain completely different although complementary base sequences, and are held together by hydrogen bonding between complementary bases. Adenine pairs with thymine by the formation of two hydrogen bonds, and guanine pairs with cytosine by the formation of three hydrogen bonds. The sequence of amino acids in a protein is encoded in the sequence of bases in DNA. An amino acid is specified by a contiguous sequence of three bases, or a codon. There are 64 possible combinations of the four nucleic acid bases taken three at a time, but only 20 amino acids occur in proteins; the genetic code is redundant and most amino acids are represented by more than one codon.

Plant nuclear genomes typically are large, although there is a great deal of variation in genome size among the different angiosperm species. Maize, for example, has 5.5 pg of DNA per haploid nucleus and *Fritillaria davisii* has 98.35 pg of DNA per haploid nucleus. *Arabidopsis thaliana*, with 0.07 pg of DNA per haploid cell, has one of the smallest genomes of any angiosperm, and this is one of the reasons this plant was selected as a model system for the molecular–genetic analysis of plant development. It is estimated that the *Arabidopsis* genome, although small, could encode more than 15,000 different proteins. Other plants with much larger genomes undoubtedly contain many more genes than *Arabidopsis,* but probably only two to three times more. Many important proteins are encoded by small multigene families. That is, there is a substantial degree of redundancy in the genetic information of plants. Plants with small genomes tend to have fewer members in their multigene families than plants with large genomes.

A substantial fraction of the DNA of most plant genomes does not encode cellular proteins or has no coding function. Frequently this DNA has a repetitive base sequence. The amount of repetitive-sequence DNA varies greatly; 60% of the *Nicotiana tabacum* genome is repetitive but only 10% of the *Arabidopsis* nuclear DNA is repetitive in sequence. Repetitive-sequence DNA consists of a number of families with similar but not identical base sequences. One class of repetitive-sequence DNA may behave as a satellite when nuclear DNA is analyzed by centrifugation. That is, the DNA has either a greater or a lesser buoyant density than the bulk of the nuclear DNA. This is because the repeats are clustered together in chromosomes, often forming bands with special staining properties, known as heterochromatin, or clustered in specific regions of chromosomes, such as the centromeres. A significant fraction of the repetitive DNA has the repeats dispersed throughout the genome. Transposons, retrotransposons, and viruses, all of which are present in multiple copies in the genome, are included in this category. Most genomes contain several kinds of transposable elements. Transposons have an inverted repeat sequence at each end. The inverted repeats are relatively short DNA regions at either end of the element that have the same nucleotide sequence, but in inverted order. Although they do not encode typical cellular proteins, transposons often encode an enzyme known as a transposase that is capable of bringing about the movement of the transposon. The transposase recognizes and binds to the inverted repeat of the transposon. In the simplest case, it breaks the DNA strand at the inverted repeat and inserts the element at another location. An autonomous transposon contains a functional transposase gene, whereas a nonautonomous transposon either does not have a transposase gene or has a nonfunctional one. The maize genome contains at least six families of transposons. The maize transposon *Ac* is autonomous; *Ds* transposons are nonautonomous. *Ds* transposons, however, can transpose in the presence of an autono-

*continues*

mous transposon. Transposons cause mutations when they insert into genes. In some cases, the function of the gene is restored when the transposon is again mobilized and moves back out of the gene.

Some viruses are similar to transposons. Both contain information facilitating their replication within the host cell and they can move into and out of the genome. Viruses differ from transposons chiefly in that at some stage in their life cycle, viruses are enclosed in a protein coat. Retroviruses have single-stranded RNA genomes, but replicate via a DNA intermediate. A virally encoded reverse transcriptase makes a DNA copy of its RNA genome after it infects a cell. In some cases the proviral DNA is integrated into the host cell nuclear DNA and is replicated along with the rest of the genomic DNA. The proviral DNA is flanked by long terminal repeats and contains three genes, *gag, pol,* and *env,* encoding viral proteins. The proviral genes are transcribed into mRNA and translated into viral proteins by the host cell machinery. Retrotransposons are a class of mobile genetic elements related to retroviruses that is found in the genomes of all classes of organisms. Retrotransposons differ from transposons in that they transpose via an RNA intermediate. Retrotransposons are flanked by long terminal repeats and carry a *pol* gene encoding a reverse transcriptase. No retroviruses are known to infect angiosperms, but angiosperms do have retrotransposons. It is not clear how retrotransposons entered plant genomes. Cauliflower mosaic virus (CaMV) has an RNA genome and replicates via a DNA intermediate. Although the CaMV genome encodes a reverse transcriptase, the DNA intermediate is not incorporated into the plant nuclear genome. The *Arabidopsis* retrotransposon family Ta is similar to the *Drosophila* retrotransposon known as *copia,* suggesting that they may have been transmitted laterally from insects to plants, possibly from a group of plant-eating insects known as leaf hoppers.

Chloroplasts and mitochondria contain their own separate genomes. Both are small in comparison with the nuclear genomes. Most chloroplast genomes range in size from 120 to 217 kb, whereas mitochondrial genomes are more variable, ranging in size from 150 to 2500 kb. These organellar genomes are smaller than most bacterial genomes, although they are organized and transcribed like bacterial genomes. Both mitochondria and chloroplasts also contain their own protein synthetic machinery. Because of this, chloroplasts and mitochondria are believed to be the descendants of prokaryotic organisms that established symbiotic relationships with the ancestors of modern eukaryotic organisms. The chloroplast genome is a circular DNA molecule. Most angiosperm chloroplast genomes contain two copies of a 20- to 50-kb sequence, one of which is arranged in an inverted order. Except for the genes found in the inverted repeat, the chloroplast genome consists entirely of unique-sequence DNA, so there is either a single copy or two copies of each chloroplast gene. The chloroplast genome of tobacco contains about 90 different genes. These include all four genes encoding chloroplast ribosomal RNAs, 20 genes encoding chloroplast ribosomal proteins, and 30 genes encoding tRNAs. A few of the photosynthetic enzymes also are encoded in the chloroplast genome. These include several of the subunits of the chloroplast ATP synthase enzyme and the large subunit of RUBISCO. These chloroplast proteins are synthesized in the chloroplast by the chloroplast protein synthetic machinery. Most of the genes encoding important photosynthetic enzymes, however, are encoded in the nuclear genome. The proteins encoded by these genes are synthesized in the cytosol and transported into the chloroplast. Mitochondrial genomes in angiosperms are highly variable in size, even in the same species. Maize mitochondria contain a 570-kb circular molecule and several smaller circular molecules which are derived from the larger molecule by homologous recombination. The large circle contains five direct repeat sequences and one inverted repeat sequence.

## Questions for Study and Review

1. Using information in this chapter, calculate the amount of DNA in grams that would be present in your body (assume that your body contains 1 billion cells). How many meters of DNA would this be if it were in a single long molecule? How many molecules of DNA would be found in a single human somatic cell?

2. Compare the DNA contents of angiosperms and mammals, using the information in Table 3.2. What conclusions can you draw from this comparison?

3. Study Fig. 3.12. What is depicted here? What can you conclude about the nuclear genomes of these two species?

4. What is the difference between an allopolyploid and an autopolyploid species? What must occur before an allopolyploid plant can be fertile? Why?

5. What is the $c$-value paradox? How does the knowledge that many plant species are polyploid help us explain the $c$-value paradox? Can plant polyploidy completely account for the $c$-value paradox?

6. What does the term $C_0t$ mean with reference to the kinetics of renaturation after DNA has been denatured? How is a $C_0t$ curve generated? How can the haploid genome size of a plant such as *Nicotiana tabacum* be calculated from a $C_0t$ curve? How would the $C_0t$ curves of *Allium cepa* and *Allium fistulosum* differ?

7. What is satellite DNA? Does satellite DNA differ from repetitive-sequence DNA? How can the technique of *in situ* hybridization be used to determine the position of satellite DNA in an organism's genome? How do you account for the fact that the satellite DNAs of

two closely related species, *Scilla siberica* and *Scilla ingredae,* are located in very different positions in these two species?

8. What are the essential features of the DNA elements known as transposons? How do autonomous and nonautonomous transposons differ? How do transposons cause mutations? How do the maize transposons *Ac* and *Ds* differ?

9. How do tobacco mosaic virus and the cauliflower mosaic virus differ? How are retroviruses different from plant plus-sense RNA viruses? How are they different from the reoviruses that infect plants? How is cauliflower mosaic virus similar to a retrovirus?

10. What are retrotransposons and how do they differ from transposons? How do they differ from retroviruses? If retroviruses do not parasitize angiosperms, as appears to be the case, how is it possible that their genomes contain retrotransposons?

11. What is the probable origin of the nonviral retrotransposons found in the genomes of many angiosperms? What is the enzyme reverse transcriptase and why would plant genomes contain genes encoding reverse transcriptase? How could nonviral retrotransposons affect the course of the evolution of angiosperms?

12. Compare the tobacco chloroplast genome and the nuclear genome of *Arabidopsis* with respect to size. How many genes could each of these genomes encode? In what ways is the chloroplast genome similar to prokaryotic genomes?

13. Compare the genomes of the chloro-plast and the amyloplast in the same species in terms of the nature, number, and order of the genes. In what way do they differ?

14. What is meant by the term *multigene family?* How could multigene families have arisen during the evolution of the angiosperms?

15. What do the codon assignments for the different amino acids tell us about the relative abundance of the different amino acids in proteins?

16. Over evolutionary time, gene transfer has occured between the chloroplast genome and the nuclear genome. What mechanism could account for the transfer of genes from the chloroplast to the nucleus and for the loss of chloroplast genes? How is it possible for chloroplast genes to be transcribed in the nucleus?

17. What is meant by the term *cytoplasmic male sterility?* Why are some maize plants highly susceptible to the T-toxin of the fungus *Bipolaris maydis* while others are completely resistant?

18. What are oncogenes and how do viruses acquire them? What are the phytooncogenes and how are they similar to and different from the oncogenes of animal retroviruses?

19. What are protoplasts and how are they produced? How are they used for plant genetic engineering?

20. What is a reporter gene and how are reporter genes used in studies of plant gene expression? How are reporter genes constructed and how might they be introduced into plants?

## Further Reading

### General References

Alberts, B., Bray, D., Lewis, J., Raff, M., Roberts, K., and Watson, J. D. (1989). *The Molecular Biology of the Cell,* 2nd ed. Garland, New York.

Darnell, J., Lodish, H., and Baltimore, D. (1990). *Molecular Cell Biology,* 2nd ed. Scientific American Books, New York.

Singer, M., and Berg, P. (1991). *Genes and Genomes: A Changing Perspective.* University Science Books, Mill Valley, CA.

Stebbins, G. L. (1971). *Chromosome Evolution in Higher Plants.* Addison-Wesley, Reading, MA

Stryer, L. (1988). *Biochemistry,* 3rd ed. W. H. Freeman, San Francisco.

Watson, J. D., Hopkins, N. H., Roberts, J. W., Steiz, J. A., and Weiner, A. M. (1987). *Molecular Biology of the Gene,* 4th ed. Benjamin/Cummings, Menlo Park, CA.

## Molecular Methods for the Analysis of Genomes and Gene Expression

Ausubel, F. M., Brent, R., Kingston, R. E., Moore, R. E., Seidman, D. D., Smith, J. G., Struhl, J. A., and Struhl, K. (1987). *Current Protocols in Molecular Biology.* Wiley, New York.

Gelvin, S. B., Schilperoort, R. A., and Verma, D. P. S. (1991). *Plant Molecular Biology Manual.* Kluwer Academic, Dordrecht.

Shaw, C. H. (Ed.) (1988). *Plant Molecular Biology: A Practical Approach.* IRL Press, Oxford.

## Specific References

Aggarwal, A. K., Rodgers, D. W., Drottar, M., Ptashne, M. and Harrison, S. C. (1988). Recognition of a DNA operator by the repressor of phage 434: A view at high resolution. *Science* **242,** 899–907.

Britten, R. J., Stout, D. B., and Davidson, E. H. (1989). The current source of human Alu retroposon is a conserved gene shared with Old World monkey. *Proc. Natl. Acad. Sci. USA* **86,** 3718–3722.

Burr, B., and Burr, F. A. (1982). *Ds* controlling elements of maize at the shrunken locus are large and dissimilar insertions. *Cell* **29,** 977–986.

Deumling, B. (1981). Sequence arrangement of a highly methylated satellite DNA of a plant, *Scilla:* A tandemly repeated inverted repeat. *Proc. Natl. Acad. Sci. USA* **78,** 338–342.

Deumling, B., and Greilhuber, J. (1982). Characterization of heterochromatin in different species of the *Scilla siberica* group (Liliaceae) by *in situ* hybridization of satellite DNAs and fluorochrome banding. *Chromosoma* **84,** 535–555.

Federoff, N. V. (1989). About maize transposable elements and development. *Cell* **56,** 181–191.

Goldberg, R. B., Hoschek, G., and Kamalay, J. C. (1978). Sequence complexity of nuclear and polysomal RNA in leaves of the tobacco plant. *Cell* **14,** 123–131.

Grandbastien, M.-A., Spielmann, A., and Caboche, M. (1989). *Tnt*1, a mobile retroviral-like transposable element of tobacco isolated by plant cell genetics. *Nature* **337,** 376–380.

Joseph, J. L., Sentry, J. W., and Smyth, D. R. (1990). Interspecies distribution of abundant DNA sequences in *Lilium. J. Mol. Evol.* **30,** 146–154.

Konieczny, A., Voytas, D. F., Cummings, M. P., and Ausubel, F. M. (1991). A superfamily of *Arabidopsis thaliana* retrotransposons. *Genetics* **127,** 801–809.

Levings, C. S., III, and Brown, G. G. (1989). Molecular biology of plant mitochondria *Cell* **56,** 171–179.

Nuss, D. L., and Dall, D. J. (1990). Structure and functional properties of plant reovirus genomes. *Adv. Virus Res.* **38,** 249–306.

Pfeiffer, P., Gordon, K., Futterer, J., and Hohn, T. (1987). The life cycle of cauliflower mosaic virus. In *Plant Molecular Biology* (D. von Wettstein and N.-H. Chua, Eds.), pp. 443–458. Plenum Press, New York.

Rees, H., and Jones, R. N. (1977). *Chromosome Genetics.* University Park Press, Baltimore, MD.

Schwarz-Sommer, Z., Leclercq, L., Gobel, E., and Saedler, H. (1987). *Cin*4, an insert altering the structure of the A1 gene in *Zea mays*, exhibits properties of nonviral retrotransposons. *EMBO J.* **6,** 3873–3880.

Shinozaki, K., *et al.* (1986). The complete nucleotide sequence of the tobacco chloroplast genome: Its gene organization and expression. *EMBO J.* **5,** 2034–2049.

Walbot, V., and Goldberg, R. (1979). Plant genome organization and its relationship to classical plant genetics. In *Nucleic Acids in Plants*, (T. C. Hall and J. W. Davies, Eds.). Vol. I., pp. 3–40, CRC Press, Boca Raton, FL.

Wilde, C. D., Crowther, C. E., Cripe, T. P., Lee, M. G.-S., and Cowan, J. J. (1982). Evidence that a human $\beta$-tubulin pseudogene is derived from its corresponding mRNA. *Nature* **297,** 83–84.

Zimmerman, J. L., and Goldberg, R. B. (1977). DNA sequence organization in the genome of *Nicotiana tabacum. Chromosoma* **59,** 227–252.

# 4

# **Regulation of Gene Expression**

**D**ifferent cells and tissues contain different proteins. This is another way of saying that the various cells and tissues of a plant are different because they have expressed different sets of genes. Despite this, when different organs of the same plant are compared, only a small percentage of their total proteins are found to be unique. The vast majority of the proteins are the same in all organs. Nevertheless, as there may be 20,000 or more different proteins in an organ such as a leaf, there are hundreds of proteins that are organ specific. Even this may be a misleadingly low estimate for the number of proteins unique to a particular cell type, however. Most plant organs contain many of the same tissue and cell types. Epidermal tissue surrounds most organs and vascular tissue runs through them. Some of the similarity in the proteins found in different organs probably results from this overlap in cell and tissue composition. A leaf mesophyll cell and an epidermal cell differ because each contains proteins that are unique to that cell type. For example, leaf mesophyll cells have high levels of proteins that are essential parts of the photosynthetic apparatus, such as chlorophyll a/b-binding proteins and ribulose bisphosphate carboxylase (RUBISCO), whereas most epidermal cells do not have these proteins. In any event, it is clear that in any given tissue or organ, there are many genes whose expression is unique to that organ, or at least to some cell type within that organ. Some of these may be important in regulating the expression of other genes in that organ; others may carry out functions unique to that organ, such as the chlorophyll a/b-binding protein in photosynthetic cells. The development of any organ is determined by the regulated expression of the genes encoding the proteins unique to that organ. In this chapter we consider the mechanisms regulating gene expression, with an emphasis on how gene transcription is controlled during plant development.

## Control of Gene Expression

Genes are expressed when the protein product they encode appears in the cell. Gene expression is a complex process that can be regulated on many levels. The steps required for gene expression include the following:

*Gene transcription.* The sequence of DNA bases encoding the protein must be transcribed into a complementary RNA transcript by an RNA polymerase.

*RNA processing and transport to the cytoplasm.* The primary RNA transcripts must be processed to produce a functional mRNA, which then must be transported out of the nucleus and into the cytoplasm where protein synthesis occurs.

*Translation.* Messenger RNAs must be translated by the cytoplasmic protein synthetic machinery to produce the protein specified by the gene.

*Posttranslational protein modification.* The function of many proteins in the cell can be drastically modified by altering certain amino acids after translation or by attaching other molecules to the proteins.

All of these steps may be necessary for a protein to appear at an appropriate place in the cell where it is to function. Any one of these can be used as a mechanism to regulate the appearance and/or function of a protein in that cell type. Each of these levels of regulation could play an important role in regulating the expression of any particular gene or group of genes. For example, the regulation of gene expression in chloroplasts occurs primarily at the posttranscriptional level. Here mRNA processing and factors regulating mRNA stability play well-documented roles in making specific mRNAs available for translation in the chloroplast; however, the expression of many nuclear genes is controlled primarily, if not exclusively, at the transcriptional level. As a result, most of this chapter is devoted to the mechanisms regulating gene transcription.

# Gene Structure

Eukaryotic genes consist of a sequence of DNA bases that, after transcription into RNA bases, can be translated into the amino acid sequence of a protein. The **coding sequence** of the gene determines the amino acid sequence of the protein. The coding sequence of most eukaryotic genes is interrupted by noncoding regions, which initially are transcribed into RNA but are not represented in the final mRNA product. These noncoding, intervening sequences are known as **introns;** the coding parts of the gene are called **exons** (Fig. 4.1). During transcription, the introns and exons are transcribed to produce a continuous RNA molecule, called the **primary transcript,** that is much larger than the final mRNA. Transcription of a gene into a sequence of complementary RNA bases begins before the ATG codon signifying the start of translation and it continues through all the exons and introns, ending beyond the stop codon in the last exon. The primary transcript is then processed in the nucleus to produce the final mRNA which is transported into the cytoplasm for translation into a protein.

A typical gene also contains sequences both upstream (before the 5′ end of the coding sequence) and downstream (after the 3′ terminus of the coding sequence) that are part of the gene and are important for its activity. These regions contain specific DNA sequences which act as signals to specify when and where the gene is to be transcribed and translated. The 5′ upstream region contains the sequences necessary for transcription, known as the **promoter** (see Fig. 4.1). Certain sequences stand out as consistent features of the promoter region when the nucleotide sequences of cloned genes are compared. One very common element in eukaryotic promoters is the sequence TATAA, or some variant of this, about 30 bases up from the cap site. This is known as the TATA, or Goldberg–Hogness box. Many genes have another sequence, known as the CAAT box, which may be found 30 to 60 bases upstream from the TATA box. Some genes lack TATA boxes, particularly those encoding the so-called "housekeeping" enzymes, such as the enzymes of the Kreb's cycle. Instead, these genes may have clusters of GC-rich regions which regulate their transcription.

The 5′ upstream regions of most eukaryotic genes contain several additional sequences that are necessary for the correct transcription of the gene. These sequences have been termed **regulatory elements.** Regulatory elements are tar-

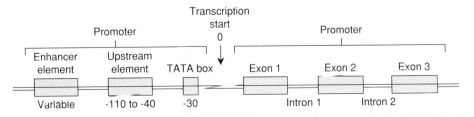

**Figure 4.1   Schematic representation of a eukaryotic nuclear gene**
Redrawn with permission from Murphy and Thompson (1988).

gets for specific DNA-binding proteins that are required for transcription. These DNA-binding proteins are called **transcription factors.** Transcription factors are proteins that bind to a specific 5- to 10-bp DNA sequence. A gene may contain additional sequences that will modify its expression. These modifiers might include elements known as **enhancers,** which stimulate transcriptional activity, or **silencers,** which repress transcription. Frequently, enhancers and/or silencers act only in certain tissues or only under certain conditions. It is rare that eukaryotic gene transcription is regulated by a single transcription factor, however. In most cases, transcription is regulated by many factors which interact with the 5′ upstream regulatory elements in a complex way.

## Structure of Messenger RNA

A messenger RNA typically is a unique linear sequence of RNA bases about 1500 nucleotides long, with distinct polarity. At its 5′ end are a number of bases which are important for translation, but which do not actually encode any of the protein's amino acids. This is known as the **5′ untranslated region.** A number of additional bases beyond the end of the coding sequence constitute the **3′ untranslated region.** Additionally, there is a unique structure at the extreme 5′ terminus known as the **cap** (Fig. 4.2A). The cap contains 7-methyl guanosine linked to the 5′ terminal nucleotide by a triphosphate group (Fig. 4.2B). The cap structure is necessary for the ribosome to recognize and bind to the mRNA, as are some of the bases in the 5′ untranslated sequence. A ribosome will bind to a capped mRNA in the 5′ untranslated region and move along it, but it does not begin translating the RNA bases into a protein until it encounters a start codon, AUG, which may be some distance from the cap. Translation then proceeds until the ribosome encounters one of the stop codons—UAA, UAG, or UGA—whereupon the ribosome releases the mRNA and the completed protein.

## RNA Processing

Transcription of a gene results in the formation of an RNA molecule that has an RNA base sequence complementary to one strand of the DNA double helix. This primary transcript is much larger than the mRNA that is produced from it. The mRNA is formed from this primary transcript as a result of a number of processing reactions (Fig. 4.2C). These include capping, polyadenylation, and splicing. **Capping** is the formation of the cap structure described earlier at the 5′ end of the primary transcript. **Polyadenylation** is the addition of a long chain of adenosine nucleotides to the 3′ terminus of the transcript. A poly(A) tail is attached to the extreme 3′ end of most, but not all, mRNAs. Typically plant mRNAs have a poly(A) tail consisting of about 100 adenine residues. The poly(A)

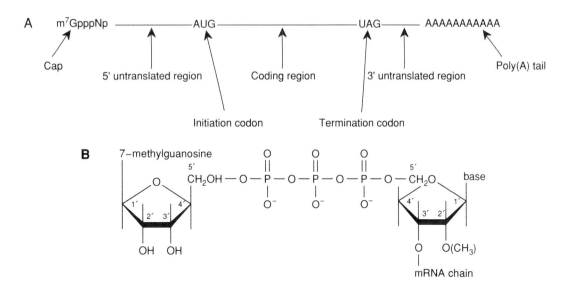

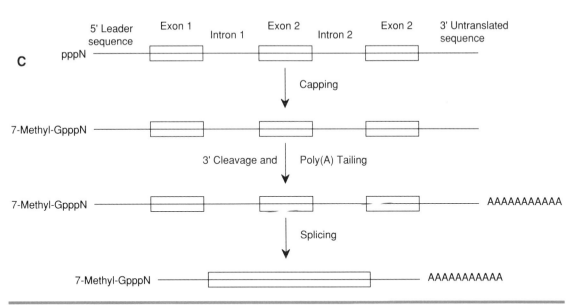

Figure 4.2    **Formation of mRNA by processing a primary gene transcript**

(A) Structure of a typical mRNA. (B) The cap structure in greater detail. (C) Gene transcription produces an RNA molecule with an RNA base sequence complementary to that of the coding strand of the gene. Although the order in which the processing reactions occur has not been determined in plants, processing includes the three main steps indicated and results in the formation of an mRNA with the characteristics shown in (A). Redrawn with permission from Murphy and Thompson (1988).

addition signal is the sequence AATAAA, which occurs in the 3′ transcribed, untranslated region of the precursor RNA, at a site 10 to 30 bases from the site of poly(A) addition. The poly(A) addition signal is rather similar in plants and animals. Finally, **splicing** is the process by which the introns are removed and the exons are spliced together to form the functional mRNA. All of these processing reactions occur in the nucleus. The completed mRNA must then be transported into the cytoplasm where protein synthesis can occur.

## Transcription of Genes by RNA Polymerases

The information encoded in nuclear DNA is transcribed into a sequence of RNA bases by means of enzymes known as RNA polymerases. Transcription begins when RNA polymerase binds to the promoter of the gene. Polymerase binding separates the two strands of the DNA double helix, exposing the bases on each strand. One strand then acts as a template, specifying the order of assembly of RNA bases with a sequence complementary to that of the template DNA strand. The polymerase connects the RNA bases with phosphodiester bonds. As the polymerase moves along the DNA, the helix is further opened up and the RNA chain is extended in a 5′ to 3′ direction (Fig. 4.3). Gene transcription is initiated when RNA polymerase binds to the DNA in the promoter region and begins to move along it, reading the base sequence of one strand of the DNA and transcribing it into a linear sequence of RNA bases. Regulating gene transcription is largely a matter of regulating the binding of the RNA polymerase to the promoter of a given gene. The configuration of the DNA within the chromatin and protein transcription factors determine whether or not RNA polymerase can bind to the promoter of a particular gene and initiate transcription. Bacterial RNA polymerase is a large enzyme that consists of five different polypeptides. Eukaryotic RNA polymerases are more complex and, in fact, all eukaryotic cells have three different RNA polymerases (described later), each of which transcribes a different class of genes. The bacterial polymerase usually is able to bind to promoters directly; however, eukaryotic RNA polymerase is not able to bind to the promoters of genes in the absence of one or more protein transcriptional factors. Two different kinds of regulation can be envisioned in simple models: positive and negative.

### Negative Control

In this model, a specific protein competes with the RNA polymerase for binding to the promoter of a given gene. This protein is called a **repressor.** If the repressor binds to a special DNA sequence in or near the promoter before the polymerase, then the polymerase cannot bind to the promoter and the gene is turned off. When the repressor is not present, the polymerase is able to bind to the promoter and the gene is expressed. In some cases, a different regulatory protein is able to bind to the repressor and loosen its affinity for the promoter. This enables the polymerase to bind to the promoter and the gene is turned on.

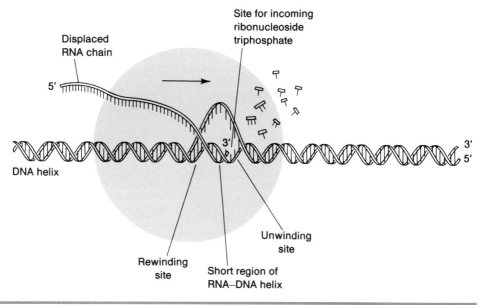

**Figure 4.3   Transcription of a gene by RNA polymerase**
RNA polymerase binds to a gene, opening up the DNA helix and allowing RNA bases to pair with the complementary DNA bases. The RNA bases are chemically linked by the polymerase. Movement of the polymerase along the DNA unwinds the helix further, allowing the RNA molecule to be extended further. Redrawn with permission from Alberts *et al.* (1989).

### Positive Control

In this case, the RNA polymerase may bind very weakly or not at all to the promoter of a given gene. Before the polymerase can bind to the promoter, a transcription factor must bind to a controlling element within or near the promoter. In some cases, the transcriptional factor must be activated before it can bind to the controlling elements in the promoter. The ability of the transcription factor to bind to a controlling element may be regulated through its interaction with some small regulatory molecule, perhaps a metabolite or a hormone, or by a posttranslational modification of the transcription factor. Most eukaryotic genes are under positive control. This mode of regulation is less common in bacteria, but it does occur, as we shall see.

### Control of Gene Transcription in Bacteria

Many important aspects of the regulation of gene expression are different in eukaryotes and prokaryotes. Nevertheless, it is instructive to consider how this process is regulated in bacteria. One of the best-studied examples of the negative control of gene expression in bacteria is the regulation of the transcription of genes involved in lactose metabolism in *Escherichia coli*. Most of the genes that

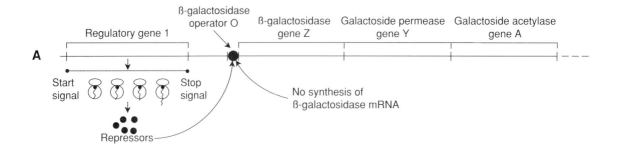

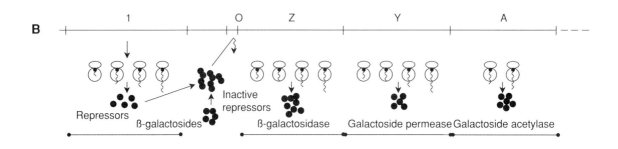

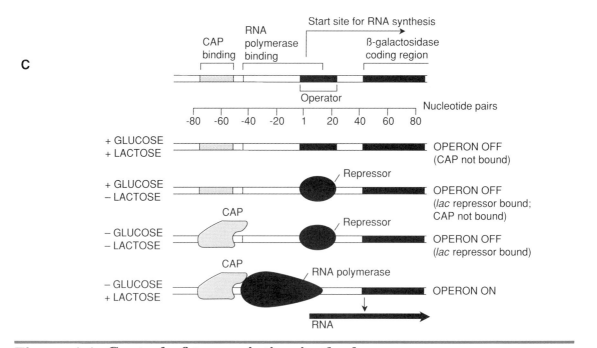

**Figure 4.4    Control of transcription in the *lac* operon**
(A) The *lac* operon consists of three genes, A, Y, and Z, as well as the promoter and operator that regulate the transcription of this operon. When regulatory gene I is expressed in the absence of lactose, its product, the repressor protein, will bind to the

must be expressed for lactose utilization are part of the *lac* operon (Fig. 4.4). *E. coli* does not express the genes of the *lac* operon when the bacterium is growing on a medium containing glucose. In the presence of lactose and the absence of glucose, however, the *lac* operon is expressed, leading to the synthesis of the enzyme β-galactosidase, which breaks the bond between glucose and galactose in the lactose molecule, enabling the bacterium to use these sugars for its metabolism.

Transcription of the *lac* operon is regulated by two DNA-binding proteins, the *lac* repressor and catabolite activator protein (CAP). The bacterium does not produce β-galactosidase in the absence of lactose, or another inducer, because the *lac* repressor protein binds to the *lac* operon, preventing transcription by the RNA polymerase (Fig. 4.4A). The *lac* repressor is a negative regulator of transcription. It turns off transcription by binding to a specific DNA sequence, known as the **operator,** thereby blocking the access of RNA polymerase to the operon. The operator is only 21 bp long and is present only once in the *E. coli* genome. It is located so that it partially overlaps the RNA polymerase binding site, which is why the RNA polymerase cannot bind to the operator when the repressor is bound to it. The *lac* repressor is a specific gene repressor. It regulates only the expression of the genes of the *lac* operon. The binding of the *lac* repressor protein to the operator is in turn regulated by small molecules that are *inducers* of lactose utilization (Fig. 4.4B). The inducers initiate gene expression by binding directly to the *lac* repressor. This reduces the affinity of the repressor for the DNA sequence of the operator. The binding of the inducer to the *lac* repressor brings about a conformational change in its structure so that its DNA binding site is no longer exposed on the surface of the protein. The repressor protein is the product of a specific gene, so the appearance of the repressor in the cell is the consequence of the expression of its gene.

Although the absence of repressor binding to the operator is necessary for polymerase binding and *lac* operon transcription, it is not sufficient. The *lac* operon also is regulated by another DNA-binding protein, CAP. RNA polymerase does not bind efficiently to the *lac* promoter by itself. If, however, CAP is bound to a site near the RNA polymerase binding site, it strengthens polymerase binding to the *lac* promoter and ensures *lac* operon transcription. CAP is a positive regulator of *lac* operon transcription. CAP can bind to the CAP binding site near the polymerase binding site only in the absence of glucose (Fig. 4.4C).

operator for the *lac* operon and prevent RNA polymerase from binding to the promoter. As a result, the operon is not transcribed. (B) β-galactosides, such as allolactose, are able to bind to the repressor. This inactivates it, preventing its binding to the operator. (C) Polymerase will bind to the operator only in the presence of catabolite activator protein (CAP), however, and CAP can bind to the CAP binding site only in the absence of glucose. As a result, transcription of the *lac* operon is regulated by both glucose and lactose. Redrawn with permission from Alberts *et al.* (1989).

# Regulation of Bacteriophage Gene Expression by DNA-Binding Proteins

A great deal of valuable information about the mechanism regulating gene transcription in an organism can be obtained by studying the expression of viral genes. Bacteriophage are virus that infect bacteria. The *E. coli* phage known as lambda has been extensively studied to determine how the expression of its genes is regulated during its infection of this bacterium. Lambda has a tadpolelike shape (Fig. 4.5A). It has a 50,000-bp circular DNA genome that is packaged into a head composed of specific phage proteins. The head is attached to a proteinaceous tail that has an affinity for the bacterial cell wall. On binding to the bacterium, the phage injects its genome into the bacterium and it begins one of two alternative life cycles (Fig. 4.5B).

In the **lytic life cycle,** phage genes are expressed, leading to replication of the lambda chromosome and synthesis of head and tail proteins. New phage particles are assembled from these components. About 45 minutes after injection the host bacteria lyses, releasing approximately 100 progeny phage. Alternatively, the phage may undergo a **lysogenic life cycle.** During the lysogenic cycle, the phage chromosome is incorporated into the bacterial chromosome and all phage genes are silent, except one. The lambda chromosome now replicates as part of the bacterial chromosome and phage particle replication is repressed. Lysogenized bacteria rarely produce phage particles unless they are damaged. Ultraviolet (UV) light is one agent that will damage bacteria and induce the lysogenized phage to undergo lytic replication.

A single phage gene is responsible for this switch in the lambda developmental pathway. It is the gene encoding a DNA-binding protein known as the **lambda repressor.** The lambda repressor is both a positive and a negative regulator of gene expression. The lambda repressor protein maintains the phage in the lysogenic state by turning off all other phage genes. At the same time, the lambda repressor promotes the transcription of its own gene. UV irradiation of lysogenized bacteria inactivates the repressor protein. This allows a second gene to be expressed, encoding a different DNA-binding protein, known as Cro. Cro binds to a DNA base sequence in the promoters of different genes. The binding of Cro to DNA inhibits transcription of the lambda repressor gene and ultimately turns on the lambda genes necessary for the lytic cycle.

The genes encoding these two DNA-binding regulatory proteins, lambda repressor and Cro, are near each other on the lambda chromosome, but they are separated from each other by 80 bp. Each gene has its own promoter. The *cl* gene encodes the lambda repressor and its promoter is designated $P_{rm}$. The *cro* gene encodes the Cro protein and its promoter is known as $P_r$ (Fig. 4.6). The binding of RNA polymerase to $P_{rm}$ orients it for transcription to the left, whereas the binding of RNA polymerase to $P_r$ orients it for transcription to the right. RNA polymerase can bind to either $P_{rm}$ or $P_r$, but not both at the same time.

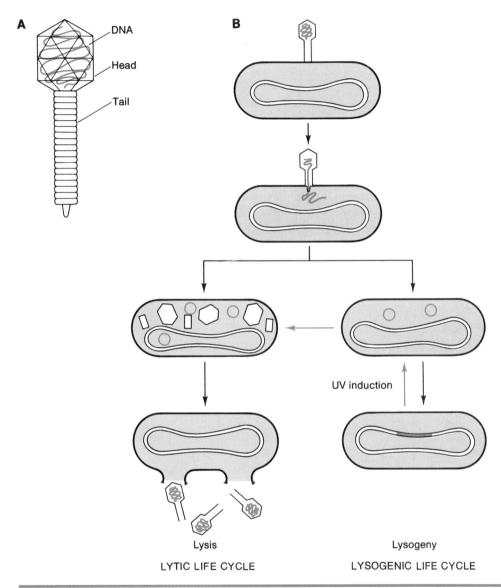

**A** DNA, Head, Tail

**B**

UV induction

Lysis

Lysogeny

LYTIC LIFE CYCLE                    LYSOGENIC LIFE CYCLE

**Figure 4.5    The bacteriophage lambda**
(A) Structure of the phage particle. (B) After the lambda chromosome is injected into a bacterium, it may either lyse or lysogenize the host. Redrawn with permission from Ptashne (1992).

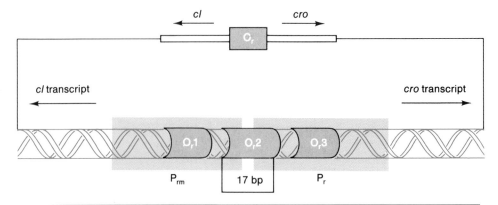

**Figure 4.6**  **The right operator of bacteriophage lambda**
The promoters for the two genes *cro* and *cl* lie next to each other on the lambda
chromosome but are oriented in opposite directions. The right operator, which consists
of three 17-bp DNA sequences, overlaps these two promoters. Redrawn with
permission from Ptashne (1992).

RNA polymerase can bind to $P_r$ without the aid of a regulatory protein, but
effective binding to $P_{rm}$ requires the participation of another positive activating
protein, namely, lambda repressor. These promoters share a controlling element
known as the lambda right operator, or $O_r$. The right operator is composed of
three separate elements: $O_r1$, $O_r2$, and $O_r3$. Each of these elements consists of
17 bp of DNA to which either lambda repressor or Cro protein can bind.

### The Lambda Repressor

The repressor protein was isolated from *E. coli* that had been infected with wild-
type lambda and it has been extensively characterized. It is a dimer in its active
form. The two monomers are identical and consist of a 236-amino-acid polypeptide. The dissociation constant for the intact repressor dimer is $2 \times 10^{-8} M$.
This means that at repressor concentrations above $10^{-7} M$, the repressor exists
as a dimer, whereas at concentrations below $10^{-9} M$, it does not dimerize. There
are approximately 100 molecules of lambda repressor in a lysogenized *E. coli*.
This means that the repressor concentration in the bacterium is approximately
$4 \times 10^{-7} M$ and that it is primarily a dimer in the bacterium.

The proteinase papain readily cleaves the repressor monomer into two fragments: a 92-amino-acid fragment containing the amino-terminal domain and a
135-amino-acid fragment containing the carboxyl-terminal domain. The central
36 amino acids from residues 93 to 131 are destroyed by the protease. The
separated carboxyl-terminal domains will dimerize in the absence of the rest of
the protein, but the amino-terminal domain does not dimerize. As a result, we
can consider the carboxyl-terminal domain to be responsible for the formation
of the dimer.

## Study of DNA–Protein Interaction by "Footprinting"

The enzyme DNase I is highly effective in degrading pure DNA. If, however, a protein is bound to a DNA sequence, it will be protected from attack by DNase. This property can be used to analyze DNA–protein interactions. When it is coupled with DNA sequencing, identification of the precise sequence of DNA that is protected, and thus the sequence that interacts with a particular protein, is possible (Fig. 4.7). The DNA is first end-labeled with $^{32}$P and then incubated with either a specific protein or a mixture of proteins, such as a cell or nuclear extract. The mixture is treated with DNase I, which will attack unprotected DNA at random, and the digested products are separated by gel electrophoresis. The position of the DNA fragments in the gel is detected by autoradiography. When the DNase conditions are set up such that, on the average, each DNA molecule is cleaved at random only once, the result is a ladder of labeled DNA fragments arranged according to size. Each band in the gel represents DNA molecules with one more base than the fragments just above it. If there is a gap in this ladder of labeled DNA fragments, it occurs because the DNA in that region was protected from DNase attack by proteins that bound to it. The DNA-binding protein leaves its "footprint" on the DNA by protecting the sequence it binds to from DNase I digestion. When a DNA sequencing gel is run alongside this, the sequence of bases to which the protein was bound can be determined.

## Binding of Lambda Repressor to 17-bp Sequence within Operator Region

Footprinting experiments with the $O_r$ region and the lambda repressor show that the repressor protects three similar 17-bp sequences in the operator region. These are the elements designated $O_r1$, $O_r2$, and $O_r3$ within the operator. Binding of lambda repressor to $O_r2$ prevents RNA polymerase from binding to $P_r$ (Fig. 4.8). As a result, the *cro* gene is not expressed and no Cro protein is made by the cells. At the same time, lambda repressor binding to $O_r2$ greatly increases

---

**Table 4.1  Base sequences of the three lambda repressor binding sites in the right operator**

| Binding site | Base position | | | | | | | | | | | | | | | | |
|---|---|---|---|---|---|---|---|---|---|---|---|---|---|---|---|---|---|
| | 1 | 2 | 3 | 4 | 5 | 6 | 7 | 8 | 9 | 10 | 11 | 12 | 13 | 14 | 15 | 16 | 17 |
| $O_r1$ | T | A | T | C | A | C | C | G | C | C | A | G | T | G | G | T | A |
| | A | T | A | G | T | G | G | C | G̅ | G | T | C | A | C | C | A | T |
| $O_r2$ | T | A | A | C | A | C | C | G | T̲ | G | C | G | T | G | T | T | G |
| | A | T | T | G | T | G | G | C | A̲ | C | G | C | A | C | A | A | C |
| $O_r3$ | T | A | T | C | A | C | C | G | C̲ | A | A | G | G | G | A | T | A |
| | A | T | A | G | T | G | G | C | G̲ | T | T | C | C | C | T | A | T |

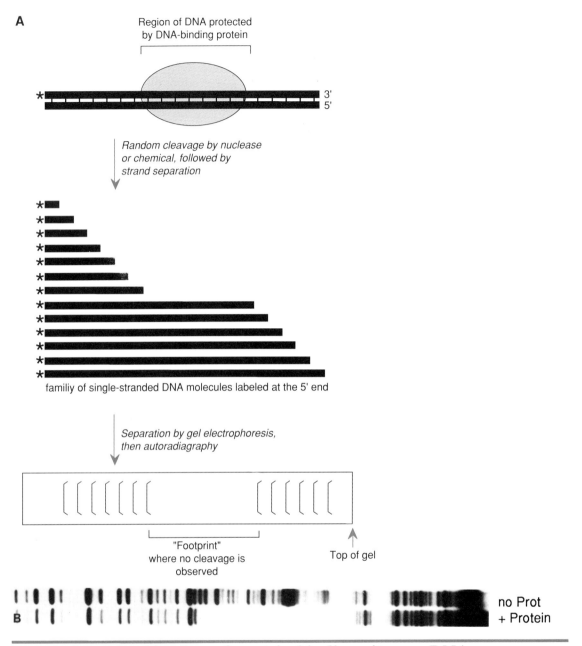

**A**

Region of DNA protected
by DNA-binding protein

*  3'
   5'

*Random cleavage by nuclease
or chemical, followed by
strand separation*

familiy of single-stranded DNA molecules labeled at the 5' end

*Separation by gel electrophoresis,
then autoradiagraphy*

"Footprint"
where no cleavage is
observed

Top of gel

**B**  no Prot
+ Protein

**Figure 4.7** **Identification of protein binding sites on DNA
by footprinting**
(A) Binding of a protein to a particular sequence within a larger DNA molecule will
protect that sequence from being cleaved by the enzyme DNase I. Random cleavage of
the DNA molecule will generate a whole range of smaller fragments. All possible
fragment sizes will be produced, except for those protected from attack by the DNA-
binding protein, and these can be separated from each other by gel electrophoresis. If
the DNA molecule is end labeled with $^{32}$P, the distribution of the fragments in the gel

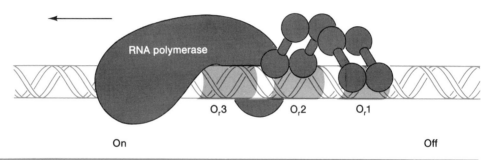

Figure 4.8 **Lambda gene transcription in a lysogenic bacterium**
Repressor bound to $O_41$ and $O_42$ prevents the binding of RNA polymerase to the *cro*
gene promoter, whereas it stimulates binding of polymerase to the $P_{rm}$ promoter. As a
result, the *cl* gene is transcribed and lambda repressor protein is made. Redrawn with
permission from Ptashne (1992).

the binding of polymerase to $P_{rm}$, with a concomitant 10-fold increase in the rate
of transcription of the *cl* gene. This is why lambda repressor is a positive regulator
of its own synthesis. In short, lambda repressor binding to $O_r2$ blocks transcription
to the right and increases transcription to the left.

Lambda repressor binding to $O_r1$ also blocks RNA polymerase binding to $P_r$
and transcription of *cro*, but repressor binding to this element has no effect on
polymerase binding to $P_{rm}$. When lambda binds to $O_r3$, the binding of polymerase
to $P_{rm}$ is blocked and the gene is not transcribed. The sequences of the three $O_r$
elements are similar, but not identical (Table 4.1). Because they are not identical,
it is not surprising that the $O_r$ regulatory elements have different affinities for
the lambda repressor (Fig. 4.9). $O_r1$ and $O_r2$ have the highest affinities, and $O_r3$,
a lesser affinity. As a result, repressor dimer is almost always bound to $O_r1$ and
$O_r2$, but usually not to $O_r3$, and the *cro* gene usually is repressed, whereas the
*cl* gene is transcribed.

Two factors determine the interaction of repressor dimer with the three
operator elements. The first is the affinity of the dimer for each of the sites;
the second is the interaction of the repressors bound to adjacent $O_r$ regulatory
elements. Footprinting has been used to determine the relative affinities of
lambda repressor for the three operator elements. One determines the amount
of repressor necessary for half-maximal protection from nuclease attack. The
affinities are different for the three regulatory elements when they are tested

can be visualized by autoradiography. (B) Autoradiogram of a footprinting experiment
in which the binding of a transcription factor to a portion of a DNA sequence has
prevented the digestion of that portion of the molecule by DNase I. (A) Redrawn with
permission from Alberts *et al.* (1989). (B) Reprinted with permission from Calzone *et
al.* (1991).

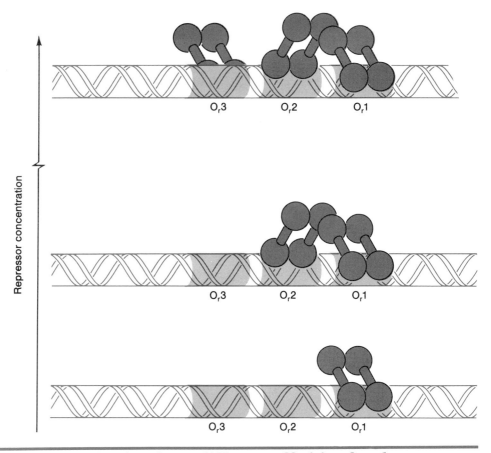

**Figure 4.9**  **The three $O_r$ sites have different affinities for the lambda repressor**
Repressor binds most tightly to $O_41$ and has the lowest affinity for $O_r3$. As a result, as the concentration of repressor increases, $O_r1$ is the first site occupied, followed by $O_r2$. $O_43$ is occupied only at high repressor concentration, after the other two sites are occupied. Redrawn with permission from Ptashne (1992).

separately than when they are tested together. When they are tested together, $O_r1$ and $O_r2$ are equally effective and both are 25 times more effective than $O_r3$. When they are tested separately, $O_r1$ is only half as effective as it is when tested in the presence of $O_r2$, and $O_r2$ is only about 1/25th as effective as it is in the presence of $O_r1$. This is because there is a cooperative effect of repressor binding to $O_r1$ and $O_r2$. The binding of repressor to $O_r2$ is greatly aided if repressor is bound to $O_r1$.

If the carboxyl-terminal domain is cleaved from the repressor with papain, the amino-terminal domain still binds to the operator; however, the repressor now binds to the three regulatory elements without cooperativity. Removal of the carboxyl-terminal domain eliminates the cooperative interaction between the repressors bound to adjacent elements. This shows that the repressor consists of two different functional domains within the same protein; the dimerization domain in the carboxyl terminus and the DNA-binding domain nearer the amino terminus. Both domains participate in the regulation of transcription. As we shall see, this is a common feature of proteins regulating transcription. This also explains how UV light induces lambda in lysogenized bacteria to undergo the lytic cycle. UV irradiation changes the behavior of a bacterial protein known as RecA so that it becomes a protease that cleaves lambda repressor in the flexible linker between the carboxyl- and amino-terminal domains (Fig. 4.10). As a result, the amino-terminal domain can bind to the $O_r$ but it cannot dimerize, and the carboxyl-terminal domain alone cannot bind to the operator regulatory elements. This means that $O_r2$ will not be occupied by repressor and the rate of *cl* gene transcription will drop dramatically. The amount of repressor in the cell thus will begin to decline. At some point the concentration of repressor will be too low to for efficient binding to $O_r1$. When this occurs, RNA polymerase will bind to $P_r$ and begin to transcribe the *cro* gene to produce the mRNA for the Cro protein.

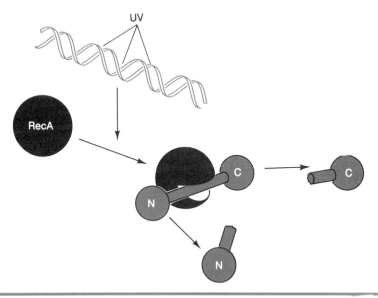

**Figure 4.10 Proteolytic cleavage of lambda repressor by irradiated RecA protein**

Ultraviolet light (UV) alters the properties of a bacterial protein known as RecA so that it cleaves the lambda repressor in the linker between the two domains. Redrawn with permission from Ptashne (1992).

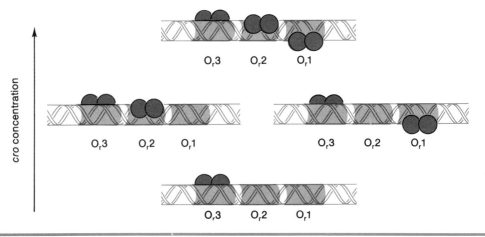

**Figure 4.11 Difference in affinities of Cro protein and lambda repressor for the three binding sites in O$_r$**

Cro has its greatest affinity for O$_4$3, and about 10-fold less affinity for either O$_r$1 or O$_r$2. Redrawn with permission from Ptashne (1992).

## Binding of Cro Protein Operator

Cro protein is a 66-amino-acid protein. It forms dimers with a very low dissociation constant so virtually all the Cro is in the dimeric state in the cell and the dimer is the active form of this protein. Cro is a strictly negative regulator. The Cro protein binds to the same 17-bp DNA sequences in the O$_r$ region as the lambda repressor. Although it binds to all three O$_r$ regulatory elements, the order of its affinity for these elements is opposite that of the lambda repressor. Cro has the greatest affinity for O$_r$3 and substantially less affinity for O$_r$1 and O$_r$2: O$_r$3 > O$_r$1 = O$_r$2 (Fig. 4.11).

## Helix–Turn–Helix Structure in Lambda Repressor and Cro

The amino-terminal domain of the lambda repressor consists of five $\alpha$-helical subdomains with precise relationships to each other (Fig. 4.12A). The third $\alpha$ helix is exposed on the surface of its protein and positioned to interact with DNA. This is known as the **recognition helix.** When the lambda repressor binds to the operator, the third helix of each repressor monomer fits into the major groove of the DNA double helix so that amino acid side chains in this helix can hydrogen bond to bases in the DNA. The second $\alpha$ helix also is important for binding. It lies across the major groove of the DNA double helix and helps position the third or DNA recognition helix (Fig. 4.12B).

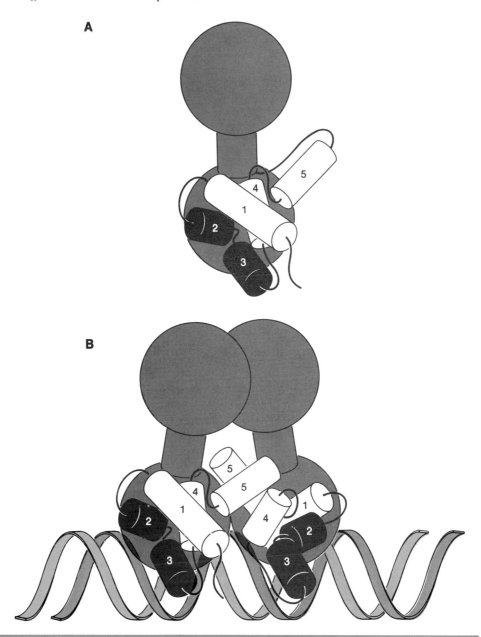

**Figure 4.12 Structure of lambda repressor**
(A) The amino-terminal domain is composed of five α-helices, represented by cylinders, connected by regions of random coil. (B) The third helix of each member of a repressor dimer fits into the major groove of the DNA where amino acids in the helix contact bases in one of the 17-bp operator elements. Redrawn with permission from Ptashne (1992).

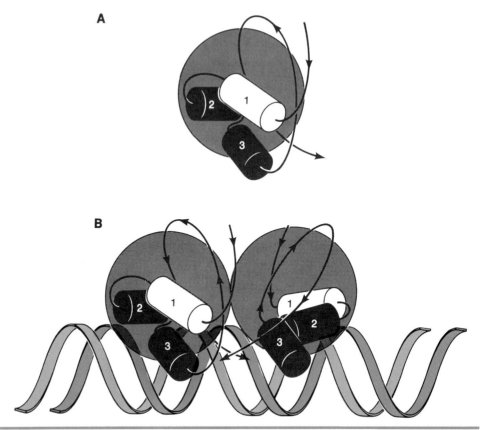

**Figure 4.13 Structure of Cro protein**
(A) Cro protein has three α-helical regions, represented by cylinders, and three regions of β-sheet structure, represented by flattened arrows, which are connected by regions of random coil. (B) The third α helix is the recognition helix that inserts into the major groove of the DNA to bind to one of the 17-bp operator elements. Redrawn with permission from Ptashne (1992).

**Figure 4.14 Interaction of Cro and lambda repressor with DNA**
(A) The recognition helixes of these two DNA-binding proteins have been pulled back from the position they occupy in the major groove of DNA to illustrate the points of contact between amino acids in the helix and bases in the DNA. The bases are numbered as designated in Table 4.1. The stars indicate base pairs that are different in $O_r1$ and $O_r3$; the diamond indicates the operator's center of symmetry. The recognition helix of the second member of the dimer would interact with bases in the other symmetrical half of the operator. (B) The binding is stabilized by hydrogen bonds that are formed between amino acids in the recognition helix and a base in the major groove of the DNA. Redrawn with permission from Ptashne (1992).

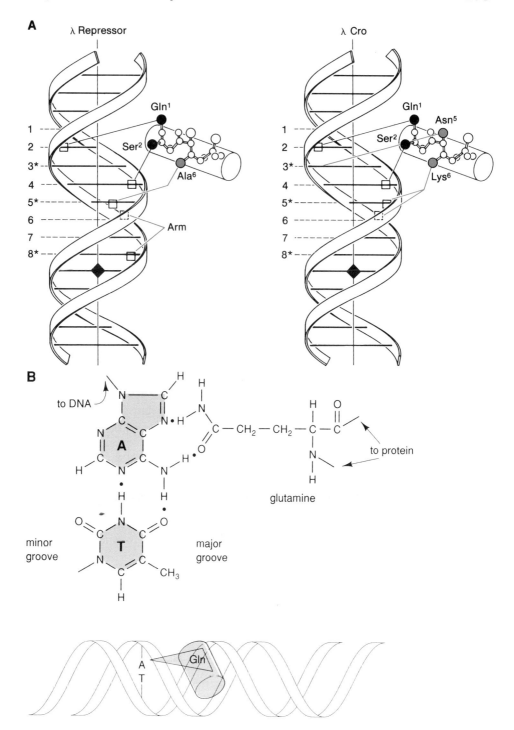

Cro has a somewhat similar structure. It consists of three $\alpha$ helices and three regions of $\beta$-sheet structures (Fig. 4.13A). The third helix of Cro is its recognition helix. That is, this helix recognizes the 17-bp regulatory elements of the $O_r$. In the Cro dimer, the recognition helices are positioned so that they will fit into the major groove of the helix (Fig. 4.13B).

In both lambda repressor and Cro, amino acids in the recognition helix (glutamine and serine at first and second positions) contact bases in the DNA of the operator (bases 2 and 4). The binding is stabilized by hydrogen bonds formed between the amino acid and a specific DNA base (Fig. 4.14). DNA-binding proteins from many different organisms have a helix–turn–helix structure, similar to that found in lambda repressor and Cro. The family of helix–turn–helix DNA-binding proteins includes many of the transcriptional factors that regulate eukaryotic gene expression.

## Control of Transcription in Eukaryotes

There are several major differences between bacterial and eukaryote transcription. In contrast to prokaryotes, eukaryotes do not have operons. Every eukaryotic gene is regulated by its own promoter. Also, it is unlikely that there are specific gene repressors or activators. That is, the DNA-binding proteins that regulate eukaryotic gene transcription probably are not specific for a particular gene, but rather activate or repress groups of related genes scattered throughout the genome. For example, all mammals develop as females unless they are exposed to the male hormone testosterone early in embryonic development. Testosterone is a steroid hormone that acts by binding to a specific cytoplasmic receptor protein. Before it binds the hormone, the steroid receptor protein is localized in the cytoplasm and cannot bind DNA. After the receptor has bound the hormone, the receptor–hormone complex acts as a transcription factor. It enters the nucleus and initiates the transcription of certain genes including many required for the development of male reproductive organs. There is a human mutation in which the gene encoding the testosterone receptor is defective so these individuals cannot respond to the hormone. The genes whose transcription is initiated by the testosterone–receptor complex remain silent in these individuals. As a result, the individuals affected by this mutation have all the secondary sexual characteristics of females, regardless of their chromosomal inheritance. This condition is called **testicular feminization syndrome.** From this we can conclude that a number of genes control the secondary sex characteristics of an individual, all of which are regulated by the testosterone receptor–hormone complex. Furthermore, these genes must be expressed early in embryonic development if the individual is to have male secondary sexual characteristics (Fig. 4.15). This example also illustrates that individual transcriptional factors in eukaryotes are involved in regulating the transcription of many genes.

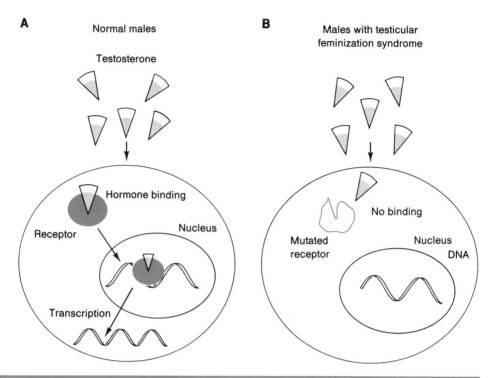

**Figure 4.15 Testicular feminization syndrome**
(A) Testosterone induces the expression of many different genes in different target cells; however, all these target cells have the same testosterone receptor, to which the hormone must bind before it is able to initiate gene expression. (B) Individuals with testicular feminization syndrome have a mutation that results in a defective testosterone receptor that is unable to bind the hormone. As a result, the target cells do not respond to the hormone and these individuals are phenotypically female even though they are genotypically male.

## Transcription of DNA in Eukaryotes by Three Different RNA Polymerases

There are three RNA polymerases in eukaryotic cells, each of which transcribes a different class of genes:

1. **Polymerase I** transcribes the genes encoding the large ribosomal RNAs.
2. **Polymerase II** transcribes the nuclear genes that encode cytoplasmic proteins.
3. **Polymerase III** transcribes transfer RNA genes and genes encoding 5 S ribosomal RNA.

These eukaryotic RNA polymerases are very large complexes, consisting of from 7 to 14 polypeptides, and they have molecular weights on the order of 500,000. Each of these polymerases recognizes a particular DNA base sequence, or group of sequences, to which it binds to initiate transcription; however, polymerase binding requires the aid of several protein *transcription factors* (TFs). The three polymerases require different transcription factors for the most part. The transcription factor is designated TFI, TFII, or TFIII, depending on which polymerase it interacts with. Each factor is additionally given a letter denoting the order in which it was discovered. TFIIA was the first transcription factor identified as required for polymerase II activity.

RNA polymerase II transcribes most of the structural genes in eukaryotes. RNA polymerase II is not competent to faithfully initiate RNA synthesis all by itself. If we place purified RNA polymerase II in a solution, along with purified DNA and other factors needed for RNA transcription, it will produce RNA molecules, but an analysis of their sequence shows that the polymerase did not faithfully transcribe the genes. Instead, it initiated transcription from a variety of false clues, such as free DNA ends, single-stranded nicks, and regions of superhelical strain. In contrast, in crude extracts of the cultured human cell line known as HeLa cells, RNA polymerase II faithfully initiates the transcription of specific genes. This demonstrates that these cells contain factors that allow correct polymerase II activity. Many of these transcriptional factors have been identified and characterized. Some of them are general transcription factors required for the transcription of most, if not all, genes, whereas others are more limited in their effects.

### TATA-Binding Protein and the TATA Box

The sequence TATAAA is found approximately 30 bp upstream from the start of transcription in most eukaryotic genes. This sequence, known as the TATA box, plays an essential role in the initiation of transcription in those genes in which it occurs. It is present in approximately 85% of the plant genes that have been sequenced thus far. Nearly all of the genes whose transcription is regulated and confined to certain tissues or organs contain a TATA box. It is less likely to be found in genes encoding metabolic enzymes. These genes may be considered "constitutive" in their expression and have sometimes been termed **housekeeping genes,** encoding housekeeping proteins.

Some modifications of the TATAAA sequence abolish the transcription of the gene, such as the mutation of the second T to a G; however, some variation in the TATA box sequence does occur when we compare different genes. Although the majority of plant genes contain a TATA box, the sequence is not identical in all cases. Most of the genes encoding photosynthetic proteins have TATA boxes with the sequence TATATA, whereas in genes encoding seed storage proteins the sequence is usually TATAAA. We do not know if these differences in plant TATA box sequences are important in the transcription of the genes that contain them, but the variation in TATA box sequence in some mammalian genes does affect their transcription. For example, the myoglobin gene contains the sequence TATAAA, whereas the simian virus 40 early promoter contains the sequence TATTAT. When the myoglobin gene TATA box sequence is changed to TATTAT, the expression of the gene is no longer stimulated by the muscle-specific enhancer present in this and other genes normally expressed in muscle tissue.

TATA boxes constitute a key component of the transcriptional machinery. The TATA box acts as a recognition signal for the binding of TFIID. TFIID is required for the transcription of genes by polymerases I and III, as well as polymerase II. Genes lacking the conserved TATA box contain a different sequence in approximately the same position to which TFIID binds. TFIID binding to either the TATA box or some other element nucleates the binding of the proteins essential for the initiation of transcription; TFIID itself is a complex of several proteins, one of which is **TATA-binding protein** (TBP). All eukaryotes use TFIID for the initiation of transcription and their genomes have one or more genes encoding TBP. Deletion or mutation of the gene encoding TBP in yeast is lethal, presumably because this blocks the transcription of most genes. *Arabidopsis* has two genes encoding TBP. Both *Arabidopsis* TBPs are very similar to yeast and mammalian TBPs with respect to their DNA binding domains. The binding of TBP to the TATA box permits the binding of another general transcription factor, TFIIB, to the promoter just downstream from the TATA box (Fig. 4.16). When both TPB and TFIIB are bound to the DNA, RNA polymerase II can initiate transcription by binding to the complex consisting of both of these proteins and the DNA to which they are bound. Although the binding of TPB and TFIIB is necessary for transcription, in most cases it is not sufficient to

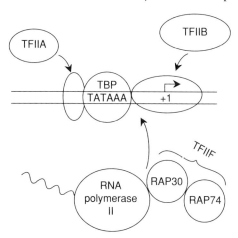

**Figure 4.16 Possible role of TATA-binding protein (TBP) and other general transcription factors in the initiation of transcription**

TBP first binds to the TATA box. Another transcription factor, TFIIB, then binds to the promoter–TBP complex. This permits the binding of RNA polymerase II to the gene so that it can initiate transcription. In some cases additional transcription factors are necessary to stabilize this complex. Redrawn with permission from Greenblatt (1991).

actually initiate transcription at a high level. Most genes have additional regulatory elements upstream from the TATA box and transcription factors must bind to these as well to initiate efficient transcription. Also, other transcription factors may interact with the TATA box–transcription factor complex to enhance or further regulate transcription.

### Viral Transcription

We have learned a great deal about the mechanism regulating transcription in eukaryotes by investigating the expression of viral genes. Viral gene promoters function in eukaryotic cells much the same way as the promoters of the host's own genes. They are transcribed by the host cell RNA polymerases and this transcription is mediated by host cell transcription factors. In some cases, viral genes can be transcribed in a wide spectrum of cells. A number of DNA viruses infect mammalian cells. The DNA tumor virus known as simian virus 40 (SV40), which infects the cells of primates, is one example. The region controlling the early expression of SV40 genes contains both promoter and *enhancer* elements. This regulatory region overlaps the region regulating the replication of the SV40 viral genome (Fig. 4.17). Although the promoter is necessary for the initiation of transcription, the enhancer greatly increases the transcriptional rate. Early gene expression of SV40 involves the transcription of the viral gene for a protein known as T-antigen. Transcription begins about 20 bp from the TATA box. There are three 21-bp repeats upstream from the TATA box which are essential parts of the promoter. The first of these is an imperfect repeat; the other two are perfect repeats. Each contains two GC-rich hexanucleotide sequences, CCGCCC. The 21-bp repeats are then followed by a different 72-bp sequence which also is repeated once.

A protein factor, known as Sp1, regulates the transcription of the SV40 early genes. Footprinting with Sp1 demonstrated that it binds to the GC-rich hexanucleotide 21-bp repeats in SV40 DNA. Many genes that lack a TATA box contain GC-rich sequences in their upstream regulatory regions, similar to the GC repeats in the SV40 early promoter. Sp1 is required for transcription of the SV40 early promoter, and it may regulate the transcription of other genes lacking a TATA box as well. Certainly it contains a structural feature that is common to a major class of DNA-binding transcriptional regulators.

### Role of Enhancers in Regulating
### Gene Expression

The 72-bp repeat also is important for transcription from the SV40 early promoter. One copy of the 72-bp repeat can be deleted without affecting transcription, but if both copies are deleted, the SV40 early promoter is inactive either *in vivo* or *in vitro*. Various modifications of SV40 DNA have been made to examine the significance of different parts of the SV40 early control region. For example,

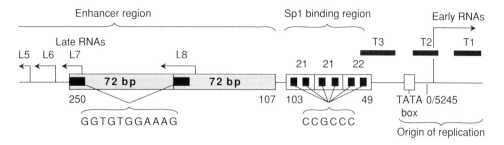

**Figure 4.17** **Diagram of the region of the SV40 chromosome that regulates early gene expression**

This region contains the TATA box and three 21-bp GC-rich repeats that are essential components of the promoter for early gene expression. The 21-bp repeats represent binding sites for the transcriptional factor known as Sp1. The 72 bp repeat is the major part of the SV40 enhancer, which contains numerous targets for transcription factor binding. T1–T3 represent binding sites for T antigen, which acts as a transcriptional repressor. L1–L8 indicate the start of transcription for individual late genes. Redrawn with permission from Herr and Clarke (1986).

a hybrid plasmid has been constructed in which the globin gene was placed under the control of the SV40 promoter. The hybrid plasmid transcribes the β-globin gene when this construct is transfected into cultured mammalian cells. Deletion of the 21-bp repeat abolishes the activity of the promoter, as does moving it elsewhere in the plasmid. Similarly, the 72-bp repeat also is necessary for promoter activity. Its role in transcription, however, is different from that of the 21-bp repeat. It could be placed anywhere on the plasmid and in either orientation with respect to the promoter and it would still stimulate transcription. This is because the 72-bp repeat is an **enhancer** or, more specifically, the SV40 enhancer. Enhancers are position- and orientation-independent stimulators of transcription. This is in marked contrast to the regulatory elements within promoters that lose activity when inverted or moved elsewhere in relation to the start of transcription. Many viral and cellular genes have enhancer sequences.

To determine the mechanism by which the SV40 enhancer stimulates transcription, different parts of its sequence have been deleted or moved around within the 72-bp sequence to determine their role in enhancing transcription. This work demonstrated that the SV40 enhancer contains several sequences, or motifs that are important for its activity (Fig. 4.18). For example, when the sequence GGGTGTGGAAAG was deleted or changed by mutation, the activity of the enhancer was reduced. This was designated the GT1 motif and it is a binding site for a HeLa cell transcriptional factor called AP3. Several other motifs important for the overall activity of this enhancer have been identified using this approach. Each of these is a binding site for a particular transcription factor (see Fig. 4.18). This illustrates a very important point about the nature of enhancers. Their activity results from the combination of several regulatory elements, or

promoter has been cloned and sequenced, and its ability to drive the transcription of other plant genes has been demonstrated. In fact, the 35 S promoter has been used extensively in biotechnology to express foreign genes in plants because it is a strong and general promoter. Why is the 35 S promoter able to be expressed at high levels and in many different tissues? This question has been investigated by analyzing the ability of the 35 S promoter to initiate the transcription of reporter genes in transgenic plants.

To investigate the components of the 5′ upstream region of the 35 S promoter that are responsible for its transcriptional activity, this region of the CaMV viral genome was cloned and fused to a *CAT* or *GUS* reporter gene lacking its own promoter. The ability of the promoter to drive the transcription of the *CAT* or *GUS* gene has been examined in several ways. In some experiments, the plasmid containing the chimeric CaMV 35 S promoter–reporter gene was introduced into protoplasts of plant cells by electroporation. In other cases, the chimeric gene has been introduced into plants by means of the *Agrobacterium* Ti vector. In either case the ability of the 35 S 5′ upstream region to drive the transcription of the reporter gene was determined by measuring enzyme activity in the transfected protoplasts or in the various tissues and organs of the transformed plants.

### Regulatory Sequences of the 5′ Upstream Sequence of the 35 S CaMV Promoter

The 5′ upstream region of the 35 S promoter is sufficient to give strong expression of a *GUS* or *CAT* reporter gene in a wide variety of plant tissues, organs, and cells that have been transformed or transfected with these chimeric genes. The signals regulating the initiation of 35 S transcription lie within the DNA sequence that extends approximately 400 bp upstream from the start of transcription (Fig. 4.20A). The location of these signals has been determined by deleting and/or manipulating portions of this sequence in the chimeric construct. The 35 S promoter is composed of two domains, designated A and B. Both of these can be further divided into seven subdomains, five within the B domain and two within the A domain (Fig. 4.20B). Domain B is the region from $-90$ to $-343$, whereas domain A is the sequence from $-90$ to the transcriptional start site. Each of these domains and subdomains has been tested for its ability to drive the expression of the *GUS* reporter gene in different tissues and organs of tobacco plants that have been transformed with these constructs. The region of the A domain from $-46$ to the transcriptional start site contains the TATA box. This is the minimal 35 S promoter. There is no *GUS* expression if it is removed from the chimeric construct. The B domain together with the A1 subdomain constitute the 35 S enhancer. Elements within the enhancer stimulate transcription and affect the ability of the promoter to be transcribed in different tissues and organs. As is the case with other enhancers, the A1 subdomain and the B domain stimulate transcription in a position- and orientation-independent manner. That is, their positions can be changed relative to the minimal promoter or they can be inverted

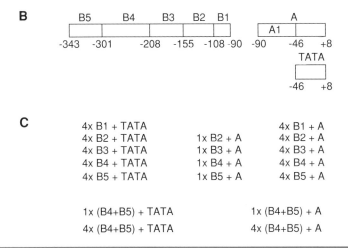

**A**

▼ -343                                                                                                    ▼ -301
TGAGACTTTT CAACAAAGGG TAATATCCGG AAACCTCCTC GGATTCCATT

GCCCAGCTAT CTGTCACTTT ATTGTGAAGA TAGTGGAAAA GGAAGGTGGC
                                                                  ▼ -208
TCCTACAAAT GCCATCATTG CGATAAAGGA AAGGCCATCG TTGAAGATGC
                                                                         ▼ -155
CTCTGCCGAC AGTGGTCCCA AAGATGGACC CCCACCCACG AGGAGCATCG
                                                                  ▼ -108
TGGAAAAAGA AGACGTTCCA ACCACGTCTT CAAAGCAAGT GGAT **TGA** TGT
       ▼ -90                                                                                       ▼ -46
GATATCTCCA C**TGACG** TAAG GGA**TGACG** CA CAATCCCACT ATCCTTCGCA
                                                                                       ⌐→
AGACCCTTCC TC**TATATAA**G GAAGTTCATT TCATTTGGAG AGGACACGCTG

**B**

|  B5   |   B4   |   B3   | B2  |  B1   |      |   A   |     |
|-------|--------|--------|-----|-------|------|-------|-----|
| -343  | -301   |  -208  | -155 | -108 -90 | -90 | -46 | +8 |

A1

TATA
-46     +8

**C**

| 4x B1 + TATA |              | 4x B1 + A |
| 4x B2 + TATA | 1x B2 + A    | 4x B2 + A |
| 4x B3 + TATA | 1x B3 + A    | 4x B3 + A |
| 4x B4 + TATA | 1x B4 + A    | 4x B4 + A |
| 4x B5 + TATA | 1x B5 + A    | 4x B5 + A |

1x (B4+B5) + TATA              1x (B4+B5) + A

4x (B4+B5) + TATA              4x (B4+B5) + A

**Figure 4.20 Sequences of the 35 S promoter and enhancer regions**
(A) Sequence of the cauliflower mosaic virus region containing the 35 S promoter and enhancer. The horizontal arrow indicates the start of transcription; the TATA sequenced is shaded. (B) To examine the function of various elements within this sequence, it was divided into the subdomains. (C) These were inserted into plants in various combinations. The effects of these combinations on the transcription of the reporter gene are shown in Fig. 4.21. From Benfey and Chua (1990) with permission.

and they still stimulate transcription. When the B domain is deleted from the 35 S promoter, the A domain alone will drive the transcription of the reporter gene, but its activity is reduced by approximately one-third. All tissues and organs are not equally affected by the loss of the B domain, however. The A domain alone supports strong expression in roots and some other meristematic tissues, although activity is lost from leaves and most other organs of the shoot. When the B domain is fused to the minimal promoter, strong expression is restored in

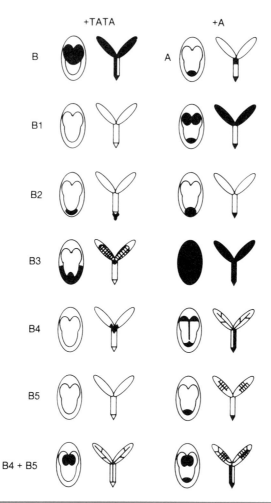

**Figure 4.21 Expression of a reporter gene in transgenic tobacco seeds and seedlings under the control of various elements of the cauliflower mosaic virus 35 S promoter and enhancer**
The two columns on the left tested the pattern of expression of the reporter gene when under the control of the element containing the TATA box combined with the B elements indicated. The two columns on the right portray the expression of the reporter gene under the control of the A element alone (top drawings). Redrawn with permission from Benfey and Chua (1990).

leaves and other organs of the shoot. The ability of the B domain to promote strong transcription in a wide variety of shoot tissues and organs results from a combination of positive regulatory signals contained within it. The individual B subdomains each have different effects on the pattern of expression of the *GUS* chimeric gene containing either the minimal promoter (TATA box) or the complete A domain, when tested in transgenic tobacco plants (Fig. 4.21). Each B subdomain contains one or more regulatory elements, and the transcriptional activity of the 35 S enhancer is a result of the synergistic interaction of these elements with the elements in the A1 region and the minimal promoter.

The A1 subdomain of the 35 S promoter contains a regulatory sequence between −85 and −64. This has been termed the **activation sequence** (*as*-1) and it is this element that is responsible for expression in root tissues. When it is inserted into the 5′ upstream region of a gene that is normally expressed in leaves, such as the gene encoding the small subunit of RUBISCO, it causes the gene to be expressed in roots as well as leaves. The *as*-1 element contains the tandemly repeated sequence TGACG which is responsible for its activity. This same regulatory element appears in some other plant genes, including some genes encoding histone proteins. The *as*-1 regulatory element is a target for a nuclear DNA-binding protein designated ASF-1. The ASF-1 protein is much more abundant in the nuclei of root tissues, which may account for the ability of the A1 subdomain to stimulate expression in root tissues.

## Oncogenes Encoding Transcription Factors

Some types of retroviruses cause tumors in vertebrates. As presented in Chapter 3, retroviruses have RNA genomes. They are called retroviruses because they reverse the usual pattern in which genetic information encoded in DNA is transcribed to RNA. When these viruses infect cells, their RNA genomes are first transcribed into DNA by the viral-encoded enzyme reverse transcriptase.

Studies of these cancer-causing retroviruses revealed that they contain genes that are able to disrupt the normal cellular machinery by changing the pattern of gene expression. These cancer-causing genes are called **oncogenes.** Viral one genes are derived from normal nuclear genes that play important roles that regulation of cellular growth, but usually they have undergone muta giving have altered their ability to function within the cell. The endogenous de pro- rise to the oncogene is known as a **protooncogene.** Protooncogen-late gene teins that are involved in the signal transdicution pathways th be involved expression. Some of these proteins are cytoplasmic and theref ibrane. Other in transmitting or amplifying signals received at the pla uclear proteins protooncogenes, including *myc, myb, erbA, jun,* and *fos*1 sarcoma *jun* was which in many cases have been shown to control of ASV17 virus 17 discovered as an insert within a retrovirus kno lacking *jun* (ASV17) which causes a type of cancer in chicl do not cause tumors, whereas those containing *jun* do. *fos* is the oncogene of the mouse virus FBJ, which causes a bone tumor. Many transcription factors in plants have been shown to be similar to the proteins encoded by these protooncogenes at the structural or amino acid sequence level.

### Leucine Zippers

**DNA-binding domains are adjacent to leucine zippers**     Analysis of the nucleotide sequences of *jun* and *fos* revealed regions of homology with the transcriptional factors AP-1 from humans and GCN4 from yeast. These proteins contain a similar region of basic amino acids in their carboxyl-terminal regions which represents their DNA binding domain. These DNA binding domains all recognize the same DNA sequence, TGACTCA, which is called the AP-1 binding sequence. The DNA binding domains of these proteins are immediately adjacent to a structure known as a **leucine zipper.** Leucine zippers are regions of proteins in which four or five leucines occur at regular intervals within an α helix so that

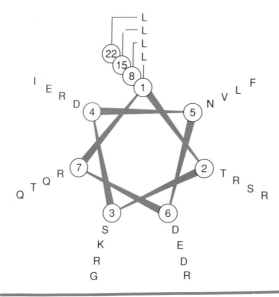

**Figure    2 Positions of amino acids in an α helix with a leucine zipper**

e amino acid sequence for residues 315–359 of the transcription factor C/EBP is
n, written as a helical wheel. The single-letter code is used to identify the amino
ote that the helix is amphipathic: acidic and polar amino acids are confined
one face of the helix; hydrophobic residues are on the other face. Redrawn
ion from Darnell *et al.* (1990).

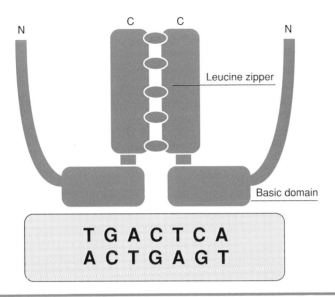

**Figure 4.23 A leucine zipper in a dimeric transcriptional factor**
Two proteins form a dimer because they interact by means of leucine zippers on similar α-helical domains. This positions their DNA binding basic domains for interaction with the two halves of a DNA sequence with dyad symmetry. Redrawn with permission from Landschulz *et al.* (1988).

they line up along the same side of the helix (Fig. 4.22). The regular occurrence of the leucines creates a hydrophobic stripe along the helix which will interact with a similar structure on another protein. As a result, proteins with leucine zippers readily form dimers (Fig. 4.23). Although *jun* will form homodimers, these do not bind to the AP-1 binding site as effectively as heterodimers composed of *fos* and *jun*. The AP-1 binding sequence has dyad symmetry, to which the *jun–fos* heterodimer will bind. There are several *jun*-like and *fos*-like proteins in plant cells. It is possible that these combine to form a series of different heterodimers, each with slightly different abilities to regulate gene transcription. Some of these *jun*-like proteins also share homology with *jun* in the amino-terminal activator domain, but others lack the amino-terminal domain. Presumably the activator domain interacts with other transcriptional factors or with RNA polymerase.

Some plant transcriptional regulators are leucine zipper proteins   The upstream regulatory regions of some plant genes contain sequences similar to the AP-1 binding site. At present it would be difficult to say how prevalent this regulatory element is in plant promoters, but it clearly occurs in the promoter of the wheat storage protein glutenin. Also, some plant regulatory genes encode proteins with leucine zippers. These include a gene known as

*Opaque 2,* which regulates the transcription of some of the seed storage protein genes in maize, and a gene that encodes a protein mediating the transcriptional effects of the plant hormone abscisic acid. Significantly, the mammalian transcriptional regulators *jun* and *fos* will initiate the transcription of genes containing AP-1 sites in tobacco cells. We can conclude that DNA-binding proteins with a leucine zipper motif play an important role in regulating plant gene transcription.

Another structural pattern found in DNA-binding proteins is the **helix–loop–helix dimer.** This is similar to the leucine zipper, except that each monomer contains two α-helical domains, separated by a region in which the protein takes on a random coil configuration and thus can loop out (Fig. 4.24). This is not to be confused with the helix–turn–helix configuration of DNA-binding proteins such as Cro and lambda repressor in which the helices form the DNA binding domains. In the helix–loop–helix proteins, the helices contain leucine zippers by which the protein dimerizes to correctly position the two DNA binding domains so that they are able to interact with the two halves of a DNA sequence with dyad symmetry. The proteins regulating the transcription of key genes in the anthocyanin biosynthetic pathway in maize have a helix–turn–helix structure.

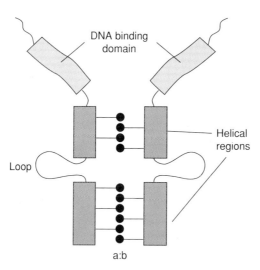

**Figure 4.24 Model of a helix–loop–helix dimeric transcription factor**
Redrawn with permission from Darnell *et al.* (1990).

## Steroid Hormones

*Steroid hormones regulate development in animals* Estrogen and progesterone regulate many aspects of animal development. Their role in the development of the chick oviduct illustrates the importance of these hormones. Estrogen is required for cytodifferentiation of glandular tissues and for the synthesis of ovalbumin, which makes up about 60% of the protein synthesized by the mature oviduct. Progesterone causes neither cytodifferentiation nor major changes in metabolic activity, but it specifically controls the synthesis of avidin, which makes up only about 0.1% of the total egg white protein. There are parallel increases in ovalbumin protein and ovalbumin mRNA after estrogen treatment of the chick oviduct. Avidin mRNA can be detected 6 hours after a single progesterone injection and it increases in amount over the course of 24 hours. In contrast to estrogen-mediated changes in ovalbumin mRNA, avidin synthesis occurs with little net change in total cellular RNA or protein, over that resulting from estrogen alone. How do steroid hormones regulate gene expression in these and other animal tissues? Extensive research on this problem in the chick oviduct and other hormonally responsive tissues has identified the mechanism by which steroid hormones regulate transcription. Although plants lack steroid hormones, many aspects of plant development are hormonally regulated and in some cases it is clear that hormonal stimulation leads to changes in gene transcription. Until we know more about the mechanism of action of plant hormones, the steroid hormones represent a useful paradigm for the action of plant hormones. In addition, analysis of the mechanism by which these hormones act has helped us understand how eukaryotic gene transcription is controlled.

*Steroid hormone receptors are potential DNA-binding proteins* There are about $10^4$ molecules of the steroid hormone receptor in each target cell. The steroid hormone receptors have a high affinity for steroid hormones ($K_d = 10^{-8}$ to $10^{-10}$ $M$) and they also can bind to specific DNA sequences within the 5′ upstream region of genes to initiate their transcription. A number of steroid hormone receptors have been characterized. Typically, their carboxyl-terminal regions represent the hormone binding domain. The central portion of the protein contains a DNA binding domain. The amino-terminal domain differs in the different steroid hormone receptors and is called the variable domain (Fig. 4.25A). The variable domain is necessary for the activity of the specific receptor. In some cases, the receptor is a dimer found free in the cytoplasm. It is unable to bind DNA or stimulate transcription because it is complexed with an inhibitor. Binding of the hormone to its receptor displaces the inhibitor and exposes its DNA binding domain (Fig. 4.25B). The DNA binding domain of the steroid hormone receptor contains two regions important for its interaction with DNA. One is rich in basic amino acids; the other is rich in cysteine. The cysteine-rich region is able to interact with the metal zinc and form what has

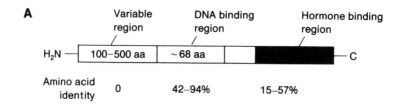

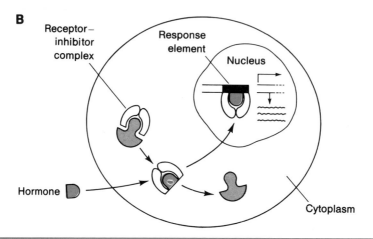

**Figure 4.25** **Steroid hormone receptors**
(A) Domain structure of steroid hormone receptors; aa, amino acids. (B) Model for the initiation of gene transcription by a steroid hormone in animal cells. The soluble, cytoplasmic hormone receptor is complexed with an inhibitor that prevents the interaction of the receptor with DNA. Binding of the hormone to the receptor displaces the inhibitor and uncovers the DNA binding site. The receptor–hormone complex enters the nucleus and binds to the response elements in the target genes to activate transcription. Redrawn with permission from Alberts *et al.* (1989).

been termed a **zinc finger.** Steroid hormone receptors have zinc fingers. Thus, analysis of the structure of steroid hormone receptors has revealed another model for DNA binding proteins.

## Summary

Gene expression is controlled on several levels, but for most genes the primary level of regulation occurs in the transcription of the gene by RNA polymerase. The coding sequence of most genes is interrupted by noncoding sequences known as introns. One or more introns divide the protein coding sequence into two or more exons. Genes also contain sequences both before the 5' end of the coding sequence (upstream) and after the 3' terminus of the coding sequence (downstream) that are important for their activity. The 5' upstream region, known as the promoter, usually contains sequences that control the transcription of the gene. One common element in eukaryotic promoters is the TATA box; other sequences also act as regulatory elements. Transcription factors bind to these regulatory elements to control the transcription of the gene by RNA polymerase. Transcription of the gene into a sequence of complementary RNA bases begins before the ATG codon, signifying the start of the protein coding sequence, and it continues through the exons and introns, ending beyond the stop codon. This primary transcript is much larger than the mRNA that will be produced from it. The mRNA is formed from the primary transcript by a number of processing reactions. These include addition of a long chain of adenosine nucleotides to the 3' terminus of the transcript (polyadenylation), addition of the 7-methyl guanosine cap structure to the 5' terminus (capping), and excision of the introns (splicing).

Control of transcription in bacteria tends to be negative. That is, a specific regulatory protein binds to a controlling element in the gene near the start of transcription and prevents the RNA polymerase from binding to the promoter and transcribing the gene. Regulation of gene transcription in eukaryotes, in contrast, usually is positive. That is, the binding of transcription factors to regulatory elements in the gene is necessary for the binding of RNA polymerase to the promoter and the subsequent transcription of the gene. It is unlikely that the transcription of any eukaryotic gene is regulated by one specific transcription factor. Rather, transcription requires a combination of transcription factors; some are common to all or at least most genes, whereas others are required for the transcription of a smaller subset of the genes, such as the genes expressed in a particular cell type. There are three RNA polymerases in eukaryotic cells: RNA polymerase I transcribes the genes encoding the large ribosomal RNAs, RNA polymerase II transcribes the nuclear genes encoding cytoplasmic proteins, and RNA polymerase III transcribes transfer RNA genes and other genes encoding small ribosomal RNAs. Each of these RNA polymerases requires a different set of transcription factors, although there is some overlap. The transcription factor TFIID is required for the transcription of all classes of genes. TFIID binding to the TATA box, or to a comparable regulatory element in genes lacking a TATA box, nucleates the binding of other transcription factors essential for binding of the polymerase to the promoter.

*continues*

This, however, usually is not sufficient to initiate transcription at a high level. The binding of several additional transcription factors to other upstream regulatory elements within the promoter may be essential to actually initiate transcription. Furthermore, the rate of transcription can be increased substantially by other regulatory elements, known as enhancers, which lie further upstream, or even in another part of the gene. Most of the regulatory elements in promoters are position dependent. They lose their activity if they are moved or inverted. In contrast, enhancers are largely position and orientation independent.

Some types of retroviruses cause tumors in vertebrates. They have tumor-inducing oncogenes in their genomes. Oncogenes were derived from normal cellular genes, which have been called protooncogenes. Several of these protooncogenes have been shown to encode transcription factors, including *myc*, *myb*, *erbA*, *jun*, and *fos*. Many transcription factors in plants have been shown to contain structural similarities to the proteins encoded by these protooncogenes. Both FOS and JUN proteins contain an α-helical region containing many basic amino acid residues that has been shown to be a DNA binding domain. The DNA binding domains of these proteins bind to a specific DNA sequence known as the AP-1 binding sequence. AP-1 binding sites often have dyad symmetry. FOS and JUN belong to a class of transcription factors known as b-ZIP proteins. These are proteins in which the DNA binding domain is immediately adjacent to a structure known as leucine zipper. The leucine zipper is an α-helical region of the protein in which hydrophobic residues, often leucines, form a hydrophobic stripe up one side of the helix. These proteins can dimerize by hydrophobic interactions through their leucine zippers. They bind more effectively as dimers than as monomers to promoters containing AP-1 binding sequences. Several plant transcription factors are b-ZIP proteins. These include the proteins encoded by the maize *Opaque* genes that control the transcription of the seed storage protein zein genes and the protein mediating the transcriptional effects of the hormone abscisic acid.

## Questions for Study and Review

1. The transcription factor Sp1 was first identified because it bound to 21-bp repeats in the early promoter of simian virus 40 (SV40). Subsequently Sp1 was found to bind to sequences in human nuclear DNA. What would you assume to be the characteristics of these

nuclear DNA sequences? Is it possible that Sp1 regulates the transcription of human genes? What role could Sp1 play in the transcription of human genes?

2. Several steroid hormone receptors have two functional domains. What are these domains and what is the role of each in the regulation of gene transcription?

3. The regulation of gene transcription can be both positive and negative. Give an example of each as it occurs in the *lac* operon of the bacterium *Escherichia coli*.

4. Many transcription factors have been identified by virtue of their ability to affect the transcription of viral genes. Why are viruses particularly useful for such studies?

5. Describe the steps required to form a mRNA from a primary gene transcript.

6. What conclusions can you draw from the fact that the genetic code is nearly identical for all organisms, including bacteria, plants, and humans?

7. What is the significance of codons that supply punctuation in the genetic code?

8. Describe three structural elements that are common features of polymers composed of many amino acids. How do these structural elements contribute to the tertiary structure of proteins?

9. Briefly outline what is required for the replication of DNA. Why is DNA synthesis discontinuous? Why is DNA replication said to be semiconservative?

10. If the guanine content of the nuclear DNA of an organism is 22%, what is its cytosine content?

11. Footprinting is a potent method for investigating regulation of transcription. What is footprinting and what types of information does it give us?

12. What features of transcription factors such as Cro and lambda repressor make it possible for them to interact with DNA? As both Cro and lambda repressor bind to the same DNA sequences, why do they regulate the transcription of different genes?

13. What are the differences between an enhancer and a promoter? Describe the position and transcriptional role of these two regulatory elements.

14. What is a peptide bond? What are the four main classes of protein amino acids, based on their chemical properties?

15. What are the chemical components of nucleotides? What are the different types of nucleotides found in nucleic acids? How do nucleotides react to form RNA? How are RNA nucleotides different from DNA nucleotides?

16. How can you show that a particular DNA sequence is involved in regulating transcription? Design an experiment to show that the 21-bp repeat of the SV40 early control region is essential for transcription.

17. What is the effect of progesterone on transcription in the chick oviduct? How does progesterone produce this effect?

18. What are the general characteristics of regulatory elements known as enhancers? Is their effect specific for a particular gene? Can you devise a model that would account for the role of enhancers in transcription?

19. M. Schena, A. M. Lloyd, and R. W. Davis (1991, *Proc. Natl. Acad. Sci. USA* **88,** 10421–10425) constructed a steroid-inducible gene expression system for plant cells and demonstrated that the steroid hormone dexamethasone could be used to induce the expression of a chimeric gene after it was electroporated into tobacco cells. What would such an expression system require to be functional in plant cells?

## Further Reading

### General References

Alberts, B., Bray, D., Lewis, J., Raff, M., Roberts, K., and Watson, J. D. (1989). *The Molecular Biology of the Cell,* 2nd ed. Garland, New York.

Darnell, J., Lodish, H., and Baltimore, D. (1990). *Molecular Cell Biology,* 2nd ed. Scientific American Books, New York.

Murphy, T. M., and Thompson, W. F. (1988). *Molecular Plant Development.* Prentice Hall, Englewood Cliffs, NJ

Ptashne, M. (1992). *A Genetic Switch: Gene Control. Phage Lambda and Higher Organisms.* Cell Press, Cambridge.

Salisbury, F. B., and Ross, C. W. (1992). *Plant Physiology,* 4th ed. Wadsworth, Belmont, CA.

Singer, M., and Berg, P. (1991). *Genes and Genomes: A Changing Perspective.* University Science Books, Mill Valley, CA.

Watson, J. D., Hopkins, N. H., Roberts, J. W., Steiz, J. A., and Weiner, A. M. (1987). *Molecular Biology of the Gene,* 4th ed. Benjamin/Cummings, Menlo Park, CA.

### Specific References

Beato, M. (1989). Gene regulation by steroid hormones. *Cell* **56,** 335–344.

Benfey, P. N., and Chua, N.-H. (1990). The cauliflower mosaic virus 35 S promoter: Combinatorial regulation of transcription in plants. *Science* **250,** 959–966.

Brennan, R. G., and Mathews, B. W. (1989). Structural basis of DNA–protein recognition. *Trends Biochem. Sci.* **14,** 286–290.

Briggs, M. R., Kadonaga, J. T., Bell, S. P., and Tijan, R. (1986). Purification and biochemical characterization of the promoter-specific transcription factor, Sp1. *Nature* **234,** 47–52.

Busch, S. J., and Sassone-Corsi, P. (1990). Dimers, leucine zippers and DNA-binding domains. *Trends Genet.* **6,** 36–40.

Calzone, F. J., Höög, C., Teplow, D. B., Cutting, A. E., Zeller, R. W., Britten, R. J., and Davidson, E. H. (1991). Gene regulatory factors of the sea urchin embryo. I. Purification by affinity chromatography and cloning of P3A2, a novel DNA-binding protein. *Development* **112,** 335–350.

Eisenmann, D. M., Dollard, C., and Winston, F. (1989). SPT15, the gene encoding the yeast TATA binding factor TFIID, is required for normal transcription initiation in vivo. *Cell* **58,** 1183–1191.

Gasch, A., Hoffmann, A., Horikoshi, M., Roeder, R. G., and Chua, N.-H. (1990). *Arabidopsis thaliana* contains two genes for TFIID. *Nature* **346,** 390–394.

Green, S., and Chambon, P. (1988). Nuclear receptors enhance our understanding of transcriptional regulation. *Trends Genet.* **4,** 309–314.

Greenblatt, J. (1991). Roles of TFIID in transcriptional initiation. *Cell* **66,** 1067–1070.

Gruissem, W. (1989). Chloroplast gene expression: How plants turn their plastids on. *Cell* **56,** 161–170.

Guiltinan, M., Marcotte, Jr., W. R., and Quatrano, R. S. (1990). A plant leucine zipper protein that recognizes an abscisic acid response element. *Science* **250,** 267–271.

Hartings, H.,, Maddaloni, M., Lazzaroni, N., Di Fonzo, N., Motto, M., Salamini, F., and Thompson, R. (1989). The O2 gene which regulates zein deposition in maize endosperm encodes a protein with structural homologies to transcriptional activators. *EMBO J.* **8,** 2795–2801.

Herr, W., and Clarke, J. (1986). The SV40 enhancer is composed of multiple functional elements that can compensate for one another. *Cell* **45,** 461–470.

Hilson, P., deFroidmont, D., Lejour, C., Hirai, S.-I., Jacquemin, J.-M., and Yaniv M. (1990). Fos and jun oncogenes transactivate chimeric or native promoters containing AP1/GCN4 binding sites in plant cells. *Plant Cell* **1,** 651–658.

Ingham, P. W. (1988). The molecular genetics of embryonic pattern formation in *Drosophila. Nature* **335,** 25–34.

Johnson, P. F., and McKnight, S. L. (1989). Eukaryotic transcriptional regulatory proteins. *Annu. Rev. Biochem.* **58,** 799–839.

Jones, N. C., Rigby, P. W. J., and Ziff, E. B. (1988). Trans-acting protein factors and the regulation of eukaryotic transcription: Lessons from studies on DNA tumor viruses. *Genes Dev.* **2,** 267–281.

Joshi, C. P. (1987). An inspection of the domain between putative TATA box and translation start site in 79 plant genes. *Nucleic Acids Res.* **15,** 6643–6653.

Kouzarides, T., and Ziff, E. (1988). The role of the leucine zipper in the Fos–Jun interaction. *Nature* **336,** 646–651.

Landschulz, W. H., Johnson P. F., and McKnight, S. L. (1988). The leucine zipper: A hypothetical structure common to a new class of DNA binding proteins. *Science* **240,** 1759–1764.

Ludwig, S. R., and Wessler, S. R. (1990). Maize R gene family: Tissue-specific helix–loop–helix proteins. *Cell* **62,** 849–851.

Matthews, R. E. F. (1991). *Plant Virology,* 3rd Ed. Academic Press, San Diego.

Murre, C., McCaw, P. S., and Baltimore, D. (1989). A new DNA binding and dimerization motif in immunoglobulin enhancer binding, *daughterless, MyoD* and *myc* proteins. *Cell* **56,** 777–783.

Ptashne, M. (1988). How eukaryotic transcriptional activators work. *Nature* **335,** 683–689.

Schena, M., Freedman, L. P., and Yamamoto, K. R. (1989). Mutations in the glucocorticoid receptor zinc finger region that distinguish interdigitated DNA binding and transcriptional enhancement activities. *Genes Dev.* **3,** 1590–1601.

Sharp, P. A. (1991). Gene transcription: TFIIB or not TFIIB? *Nature* **351,** 16–18.

Struhl, K. (1989). Helix–turn–helix, zinc-finger, and leucine-zipper motifs for eukaryotic transcriptional regulatory proteins. *Trends Biochem. Sci.* **14,** 137–140.

Varmus, H. E. (1987). Oncogenes and transcriptional control. *Science* **238,** 1337–1339.

Vogt, P. K., and Bos, T. J. (1989). The oncogene *jun* and nuclear signalling. *Trends Biochem. Sci.* **14,** 172–175.

Wefald, F. C., Devlin, B. H., and Williams, R. S. (1990). Functional heterogeneity of mammalian TATA-box sequences revealed by interaction with a cell-specific enhancer. *Nature* **344,** 260–262.

# 5

# Characteristics of Plant Cells That Are Important in Development

I ndividual cells of a complex organism are alive and they may continue to carry out many of their specialized functions when they are isolated from the organism and placed in culture. The cell is a functional unit consisting of a complex, interacting system of subcellular structures. As we have come to understand more about the organelles and macromolecular complexes that make up eukaryotic cells, the similarities between plant and animal cells have become more apparent. The cells of all eukaryotic organisms have many homologous structures with very similar functions. This similarity is a reflection of the fact that plants and animals have evolved from common ancestors. Nevertheless, higher plant cells contain a number of unique structures that are not found in animal cells. These include the plant vacuole, the cell wall, and plastids (Fig. 5.1). Furthermore,

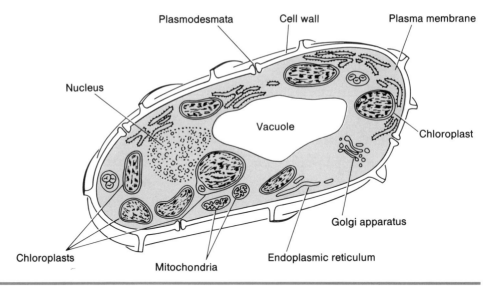

**Figure 5.1  Leaf mesophyll**
Some of the structures that might be seen in thin sections prepared for the electron microscope. Redrawn with permission from Salisbury, F. B., and Ross, C. W. (1992). *Plant Physiology*, 4th ed. Wadsworth, Belmont, CA.

the presence of a cell wall means that plants have evolved structures for cell–cell communication, namely, the plasmodesmata, that are rather different from their counterpart in animal cells. These unique structures are very important in the growth and development of plants. Furthermore, the functions of some of the structures common to plants and animal cells differ sufficiently so that the unique features of these structures in plants must be brought out if we are to understand their role in plant development. This is particularly true of the cytoskeleton, which not only functions somewhat differently in plants and animals, but also plays several key roles in plant development.

# Membranes

### Lipid Bilayer

The plasma membrane plays a vital role in cell function in that it is a selectively permeable sieve, not only keeping the cell contents in but also monitoring the entrance and exit of molecules from the cell. Often, membranes are the eyes and ears of the cell because membrane-bound receptors monitor signals coming from the external environment. Membranes are composed of a lipid bilayer, in which proteins are embedded in a mosaic pattern. The lipids found in cell membranes are **amphipathic.** That is, they tend to be linear molecules with hydrophilic groups at one end; the remainder of the molecule is hydrophobic (Fig. 5.2). The hydrophobic parts of the lipid interact with each other to form the bimolecular layer, which is oriented with the hydrophilic groups lining the surfaces of the membrane, in contact with the water of the cytoplasm or the external solution. At about 20°C the membrane lipids are in a fluid phase and have the consistency of an oil. At lower temperatures, however, the membrane lipids may solidify, preventing the movement of molecules within the bilayer.

## Membrane Proteins

Biologists recognize two groups of membrane proteins, peripheral and integral. Peripheral proteins are attached to the surface of the bimolecular lipid layer, probably by electrostatic interactions, whereas integral proteins are integrated into the lipid bilayer in whole or part (Fig. 5.3A). Integral proteins contain large hydrophobic regions, so they are held in the bilayer by hydrophobic interaction with the lipids. Frequently, integral proteins contain hydrophobic membrane-spanning domains, with one hydrophilic domain that resides within the cytoplasm and another hydrophilic domain that extends into the extracellular environment. In many cases membrane proteins are glycoprotein; that is, chains of carbohydrate molecules are covalently bonded to them. These carbohydrate chains usually are attached to the extracellular domain of the protein and they protrude into the extracellular space (Fig. 5.3B). These membrane proteins have many functions, including a very important role in cell signaling. Some membrane proteins are receptors for hormones and other molecules that provide the cell with information about its environment.

## Division of Cell into Compartments

The interior of the cell is extensively compartmentalized by membranes that surround and define several different subcellular organelles and compartments. These include the Golgi apparatus and endoplasmic reticulum (ER), which are the most important components of an interacting unit known as the endomembrane system. The membranes of this endomembrane system create a compartment that is separated from the cytosol, in which different chemical reactions can occur at different pH and in different ionic environments. Each membrane in the endomembrane system has two faces: a cytosolic face, which interacts with the cytoplasm, and a cisternal face, which interacts with the compartment interior. The ER consists of an interconnected set of membrane-bound vesicles and cisternae, which are often flattened. The ER membrane system may be physically continuous with the nuclear envelope, and the nuclear envelope is derived from the ER.

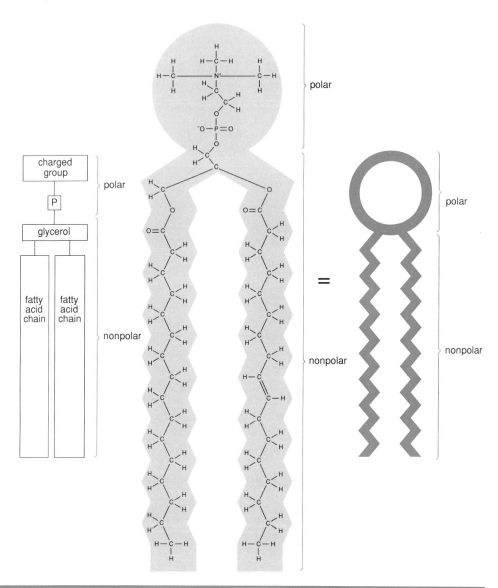

**Figure 5.2   Structure of phosphatidylcholine, a membrane phospholipid**

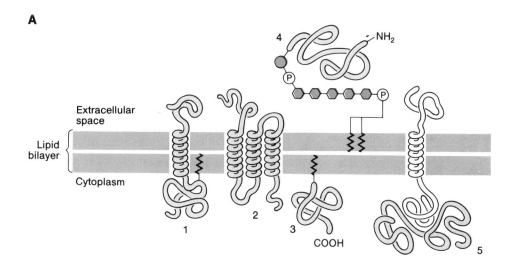

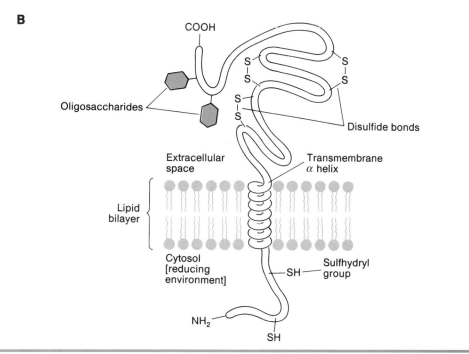

Figure 5.3 **The cell membrane**

(A) Model for the structure of the cell membrane showing the attachments of integral and peripheral proteins. Transmembrane proteins have one (1) or more (2) helices that pass through the lipid bilayer. Peripheral proteins may associate with the membrane by attachment to a lipid that is inserted in the bilayer (3), or they may be bonded to the phospholipid through an oligosaccharide (4). Proteins also may associate with membranes as a result of noncovalent interactions with integral membrane proteins (5). (B) A transmembrane protein often has oligosaccharide chains attached to its carboxy-terminal domain that can interact with cell wall polysaccharides. Intrachain disulfide bonds may occur in the extracellular domain but not the cytoplasmic domain because the cytosol is a reducing environment. Redrawn with permission from Alberts *et al.* (1989).

# Protein Synthesis and Protein Targeting

Protein synthesis is carried out by ribosomes in the cytoplasm and in mitochondria and plastids. Cytoplasmic ribosomes in plant cells are very similar in size and structure to mammalian cytoplasmic ribosomes, whereas organellar ribosomes are somewhat similar to bacterial ribosomes. In either case, the ribosome consists of a complex of RNA molecules and special ribosomal proteins. Organellar and cytoplasmic ribosomes differ in the sizes and numbers of RNA molecules they contain, as well as in the numbers of ribosomal proteins. The mechanism of protein synthesis is not covered in this book and the reader is referred to one of the general texts cited at the end of this chapter for a thorough discussion of the basic mechanisms of protein synthesis. Ribosomes initiate protein synthesis by binding to the mRNA in the 5′ transcribed, untranslated region, near the AUG codon specifying the start of translation. The ribosome then moves toward the 3′ end of the mRNA, reading the sequence of nucleotides in groups of three. Transfer RNA molecules carry amino acids to the ribosome as specified by the codon currently being read by the ribosome, and a peptide bond is formed to covalently join the amino acid to the carboxy-terminal end of the growing polypeptide chain. Before one ribosome can completely translate the mRNA, other mRNAs will initiate the translation of the same mRNA. As a result, each mRNA will be translated by, and thus attached to, many ribosomes, often 10 to 15. This complex of many ribosomes, connected by the mRNA each of them is translating, is known as a **polyribosome** or, simply, **polysome.** There are two types of cytoplasmic polysomes in most cells: free polysomes and membrane-bound polysomes. Free polysomes are not attached to any membrane surface, but appear to be freely suspended within the cytosol. In contrast, membrane-bound polysomes are attached to a membrane surface, usually the cytosolic face of the ER and, in some cases, also the cytosolic face of the nuclear envelope. These translate different messages. Proteins that function within the cytosol generally are translated by free polysomes. The membrane-bound polysomes also translate a special class of mRNAs that encode proteins that are either exported from the cell or imported into another membrane-bound compartment. Membrane proteins also are translated on rough ER. Typically, these mRNAs encode a signal peptide at the amino-terminal end of the protein that contains the information necessary for attachment of the mRNA–ribosome complex to the ER membrane. As the mRNA is translated by the ribosome, it is extruded through the ER membrane and into the ER lumen (Fig. 5.4). Once it is in the ER lumen, the signal peptide will be cleaved off and the protein may be further modified by the addition of carbohydrate residues or other posttranslational modifications.

Each protein synthesized by the cell carries out a unique structural or enzymatic function, but to do so, it must find its way to the organelle, membrane, or structure where it is to function. Proteins are targeted for specific locations within the cell. The signal sequence of proteins destined to be synthesized by membrane-

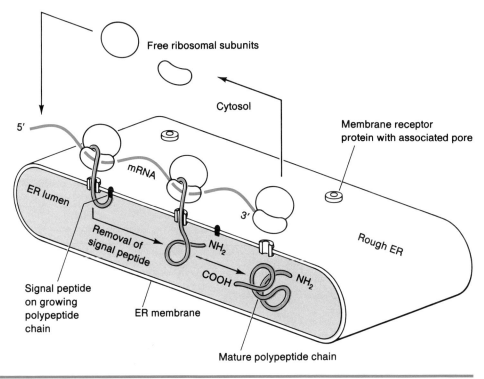

**Figure 5.4** **Model for protein synthesis on the rough endoplasmic reticulum**

Translation of the mRNA produces a signal peptide that directs the ribosome to a membrane receptor protein associated with a pore in the endoplasmic reticulum (ER) membrane. The peptide passes through the pore into the ER lumen as translation of the mRNA continues on the cytoplasmic face of the membrane. The signal peptide is removed at some point and the completed protein is released into the ER lumen. Redrawn with permission from Alberts *et al.* (1989).

bound polysomes is only one example of this kind of targeting information. Nuclear proteins are in the nucleus because they contain targeting information that will allow them to be taken up by the nuclear envelope. Similarly, specific amino acid sequences in chloroplast proteins enable these proteins to pass from the cytoplasm, where they are synthesized, into the chloroplast.

## The Golgi Apparatus

The Golgi apparatus is not physically connected with the ER, but ER-derived, membrane-bound vesicles carry materials from the ER to the Golgi apparatus.

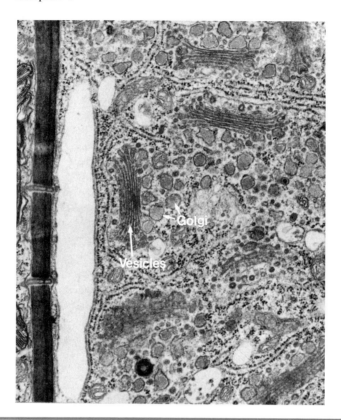

**Figure 5.5    Electron micrograph of a cell from the green alga**
*Bulbochaete* **showing numerous stacks of Golgi cisternae**
**and associated vesicles**
Reprinted with permission from Gunning and Steer (1975).

Structurally, the Golgi apparatus consists of a stack of four to six flattened sacs, called **cisternae,** surrounded by a cloud of vesicles (Fig. 5.5). The stack of cisternae is known as the **dictyosome.** Membrane-bound vesicles derived from the ER fuse with one of the dictyosome cisternae known as the forming face. Materials carried by these vesicles may be processed within the dictyosome cisternae. Finally, vesicles that are pinched from the maturing face of the dictyosome may be transported to the plasma membrane or to another membrane-bound compartment, such as the vacuole (Fig. 5.6). This is the path taken by cell wall proteins, such as extensin (see later) and some seed storage proteins. In addition, the Golgi apparatus plays an important role in the synthesis of cell wall polysaccharides.

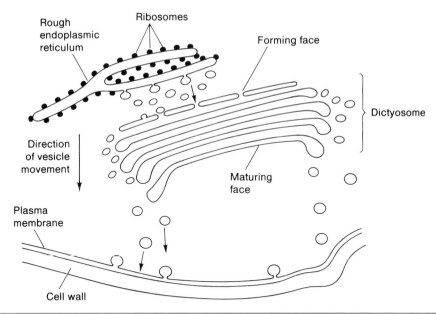

**Figure 5.6** **Flow of membranes and other materials from the endoplasmic reticulum through the Golgi to the plasma membrane**
Proteins synthesized on the endoplasmic reticulum, such as the wall protein extensin, and noncellulosic carbohydrates synthesized in the Golgi are transported to the plasma membrane. This is the source both of new cell wall components and of plasma membrane.

## Plant Vacuoles

The plant vacuole is a unique, multifunctional organelle that has no direct analog in animal cells, although some of its properties are similar to those of lysosomes. The vacuole is the largest compartment in mature plant cells and it may occupy up to 90% of the total cell volume (Fig. 5.7). It is separated from the cytoplasm by a semipermeable membrane called the **tonoplast.** Meristematic cells have numerous small vacuoles, but as plant cells grow, these fuse to form a single central vacuole. Vacuoles are derived from the membrane network at the maturing face of the Golgi apparatus. Small vacuoles formed by this system fuse to construct the large central vacuole. As the vacuole volume increases, the tonoplast expands by incorporating vesicles derived from the ER and the Golgi apparatus. Vacuoles have many different functions and properties, depending on the cell type in which they occur.

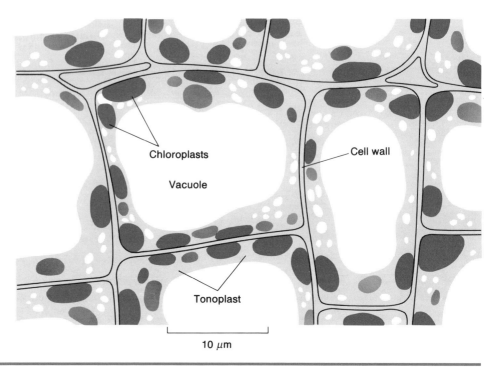

**Figure 5.7    Electron micrograph of tobacco leaf mesophyll cells with large central vacuoles**
Redrawn with permission from Gunning and Steer (1975).

### Vacuoles as Osmotically Active Compartments

Vacuoles play a dynamic role in plant growth and development. In elongating cells, organic compounds and inorganic solutes accumulate in the vacuole. These solutes create the osmotic pressure that is responsible for the turgor pressure essential to their extensive growth. Solutes may accumulate in the vacuoles by active transport against an electrochemical concentration gradient. The vacuolar sap is acidic because the tonoplast contains an active ATP-driven proton pump, in which protons are transported into the vacuole from the cytoplasm, creating a proton-motive force (PMF). This PMF, then, is the driving force for a variety of secondary active transport systems. The pH of the cytoplasm is about 7.5, and that of the vacuole typically is about 5.0; however, plants can have a broad range of vacuolar pH values. Cells with a high concentration of organic acids in their vacuoles typically have very acidic vacuolar pH values. Succulent plants store captured light energy during the day but do not fix carbon with this energy until it is dark, when the light energy is used to synthesize organic acids which are stored in the vacuole. This diurnal fluctuation of vacuolar organic acids is known as **crassulacean acid metabolism** (CAM). As a result, the vacuoles of CAM plants undergo diurnal fluctuations, changing from pH 6.0 in the light-grown plant to pH 3.5 when malic acid and protons accumulate in the vacuole during the dark cycle.

Isolated vacuoles also may have transport systems for a number of specific substances, including sugars, organic acids, and alkaloids, where these substances accumulate in vacuoles. For example, a membrane protein known as a permease is responsible for uptake of the organic acid malate by *Bryophyllum* and barley vacuoles. Often these permeases are very specific. Nicotine and vinblastine are both alkaloid secondary products and are synthesized by *Nicotiana* (tobacco) and *Catharanthus* (periwinkle), respectively. Nicotine is taken up by isolated tobacco vacuoles, but not by periwinkle vacuoles. Similarly, vinblastine is taken up by periwinkle vacuoles, but not by tobacco vacuoles.

## Vacuoles as Lysosomes

Vacuoles are similar to lysosomes. Lysosomes are characterized as unit membrane-bound organelles that are autophagic; that is, they are capable of digesting other cellular components. They contain enzymes that enable them to hydrolyze proteins and nucleic acids, as well as enzymes that rupture phosphate and other esters and glycosidic bonds (Fig. 5.8). Most of the acid hydrolases in plant vacuoles are glycoproteins, as are the hydrolases of animal lysosomes. Autophagy is a function of the small vacuoles in young cells, but mature vacuoles have not been shown to degrade cytoplasmic macromolecules. Although vacuolar proteinases are highly active in plants, their function in the dynamics of cellular protein turnover is not known. Typically, only 1 to 10% of the total protein of the mature cell is contained in the vacuole and the proteins that remain in the vacuole at maturity may be the remains of the early developmental phase of the vacuoles which were autophagic. There are, however, specialized tissues in which the vacuole is used as a storage compartment for proteins.

## Vacuoles as Storage Depots

Vacuoles also may be dynamic storage compartments in which ions, proteins, and other metabolites accumulate and later can be mobilized. Seed storage proteins in legumes and other dicots are synthesized on rough ER, glycosylated in the ER, transported to the Golgi, where further glycosylation takes place, and finally packed into the protein bodies, where they are further processed by partial deglycosylation and partial proteolysis (Fig. 5.9). During germination, a protein-ase transported into the protein bodies degrades the storage proteins. Similarly, plants that store large amounts of protein in vegetative tissues frequently do so in vacuoles. Leaves of potatoes, tomatoes, and other members of the Solanaceae contain protein inhibitors of chymotrypsin and other proteinases which are deposited in vacuoles.

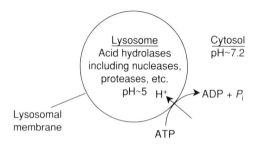

**Figure 5.8    Lysosomes**

Lysosomes are membrane-bound organelles that contain numerous hydrolytic enzymes with acid pH optimums. The acidic pH of the lysosome is maintained by a proton pump in its membrane which uses energy derived from the hydrolysis of ATP to pump $H^+$ into the organelle from the cytoplasm.

### Vacuoles as Depositories for Secondary Products

Vacuoles are depositories for secondary metabolites and metabolic waste products. Many plant cells synthesize water-soluble pigments, such as anthocyanins and betanins, which are localized in the vacuole. Other secondary products, including alkaloids, saponins, glucosinolates, cyanogenic glycosides, and coumaryl glycosides, are nearly completely sequestered in vacuoles. Many of the secondary products accumulated in vacuoles are toxic to pathogens, parasites, herbivores, and the plant itself. For example, the alkaloid nicotine is synthesized by tobacco roots and transported into the shoot, where it preferentially accumulates in the vacuole. In many cases, the vacuole is a static compartment. Substances deposited in the vacuole never return to the cytoplasm. For example, specialized cells in plants such as *Dieffenbachia* form calcium oxalate crystals in their vacuoles. These stores of calcium and oxalic acid never return to the cytoplasm. This is what we would expect from the classical view of the vacuole as a substitute for the excretion system of animals. Yet, the vacuole system also may be dynamic, as we have seen from the preceding discussion of the diurnal fluctuation of vacuolar organic acids in CAM plants. In addition, vacuoles of storage organs also exhibit seasonal changes in the accumulation and mobilization of carbohydrates, and the opening and closing of stomates involve rapid movement of solutes in and out of the guard cell vacuoles.

## The Cytoskeleton

The cytoskeleton is an interconnected network of filamentous proteins found in the cells of nearly all eukaryotes. The principal functions of the cytoskeleton are to determine cell shape, to organize the cytoplasm, to transport cellular structures

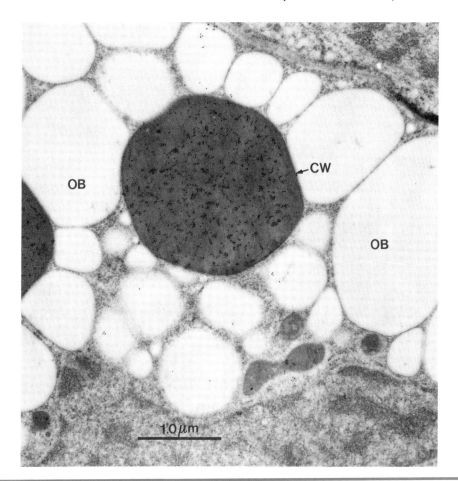

**Figure 5.9    Synthesis and vacuolar deposition of the maize storage protein zein in transgenic tobacco embryonic cells**
The gene encoding a maize 15-kDa zein storage protein was placed under the control of regulatory elements for a dicot seed storage protein and introduced into the tobacco genome by *Agrobacterium*-mediated transformation. The chimeric gene was expressed only in the developing embryo of the transgenic tobacco plants, and it was correctly targeted for the vacuole and packaged into protein body crystalloids (C). The black grains over the crystalloid are immunogold particles. This section was stained with zein-specific antibody attached to colloidal gold particles. The fact that these colloidal immunogold particles are localized only over protein body crystalloids demonstrates that the zein protein was deposited in these structures. The embryonic cells also store oil in oil bodies (OB). CW, cell wall. Reprinted with permission from Hoffman *et al.* (1987).

such as vesicles and chromosomes, and to bring about cell motility. The cytoskeleton is composed of three filamentous structures: microtubules, microfilaments, and intermediate filaments. **Microtubules** (MTs) are constructed chiefly of subunits of the heterodimeric protein tubulin, whereas microfilaments (MFs) are composed principally of subunits of the protein actin. Although plant cells contain intermediate filament proteins, their biochemical nature and their role in plant cell structure have not been determined. All three cytoskeletal components may contain additional specific proteins that contribute to their function and connect them with each other or with other cellular structures. Furthermore, the different components of the cytoskeleton interact with each other as well as with membranes and other cellular structures to carry out its functions.

## Microtubules

Microtubules are filaments that are composed principally of a protein known as tubulin; and other proteins known as **microtubule-associated proteins (MAPs).** They average about 240 Å in diameter and may be many micrometers in length. In cross section in transmission electron microscopy they look like small tubes or pipes. The electron-translucent core of the microtubules is about 100 Å in diameter and the wall of the tubule is about 70 Å thick. In a negatively stained preparation, the wall of the microtubule can be seen to be composed of roughly globular subunits, each of which is about 40 Å in diameter. The microtubule appears to be composed of 11 to 13 filaments arranged in a circle, with each filament consisting of a string of globular units (Fig. 5.10). In thin sections, microtubules usually have lateral appendages by which they seem to be connected to other microtubules or to other cellular organelles. MAPs associate with the surface of the microtubule to form these lateral projections. There are many different MAPs, and the nature of the MAPs determines which structures the microtubule may interact with and how this interaction will occur. MAPs stabilize microtubules and they determine the specific function of the microtubule, but they are not necessary for the formation of the basic microtubule structure. The tubulin dimer contains all the information necessary to self-assemble and form the microtubule. Microtubules have a distinct polarity, which may reflect the fact that the tubulin dimer is inherently asymmetrical. This polarity is easily demonstrated experimentally (Fig. 5.11). Polarity is of considerable significance to the cell, as transporter molecules move along the microtubule in one direction only.

Tubulin is a highly conserved protein    Microtubules are formed by the self-assembly of soluble tubulin protein. Tubulin is a heterodimer, composed of polypeptides designated $\alpha$- and $\beta$-tubulins. These are similar in amino acid composition, but they are sufficiently different so that antibodies raised to $\alpha$-tubulin do not react with $\beta$-tubulin, and vice versa. The tubulin dimer has a molecular weight of about 100,000, whereas the molecular weights of $\alpha$- and $\beta$-

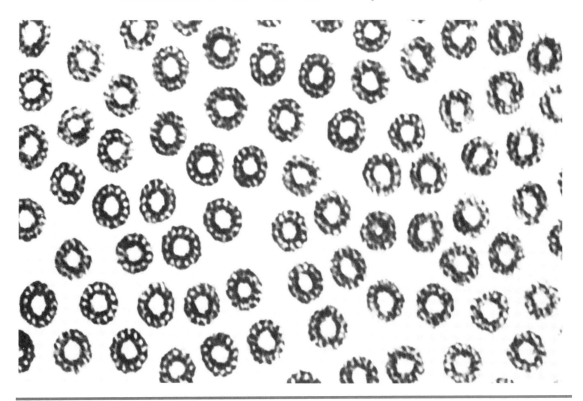

**Figure 5.10 Electron micrograph of *in vitro*-assembled yeast microtubules**
Reprinted with permission from Morejohn and Fosket (1986).

tubulin are both near 50,000. Tubulins are encoded by multigene families in all of the higher plants examined. There are at least six $\alpha$-tubulin and nine $\beta$-tubulin genes in the *Arabidopsis* genome. These encode closely related proteins, and in fact the tubulins from very divergent organisms will co-assemble to form microtubules. We do not know if these different tubulins have different functions in the plant, but it appears that they are interchangeable as far as their ability to form microtubules is concerned. If they have different functions, it may be because they interact with different MAPs.

Microtubules form by a self-assembly process   Tubulin will undergo self-assembly to form microtubules in a test tube if certain conditions are met. This self-assembly is promoted by GTP, $Mg^{2+}$ ions, and a low concentration of $Ca^{2+}$ ions, and has a pH optimum of 6.9. Under these conditions and at 20 to 37°C, tubulin spontaneously assembles into microtubules. Microtubules will

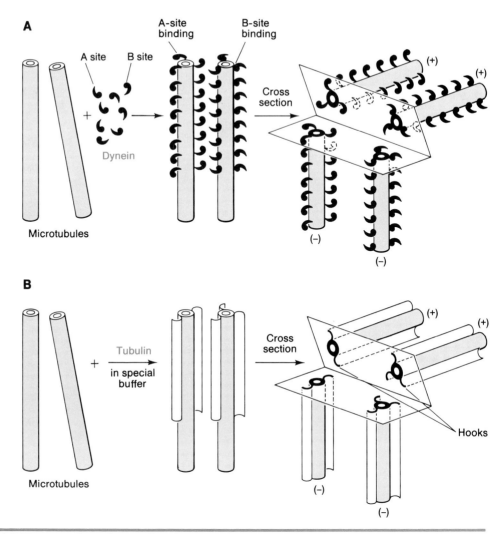

**Figure 5.11 Determination of microtubule polarity**
The polarity of microtubules is determined by incubating them in a solution containing either dynein or free tubulin dimers. Dynein, a microtubule motor of flagella, is an asymmetrical molecule and it binds to microtubules by either its A or B binding site. When viewed in thin section, the orientation of the dynein or tubulin hooks demonstrates the polarity of the microtubule. Redrawn with permission from Bershadsky and Vasilev (1988).

disaggregate to tubulin dimers if the solution is cooled to 0°C. This is a reversible reaction:

$$\text{tubulin dimers} \rightleftharpoons \text{microtubules}$$

Both of these reactions occur simultaneously. Whether the equilibrium of the reaction is shifted toward microtubule formation or toward dimer formation depends on the conditions under which the reaction occurs. Cold temperature and high $Ca^{2+}$ concentration favor dimer formation and microtubule disaggregation, whereas warm temperature and low $Ca^{2+}$ concentration favor microtubule formation. There is a critical tubulin concentration, $C_c$, below which no microtubules are formed. At concentrations above the $C_c$, the amount of polymer formed is proportional to total tubulin concentration, but all the tubulin dimers do not assemble. The microtubules behave as though they are in equilibrium with a pool of unassembled free tubulin dimers, the concentration of which equals the $C_c$ (Fig. 5.12A). Tubulin self-assembly is a two-step process in which a comparatively slow nucleation step is followed by a rapid elongation reaction. Nucleation is the process in which aggregates are formed that have a greater probability of persisting than disaggregating. The rate of subunit addition to the microtubule ends, $k^+$, is proportional to the free subunit concentration, $c$. Subunits also are lost from the ends of the microtubules at a rate, $k^-$, independent of the free subunit concentration. As a result, the net assembly rate at one end of the microtubule is

$$dn/dt = ck^+ - k^-$$

This equation also defines the $C_c$, where $dn/dt = 0$, as

$$C_c = k^-/k^+$$

Assembly results in the progressive reduction of the free tubulin dimer concentration so that the reaction should come to equilibrium and the net assembly rate should reach zero when the free tubulin dimer concentration is reduced to $C_c$.

GTP hydrolysis is important for tubulin assembly–disassembly
Tubulin binds 2 moles of GTP per mole of tubulin dimer. One of the GTPs is exchangeable with GTP in solution; the other is not exchangeable. After assembly, GTP bound at the exchangeable site is hydrolyzed to GDP by a GTPase activity of the tubulin itself. Although GTP must be bound to tubulin if it is to assemble, GTP hydrolysis is not necessary for assembly. The dissociation of GTP-tubulin occurs at a slower rate than that of GDP-tubulin, and it has been proposed that GTP-tubulin provides a kinetic cap on the microtubules (Fig. 5.12B). According to this model, GTP hydrolysis lags behind assembly, particularly at high assembly rates. GTP hydrolysis is, however, a stochastic process and there is a probability that GTP hydrolysis will occur in all or most of the subunits at the microtubule end so that the cap is lost. Once the microtubules have lost their GTP-tubulin caps, the exposed GDP-tubulin rapidly disassembles. Once the GDP-tubulin is released from the microtubule, the bound GDP is free to exchange with GTP in solution, thus regenerating GTP-tubulin which can again participate in the assembly reaction. It is assumed that there is much greater probability that the GTP-tubulin will be added to capped than uncapped microtubules. As a result,

an individual has been used to follow the behavior of the centrosome during the cell cycle of mammalian cells. The scleroderma antiserum containing antibodies to human pericentriolar material also stains the granules found at the poles of mitotic onion cells, from which the microtubules of the mitotic spindle emanate. Angiosperm cells have an extensive array of microtubules in the cortical cytoplasm. There are no obvious structures acting as MTOCs from which these cortical microtubules are assembled and no structures in the cortical cytoplasm stain with the scleroderma anticentrosome serum. Although the microtubules of the mitotic spindle may be nucleated by MTOCs with some homology to those nucleating the microtubule arrays of animal cells, other plant microtubule arrays appear to be nucleated by different MTOCs.

**Plant microtubules often are cold stable**   The microtubules of many higher plants are resistant to cold depolymerization. This is not surprising, as temperate zone plants must be able to withstand a wide range of temperatures, and some plants are capable of growing at temperatures that would depolymerize the microtubules of homothermic animals. Cold-stable microtubules may arise by two mechanisms. First, microtubules are stabilized by interaction with MAPs. A group of mammalian brain MAPs known as STOPs bind to microtubules at substoichiometric levels and increase their resistance to cold-induced depolymerization. In other cases, the tubulin itself may be resistant to cold-induced depolymerization. At present, we do not know which of these mechanisms results in the cold stability of plant microtubule arrays.

Some plants have cold-sensitive microtubules. The sensitivity of plant microtubules to cold depolymerization may be related to their cold hardiness. Interestingly, plant hormone treatments have been shown to increase both the resistance of sensitive plants to chilling injury and the resistance of their microtubules to cold-induced depolymerization.

**Some plant secondary products affect microtubule assembly**   A number of plant alkaloids have been found to bind to tubulin and to affect microtubule formation, at least in preparations of animal tubulin (Fig. 5.13). The best known of these is colchicine, an alkaloid produced by the autumn crocus, *Colchicum autumnale,* which binds to the tubulin dimer and prevents its self-assembly. As a result, colchicine brings about rapid microtubule disassembly. An alkaloid produced by the yew plant, *Taxus brevifolia,* and named **taxol** also binds to tubulin dimer; however, tubulin complexed with taxol assembles more rapidly, and the subsequent taxol microtubules cannot be disassembled by cold treatments. Other plant alkaloids that have been shown to bind to tubulin include podophyllotoxin and the *Catharanthus rosea* alkaloids vincristine and vinblastine. Many of these compounds have been used in medicine to treat specific human diseases. Both taxol and vinblastine are used in cancer chemotherapy. They are effective because they are able to block the division of cancer cells by preventing formation

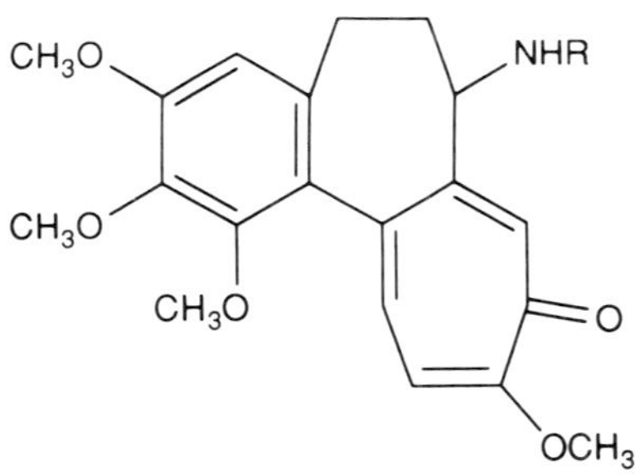

nocodazol

colchicine (R = COCH₃)
colcemid (R = CH₃)

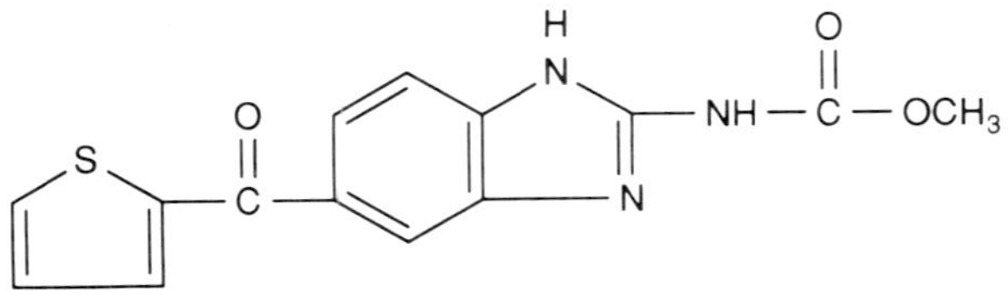

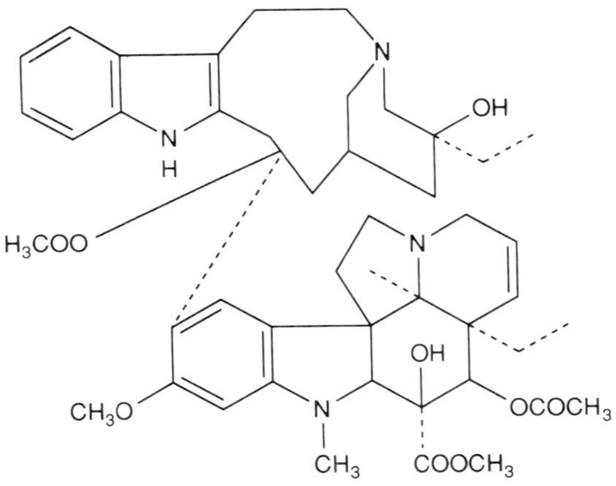

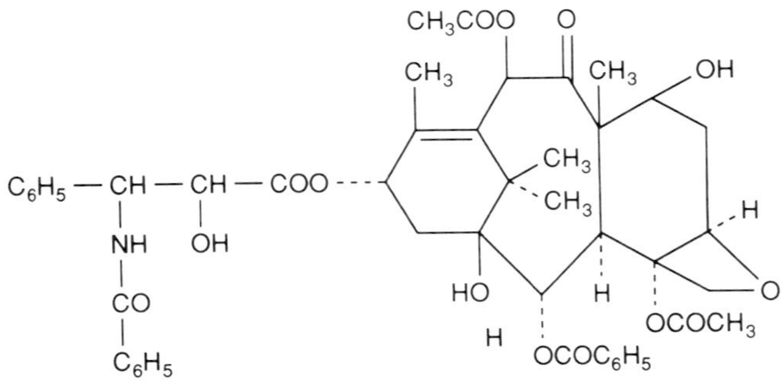

taxol

vinblastine

**Figure 5.13 Structures of some chemical compounds that affect microtubule assembly**

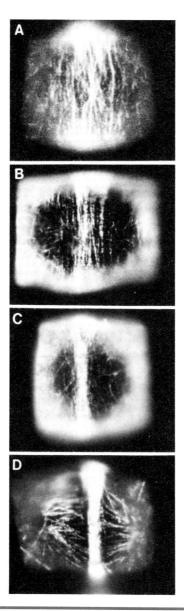

**Figure 5.26 Formation of the preprophase band**

A series of fluorescent micrographs of wheat root cells are shown after staining with a fluorescent-labeled antibody against tubulin. (A) The beginning of the preprophase band is indicated by a loss of microtubules near either end and the formation of additional microtubules in the middle of the cell. (B) The microtubules become more tightly focused in the middle region of the cortex, but retain a distinctly fibrillar appearance. (C) The preprophase band narrows and it is difficult to distinguish individual fibers. (D) As the mitotic spindle microtubules are initiated in prophase, the preprophase band begins to break down. Reprinted with permission from McCurdy and Gunning (1990).

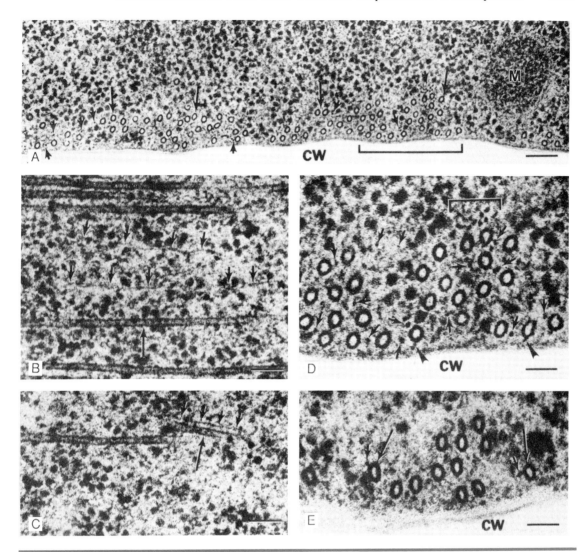

**Figure 5.27 Microtubules and microfilaments of the preprophase band in tobacco root cells**

These are high-resolution transmission electron micrographs. (A) Cross section of a preprophase band (PPB). Bar = 100 nm. (B,C) Longitudinal sections through a PPB. Bar = 80 nm. (D) Enlargement of the region of the PPB in (A) indicated by brackets. Bar = 80 nm. (E) Cross section of a PPB showing the close association between microtubules and microfilaments. Bar = 50 nm. Long arrows indicate microtubules; short arrows point to microfilaments. Bridges between microtubules and the plasma membrane are indicated with arrowheads. CW, cell wall; M, mitochondrion. Reprinted with permission from Ding *et al.* (1991).

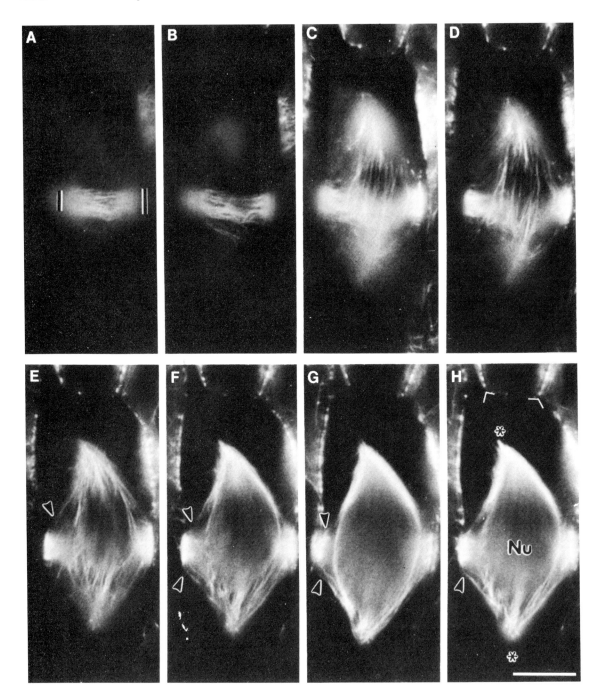

spindle develops (see Fig. 5.28). The preprophase band begins to break down after the mitotic spindle is formed, and it is largely gone by the time the metaphase spindle is fully formed. Other functions of the preprophase band are to determine the plane of cytokinesis and to determine the position in which the cell plate is inserted into the parental cell wall. Although the mechanism by which this is accomplished is not known, the preprophase band is thought to alter the cell wall adjacent to where it occurs so as to mark this position for the insertion of the cell plate. The preprophase band plays a central role in plant morphogenesis as a result of its ability to determine the plane of cell division, as we shall see later.

The mitotic spindle begins to form in prophase when the nucleus is still surrounded by radial array microtubules   Initially the microtubules of the radial array show no preferential orientation, but as the polar determinants begin to exert their effect, microtubules emanating from MTOCs in the two polar regions toward the cell equator become more prominent. The framework of the mitotic spindle is in place as a cage around the nucleus before the nuclear envelope breaks down in late prophase, but the chromosomes do not become attached to the spindle microtubules until after the nuclear envelope has disintegrated. Once the nuclear envelope has disassembled, the spindle becomes more highly ordered. Some spindle microtubules are captured by the specialized structure known as a kinetochore that is localized at the chromosomal centromere; whereas others remain free and terminate in the cell equator. The latter are **polar microtubules;** those captured by kinetochores are known as **kinetochore microtubules.** Both kinds of microtubules have the same polarity and the two halves of the mitotic spindle are mirror images of each other (Fig. 5.29). The plus end of the microtubules is toward the cell equator; the minus ends are at the poles. Each chromosome is replicated during interphase. At prometaphase

**Figure 5.28** **The preprophase band in relation to the mitotic spindle in a dividing onion cotyledon cell**

A series of optical sections were made through a single prophase cell that contained a tightly focused preprophase band (PPB) and was initiating its mitotic spindle. The cell was stained by an immunofluorescence procedure to visualize microtubules. (A,B) The microscope is focused on the microtubules nearest the plasma membrane, with the width of the PPB indicated by vertical bars. The microtubules of the spindle are not visible in this focal plane. (C,D) The microscope is focused further into the cytoplasm so some of the spindle microtubules, as well as those of the PPB, become visible. (E–H) Each micrograph is focused deeper into the cell, revealing progressively more of the mitotic spindle microtubules. The asterisk in (H) indicates the location of the spindle poles. Arrowheads indicate microtubules linking the PPB with the spindle. Nu, nucleus. Bar = 10 $\mu$m. Reprinted with permission from Mineyuki *et al.* (1991).

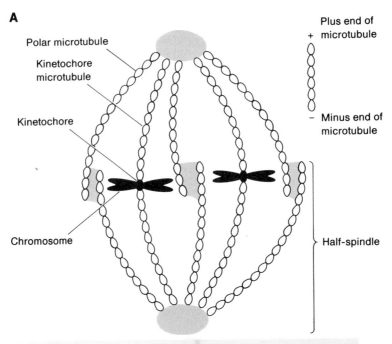

**A**

Polar microtubule

Kinetochore microtubule

Kinetochore

Chromosome

Plus end of microtubule +

− Minus end of microtubule

Half-spindle

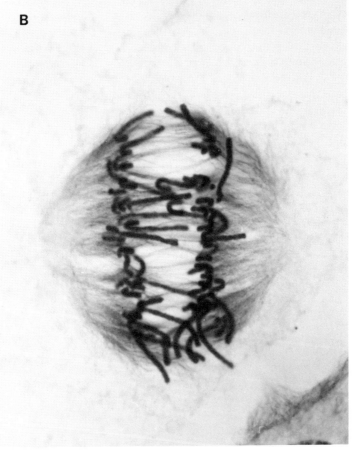

**B**

these two chromosomes are visible, but they are still held together at their centromeres and are called chromatids. Each chromatid has its own kinetochore. The two chromatids of any given chromosome have their kinetochores facing in opposite directions. The prometaphase chromosomes are pulled in various directions as they become attached to the spindle microtubules, and it is not until after microtubules have attached to both chromatids that they come to rest at the metaphase plate.

Transport of the chromosomes to the poles is a complex process that appears to involve three different microtubule behaviors. Motor proteins, possibly a type of cytoplasmic dynein, may attach the microtubules to the kinetochores. These molecular motors are able to walk along the microtubules, pulling the chromosomes toward the poles. They use energy derived from the hydrolysis of ATP to transport the chromosome along the spindle microtubules (Fig. 5.30). While this is occurring, there is a depolymerization of the kinetochore microtubules from the kinetochore end. Apparently tubulin subunits are removed from the kinetochore microtubules as the chromosomes are pulled toward the poles by the microtubule-dependent motor also localized here. These movements are known as anaphase A. Although anaphase A is a major part of the mechanism by which microtubules transport chromosomes to the poles, there is an additional mechanism that is important in separating the chromosomes. During anaphase B, which begins part way through anaphase A, the poles move further apart. This is brought about by a lengthening of the polar microtubules, and possibly their sliding past each other in the equatorial region where they overlap.

**The phragmoplast constructs the cell wall that will partition the cytoplasm and separate the two daughter nuclei of the dividing cell**   The phragmoplast forms in late telophase, as the spindle disappears. It consists of two sets of parallel microtubules, both oriented at right angles to the division plane (Fig. 5.31). The two sets of microtubules overlap at their tips and have the same polarity. Their plus ends are in the equatorial plane where they overlap; their minus ends are nearer the poles. Microfilaments also are present in the phragmoplast and connect the phragmoplast to cortical cytoplasm adjacent to the lateral walls at the site of the PPB. The phragmoplast begins to form in late telophase and it represents a new site of microtubule assembly. The microtubules of the mitotic spindle have largely disappeared by the time the

**Figure 5.29  The mitotic spindle**
(A) Diagram of the mitotic spindle as it might appear at metaphase. The microtubules in both halves of the spindle are oriented with their minus ends toward the poles. Some of the microtubules overlap in the center of the cell. (B) Immunogold-stained *Hemanthus* cell in anaphase. The fibers seen are the spindle microtubules, which are pulling the chromatids toward opposite poles. (A) Redrawn with permission from Alberts *et al.* (1989) and (B) courtesy of Andrew Bajer.

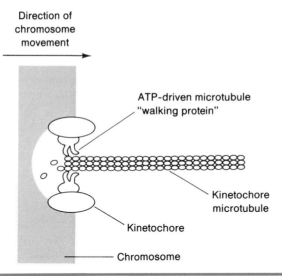

**Figure 5.30 Model for the movement of chromosomes to the poles at anaphase**
A motor protein such as kinesin may be associated with the kinetochore so that it walks along the spindle microtubules, pulling the chromosome toward the pole. The microtubule is disassembled as the motor walks toward the pole. Redrawn with permission from Alberts *et al.* (1989).

phragmoplast is formed, although some polar spindle microtubules may be recruited for the phragmoplast, and many additional microtubules are assembled to form the dense phragmoplast array. The **cell plate,** which is the new cell wall that will separate the daughter cells, is constructed in the region of the phragmoplast where the ends of the microtubules overlap. It consists largely of noncellulosic polysaccharides, which are synthesized in the Golgi and transported to the cell plate in Golgi-derived vesicles by the phragmoplast microtubules. The vesicles fuse in the equatorial plane and the noncellulosic polysaccharides they contain become the middle lamella of the cell wall that will divide the daughter cells. The vesicle membrane becomes the plasma membrane. Microfilaments radiate out from the phragmoplast to the peripheral cytoplasm, into the cortical cytoplasm (Fig. 5.32). These microfilaments probably orient the growing cell plate, ensuring that it will insert into the site occupied by the preprophase band before the initiation of mitosis.

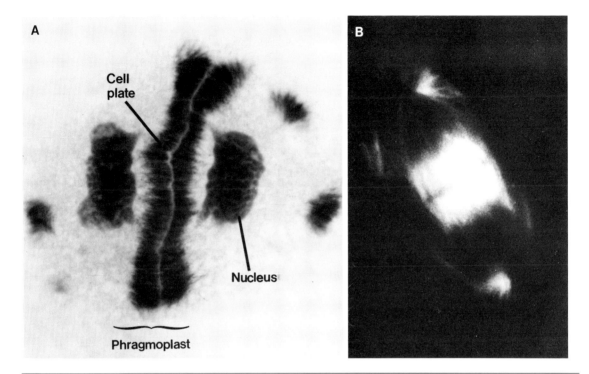

**Figure 5.31 The phragmoplast of dividing plant cells**
(A) The microtubules of the phragmoplast are visualized by an immunogold procedure using an antibody specific for tubulin. The developing cell plate is indicated. (B) The phragmoplast microtubules are visualized by an immunofluorescence procedure using a fluorescein-labeled antitubulin antibody. Both the immunofluorescence and immunogold procedures demonstrate that the phragmoplast consists of two overlapping sets of microtubules. Courtesy of (A) Andrew Bajer and (B) Susan M. Wick.

## Plant Cell Walls

One of the most significant differences between a plant cell and an animal cell is that the plant cell is surrounded by a wall in nearly all stages of its development. There are wall-less plant cells, chiefly the endosperm cells during part of seed development in some monocot species; however, these exceptions are relatively rare. The wall is an integral part of the plant cell. A growing plant cell must have its wall to survive, and if it is to grow, it must continually modify and expand the wall. To understand the mechanism of plant cell enlargement, it is first necessary to know something about the chemistry, biosynthesis, and physical

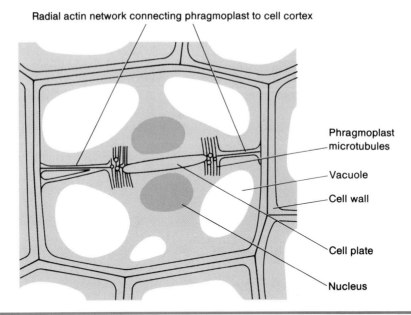

Radial actin network connecting phragmoplast to cell cortex

Phragmoplast microtubules

Vacuole

Cell wall

Cell plate

Nucleus

**Figure 5.32 Cytokinesis in plant cells**
Organization of the actin filaments and microtubules within the phragmoplast. Actin filaments extend from the periphery of the phragmoplast to the cortical cytoplasm, as well as parallel to the phragmoplast microtubules. Redrawn with permission from Alberts *et al.* (1989).

properties of cell walls. Plant anatomists recognize two kinds of cell walls, **primary** and **secondary,** based both on their composition and on when they are formed. Primary cell walls are laid down during cell growth, whereas secondary cell walls are deposited after growth has ceased. Usually the secondary wall is very much thicker than the primary wall. In some cases, secondary walls are deposited in an intricate, sculptured pattern. In other cases, secondary walls are deposited in layers (Fig. 5.33). Typically, secondary walls are formed during cellular differentiation. Cells undergoing secondary wall deposition are becoming specialized for water conduction or as mechanical support tissue.

Primary and secondary walls differ in their chemical composition, thickness, and physical properties. The differences in chemical composition of primary and secondary walls are shown in Table 5.1. The walls of particular cell types may differ substantially from this. Some walls, such as those of the pollen tube, contain almost no cellulose. The secondary walls of some cells, such as those in the tissue known as collenchyma, contain no lignin. The data presented in Table 5.1 are a generalization that is true for many cell types, but is inaccurate in depicting the wall composition of some specialized cell types or cells growing under unusual conditions.

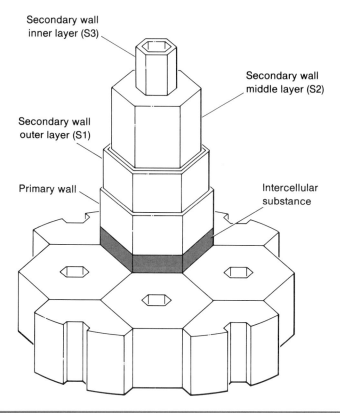

Secondary wall
inner layer (S3)

Secondary wall
middle layer (S2)

Secondary wall
outer layer (S1)

Primary wall

Intercellular
substance

**Figure 5.33 Secondary wall structure**
A cross section of a fiber cell is diagrammed to illustrate the structure of its secondary wall. Typically the secondary wall consists of three layers, which differ from each other primarily in the orientation of their cellulose microfibrils. Cellulose microfibrils are highly ordered within any given layer of the secondary wall, but the orientation is different in each layer. Redrawn with permission from Wardrop and Bland (1959). The process of lignification in woody plants. In *Proceedings, 4th International Congress of Biochemistry,* Vol. II, p. 96.

**Table 5.1    Comparison of the polymeric compositions of primary and secondary walls**

| Wall component | Primary walls | Secondary walls |
|---|---|---|
| Polysaccharides | 90% | 65–85% |
|   Cellulose | 30% | 50–80% |
|   Hemicellulose | 30% | 5–30% |
|   Pectins | 30% | — |
| Protein | 10% | — |
| Lignin | — | 15–35% |

Cell walls are composed mainly of polysaccharides    Polysaccharides are macromolecules composed of many sugar molecules or sugar derivatives, linked together by covalent bonds. The types of sugars found in cell walls include six-carbon sugars, or hexoses, such as glucose, galactose, fucose, and mannose; five-carbon sugars (pentoses), such as xylose and arabinose; and uronic acids, such as galacturonic acid (Fig. 5.34). About half of the carbonyl groups of the

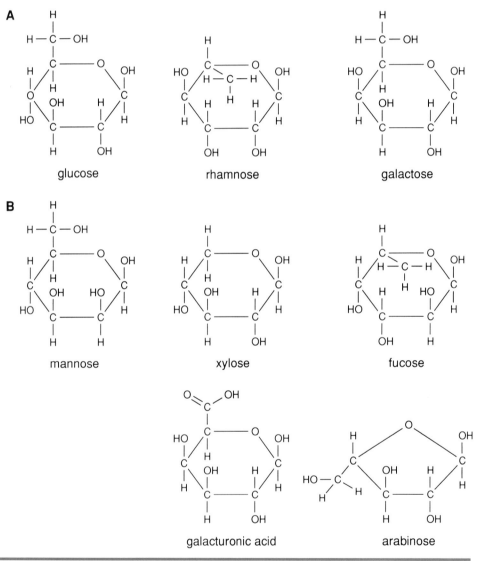

**Figure 5.34 Some of the monosaccharides found in primary cell wall polysaccharides**

galacturonic acid residues in the wall are esterified with methanol, and frequently the glucuronic acid residues have a methyl group on the number 4 carbon.

Several sugars play important roles in cell structure and metabolism inside the cell but are never found in wall polysaccharides. These include the ribose and deoxyribose of RNA and DNA. Similarly, keto sugars, such as fructose and ribulose, are found inside the cell, where they play important roles in cell metabolism, but never as part of cell wall polysaccharides. In addition, plant cell walls have no amino sugars and no sulfur-containing sugars, such as those that make up the cell walls of bacteria and the extracellular matrix of animals.

Cellulose is a simple, linear polymer of glucose subunits     Cellulose is an unbranched, linear polymer that is composed of a great many glucose subunits, linked end-to-end with a $\beta$1,4 linkage (Fig. 5.35A). The length of individual cellulose chains is difficult to determine because cellulose must be dissolved to make length measurements. As is true for all wall polymers, cellulose is not water soluble and a strong base must be used to put the cellulose into solution. The alkali used to dissolve the cellulose breaks some of the linkages, so we may underestimate the number of glucoses in a typical cellulose molecule. Nevertheless, alkali-solubilized cellulose is from 6000 to 10,000 glucose molecules long, corresponding to a chain 30,000 to 50,000 Å in length and a molecular weight of more than a million. The cellulose in cell walls does not consist of single molecules, but rather bundles in which about 40 cellulose molecules combine to form linear crystals known as **microfibrils.** The individual cellulose molecules are parallel to each other within the microfibril and are held together by a large number of hydrogen bonds, which form between the glucose subunits of the adjacent cellulose molecules (Fig. 5.35B). This arrangement of cellulose molecules makes the microfibril very strong. Each cellulose microfibril is around 35 Å thick and many micrometers long, but it is still far short of extending completely around the cell.

Noncellulosic polysaccharides are heterogeneous in composition In contrast to cellulose, the noncellulose polysaccharides are an exceedingly complex and heterogeneous group of polysaccharides known as hemicelluloses and pectins. The average noncellulosic wall polysaccharide (NCWP) has a backbone consisting of many linked monomers and numerous side chains extending from the backbone. Frequently, the side chains are composed of two or more different sugars, one of which is likely to be a uronic acid. These NCWPs are named for their predominant sugar, which usually is the one forming the backbone, and thus are called xylans, arabans, galactans, and so forth (Fig. 5.36). The cell walls of grasses have xylans as their major hemicellulose, but there is relatively little xylan in dicot cell walls. Instead, dicots contain a xyloglucan as their major hemicellulose. The xyloglucan contains a backbone of glucose units held together by $\beta$-1,4 linkages, with chains of xylose as side chains on most of the glucose subunits. Cell wall polysaccharides are hydrophilic, even if they are not soluble in water. Because of their

**A**

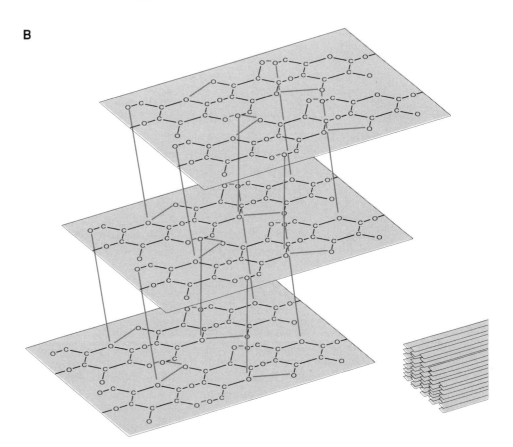

**B**

**Figure 5.35 Structure of cellulose**

(A) Cellulose is a simple unbranched polymer composed of glucose units linked by glycosidic bonds in the $\beta$ configuration. (B) The cellulose microfibril consists of approximately 40 cellulose molecules which are held together by hydrogen bonds between oxygen atoms and the hydrogens of the hydroxyl groups. Hydrogen atoms are not shown in this drawing. Redrawn with permission from Albersheim (1975).

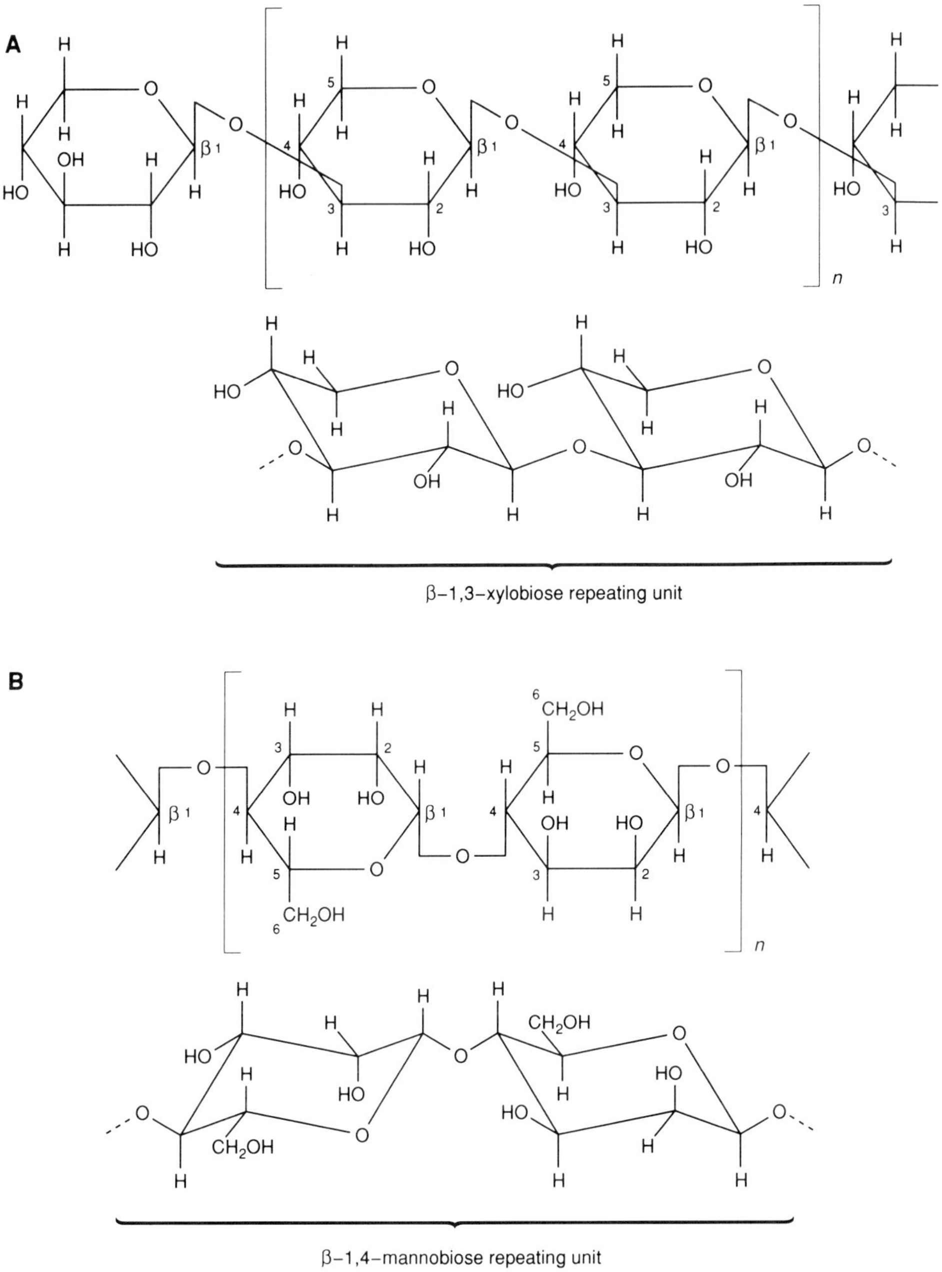

**A**

β−1,3−xylobiose repeating unit

**B**

β−1,4−mannobiose repeating unit

**Figure 5.36 Structures of some cell wall noncellulosic polysaccharides**
These might form the backbone of a more extensive polysaccharide. (A) Xylan.
(B) Mannan.

highly branched structure, the NCWPs not only attract but also trap water molecules to form gels.

The pectins are a special class of noncellulosic polysaccharides that consist of polymers with a high content of uronic acids, particularly galacturonic acid. Pectins differ from hemicelluloses in their solubility. Historically, pectins have been classified as wall polymers that are soluble in hot water, whereas the hemicelluloses are soluble only in dilute acidic or basic solutions. This is true only if the bonds that link the pectins to other cell wall noncellulosic polysaccharides have been broken. In the walls of growing plant tissues, the pectins either are much larger or are bonded to other wall polysaccharides and are not soluble in cold or hot water. During fruit ripening or as tissues become senescent, however, the chemical bonds between the pectins and other wall polymers are broken, and/or the pectin molecules themselves are partially degraded, rendering these polymers readily soluble. The different glycosidic bonds in the hemicelluloses and pectins vary in their susceptibility to acid hydrolysis. The layer between cells, known as the middle lamella, is especially rich in pectins. Enzymatic digestion of the pectin from the middle lamella releases many cells from their attachment in a tissue, so the pectin of the middle lamella may be seen as the glue that holds plant cells together. The pectins are fibrous molecules that may be crosslinked through divalent calcium ions. Reagents that chelate calcium ions also can lead to cell separation.

**Structures of several sycamore cell wall noncellulosic polysaccharides have been determined**    The cell wall polysaccharides of cultured sycamore cells have been characterized extensively. These cell walls contain substantially less cellulose and proportionally more protein and noncellulosic polysaccharides than the average primary cell wall, and in this respect they are not typical of primary cell walls. Nevertheless, the structures of individual sycamore cell wall noncellulosic polysaccharides probably are similar to the structures of those found in most plant cell walls. The structure shown in Fig. 5.37A is a rhamnogalacturonan, a pectic noncellulosic polysaccharide, whereas the molecule shown in Fig. 5.37B is an arabinogalactan and would be considered a hemicellulose. These molecules are extensively interconnected in the wall, and for the most part, it is not possible to isolate them intact without breaking bonds linking them to other molecules.

**Primary cell walls also contain proteins**    Primary cell walls contain structural proteins that, in part, determine the properties of the wall. The most thoroughly studied cell wall protein has been given the name extensin. The amino acid composition and structure of extensin are highly unusual. Forty-five percent of the amino acid residues of this protein are hydroxyproline, an amino acid that rarely occurs in cellular proteins. Extensin is a member of a family of hydroxyproline-rich glycoproteins (HRGPs). The universal genetic code does not contain a codon for hydroxyproline. Instead, the genes encoding these hydroxyproline-rich proteins contain codons specifying the amino acid proline, which is

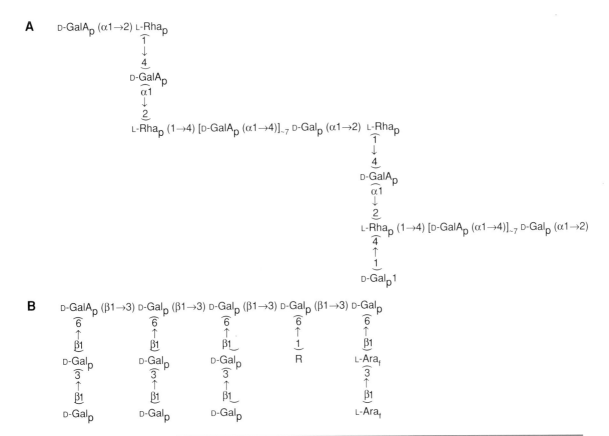

**Figure 5.37** **Partial structures of some cell wall noncellulosic polysaccharides**

(A) A rhamnogalacturonan component of pectin from sycamore primary wall. (B) An arabinogalactan component of hemicellulose from a primary wall. R = BD-Gal$_p$ or, less frequently, B-L-Ara$_f$ or D-GlcA$_p$. L-Ara$_f$ = L-arabinos; GalA$_p$ = D-galacturonic acid; L-Rha$_p$ = L-rhamnose, D-Gal$_p$ = D-galactose.

modified to hydroxyproline after the protein is synthesized. Extensin also is rich in five other amino acids: histidine, lysine, serine, valine, or tyrosine. Together, these six amino acids make up 98% of the total amino acid residues of the protein. The amino acid sequence of HRGPs contains a characteristic repeating peptide, Ser–Pro$_4$, interrupted periodically by one of these other residues. Two other kinds of cell wall proteins have been identified and characterized—the proline-rich cell wall proteins (PRPs) and the glycine-rich cell wall proteins (GRPs) both of which also are families of related proteins, encoded by small multigene families. The GRPs contain a repeating amino acid sequence: [Gly-$X$]$_n$, where $X$ often is glycine. The PRPs also contain a repeating motif consisting of the sequence Pro–Pro–Val–$X$–Lys, where $X$ is glutamic acid or tyrosine. These cell wall proteins are not found universally in primary cell walls, but rather are restricted in their distribution. For example, the GRPs are localized in provascular tissue and in the lignified cell walls of xylem in soybean stems. HRGPs occur more widely in the parenchymal tissues, but are also found in the walls of cambial cells. The individual members of the PRP family have been shown to differ very significantly in their distribution within the hypocotyl tissues of soybean. Of the three members of this family examined, one was localized in epidermal cell walls in the elongation and mature regions of the hypocotyl, another was present in cortical cells and vascular tissues, and the third was localized in the endodermis. It seems likely that as more is learned about the function of these cell wall proteins, we will find that they play an important role in growth and development.

Wall polymers are interconnected in the cell wall matrix    The cell wall matrix has often been compared with reinforced concrete. Although the physical properties of the cell wall and concrete are very different, the properties of the cell wall are a result of the interactions of the noncellulosic polysaccharides with the cellulose microfibrils, much as the properties of reinforced concrete are the result of the combination of steel and concrete. Cellulose microfibrils individually are very strong, but they are not continuous around the circumference of the cell; however, the microfibrils are embedded within the somewhat amorphous, gelatinous, pectins. Hemicelluloses, which also are fibrous, cover the cellulose microfibrils, where they are held by hydrogen bonds. The hemicelluloses coating the cellulose microfibrils are in turn chemically bonded to another class of hemicellulose that serves as a cross-bridge to pectin molecules (Fig. 5.38). This model represents the structure of the primary wall in cells in which the cell wall polysaccharides are not crosslinked with hydroxyproline-rich cell wall proteins to any significant extent. A somewhat different model has been derived from studies of the sycamore primary cell wall, and this model may represent the structure of mature walls in primary cells that contain more protein (Fig. 5.39). In this model, the protein extensin is bonded to an arabinogalactan. As a result of this interconnected nature, the wall is fairly rigid and unyielding to pressure from within the cell.

Noncellulosic cell wall polysaccharides are synthesized by the Golgi apparatus and transported to the wall in vesicles    Noncellulosic polysaccharides are synthesized in the Golgi dictyosomes. The noncellulosic polysaccharides are transported to the cell wall via Golgi-derived vesicles. These vesicles then fuse with the plasma membrane, excreting the noncellulosic polysaccharides into the wall. This mechanism is most apparent when the cell wall is first formed after cell division. The new cell wall is formed by the phragmoplast in late telophase of mitosis, as described earlier. The cell plate is the beginning of not only the cell wall, but also the plasma membranes of the two daughter cells. The membrane surrounding the Golgi vesicles fuses with the plasma membrane,

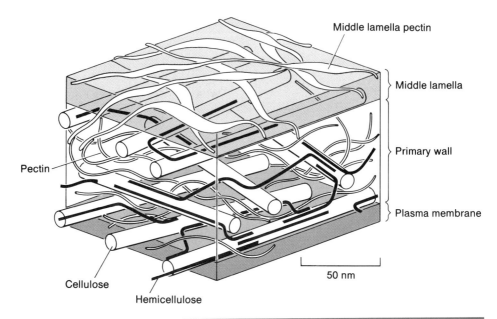

**Figure 5.38 Simplified model for the structure of the cell wall showing the interactions of the three classes of polysaccharides**
Hemicellulose xyloglucans adhere tightly to the surface of the cellulose microfibril and crosslink them. The cellulose microfibrils probably are completely coated with hemicellulose chains. The pectins are considered to form a separate network of fibrous molecules that interdigitate with the cellulose–hemicellulose network, except in the region of the middle lamella, which is composed primarily of pectin. Redrawn with permission from McCann *et al.* (1990).

becoming part of it, and the noncellulosic polysaccharides become the matrix of the cell wall. The first formed wall becomes the middle lamella which is shared by the two daughter cells and cements them together.

Secondary walls differ from primary walls in composition and structure   Secondary walls consist of both cellulose and NCWPs, but lack protein and contain lignin. Furthermore, the cellulose microfibrils often are deposited in the wall in a highly ordered manner. The secondary wall is laid down in a number of discrete layers, usually three, in which the orientation of the cellulose microfibrils differs. Within a particular layer, the cellulose microfibrils are deposited parallel to each other, but the cellulose microfibrils in adjacent layers are at markedly different angles (Fig. 5.40). The orientation is very precise. In any given layer all the cellulose microfibrils have the same orientation. Presumably this is because their orientation is not disturbed by the subsequent growth of the cell.

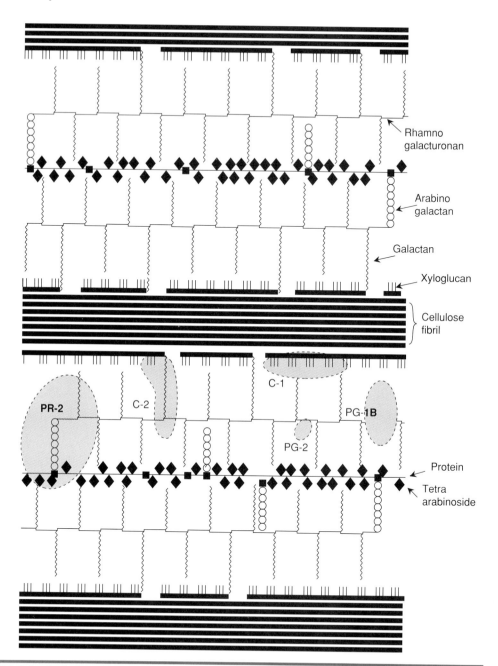

**Figure 5.39 Model for the primary wall in mature cells that have ceased elongation**

In this case, the protein extensin crosslinks the various wall polysaccharides, forming a more rigid structure. Redrawn with permission from Keegstra *et al.* (1973).

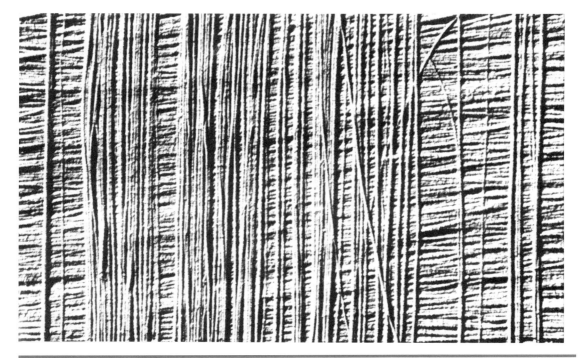

**Figure 5.40 Ordered pattern of cellulose microfibrils in two layers of the cell wall of a green alga**
Courtesy of R. D. Preston.

Secondary walls usually contain another polymer known as lignin that is not commonly found in primary cell walls. Lignin is a complex polymer that contains many chemically distinct monomers, all of which are derivatives of cinnamic acid. Cinnamic acid is formed by the deamination of the amino acid phenylalanine which is brought about by an enzyme known as phenylalanine ammonia lyase. Cinnamic acid is then further modified by hydroxylation, methylation, or reduction to form many different monomers, including ferulic acids, sinapic acid, caffeic acid, coniferyl alcohol, and syringaldehyde.

All of these chemical reactions must occur within the cytoplasm, possibly within dictyosomes. These monomers then are excreted from the cell during secondary wall deposition, where they are dehydrogenated by the enzyme peroxidase to form free radicals. Apparently, these free radicals polymerize spontaneously to form the complex structure known as lignin (Fig. 5.41). Lignification occurs throughout secondary wall deposition, but some regions of the wall become more heavily lignified than others. In conifer tracheids the middle lamella actually is the most heavily lignified region of the wall.

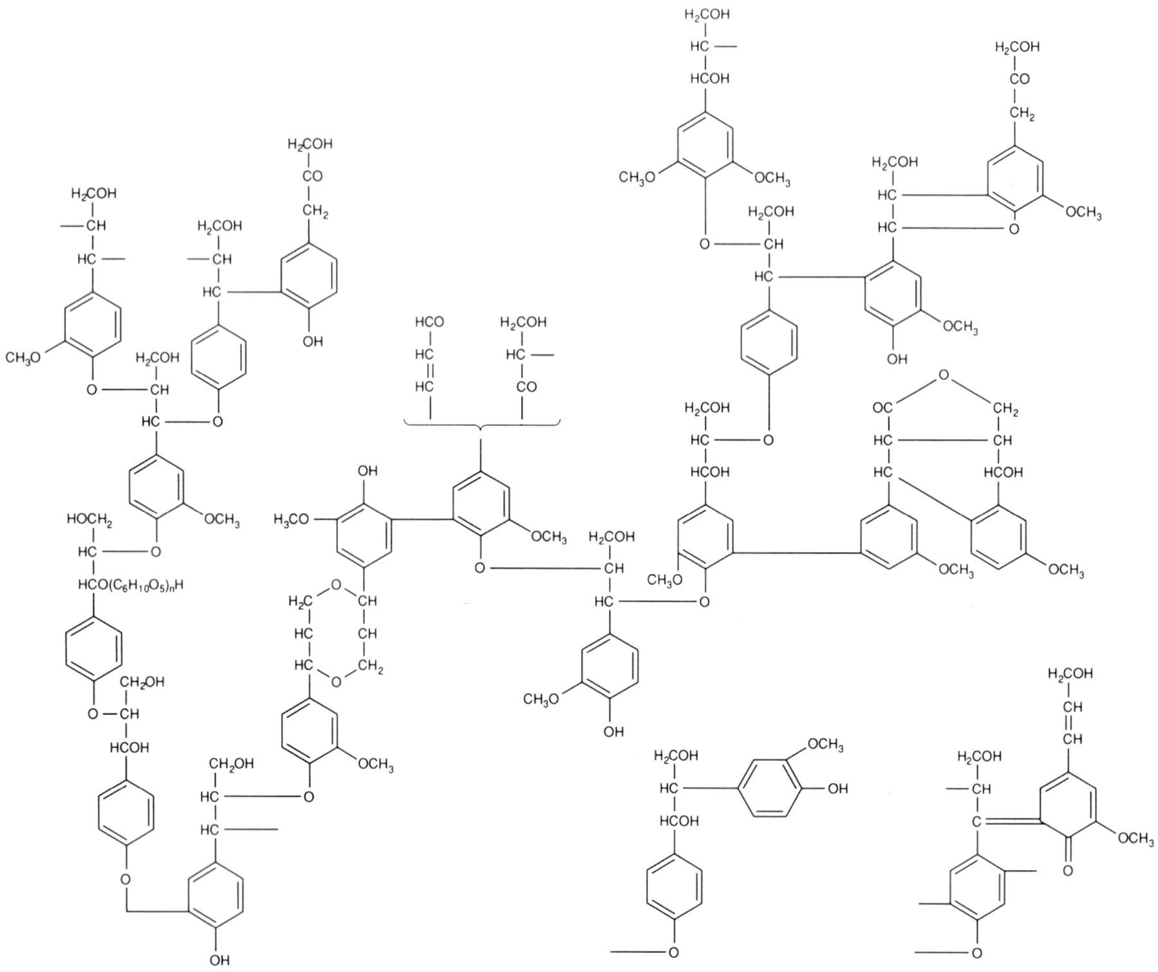

**Figure 5.41 Partial structure of spruce lignin**
From Freudenberg and Neish (1968).

Cellulose microfibrils may be oriented in primary walls  The cellulose microfibrils of primary cell walls in nonelongating cells form an intertwined network of fibers with a random orientation (Fig. 5.42). In contrast, the most recently deposited cellulose microfibrils in an elongating cell tend to be laid down at right angles to the long axis of the cell (Fig. 5.43). As the cell elongates, the older wall material is stretched and the microfibrils are pulled out of the orientation in which they were deposited to form the net. There is no apparent order in the cellulose microfibrils of the primary cell walls, except in the most recently deposited layer, where the microfibrils tend to be oriented at right angles to the direction of elongation. This most recently deposited cellulose microfibril layer is responsible for the structural anisotropy of the cell wall and, consequently, the polarity of cell elongation. The wall retains the same thickness as the cell grows. This means that new wall material is constantly deposited as the cell elongates.

How are cellulose microfibrils synthesized and deposited?     Cellulose is synthesized by an enzyme complex known as cellulose synthetase, which is embedded in the plasma membrane of higher plants. Although the cellulose synthetase complex has not been isolated and characterized biochemically, it has been seen in electron micrographs. Cellulose microfibril deposition has been extensively studied during secondary wall formation in the alga *Micrasterias denticulata,* which has the highly ordered pattern of cellulose microfibril deposition (Fig. 5.44). Here, the growing ends of the cellulose microfibrils in the developing secondary wall can readily be seen. Freeze-etch electron micrographs of the plasma membrane of these cells show an unusual pattern in which rows of rosette particles are embedded in the membrane, aligned over the growing ends of the cellulose microfibrils (Fig. 5.44B). A model for the deposition of the cellulose microfibrils by these membrane-bound rosette complexes is shown in Fig. 5.44C. Each rosette complex is believed to secrete one 5-nm-diameter cellulose microfibril, and a row of these complexes secretes a set of these 5-nm microfibrils which become hydrogen bonded to each other to form the approximately 20-nm-diameter-thick cellulose microfibril of the *Micrasterias* secondary wall. Primary walls, in which the cellulose microfibrils are smaller in diameter, probably are synthesized by single rosette particles, rather than the flotillas of particles that form the fibrils of the secondary wall.

It is not clear why it has been so difficult to isolate the cellulose synthetase complex, but perhaps this difficulty is an indication that the orientation of the rosette particles within the membrane is important in determining the enzymatic activity of the complex. The immediate precursor of cellulose is the sugar nucleotide uridine diphosphate glucose (UDPG). The UDPG precursors come to the complex on the cytoplasmic face of the membrane, whereas the product (cellulose) is excreted from the external face of the membrane. As illustrated in Fig. 5.45, the pH and ionic conditions on the two sides of the membrane are very different and this asymmetry may be essential for the synthesis of cellulose.

Usually the most recently deposited cellulose microfibrils tend to be aligned in primary cell walls     As presented earlier, cellulose microfibrils occur in a highly ordered configuration within the different layers of the secondary wall. In contrast, in the primary wall, most of the cellulose microfibrils are in a random network, with little order. Even in the primary cell wall, however, the most recently deposited cellulose microfibrils tend to be oriented at right angles to the long axis of the cell, particularly in elongating cells, and this ordered layer plays a critical part in determining the direction of cell enlargement. The loss of cellulose microfibril order in the older parts of the primary wall is explained by the fact that the cell wall increases in surface area as the cell grows. As the cell wall grows in area, the cellulose microfibrils are pulled out of the alignments

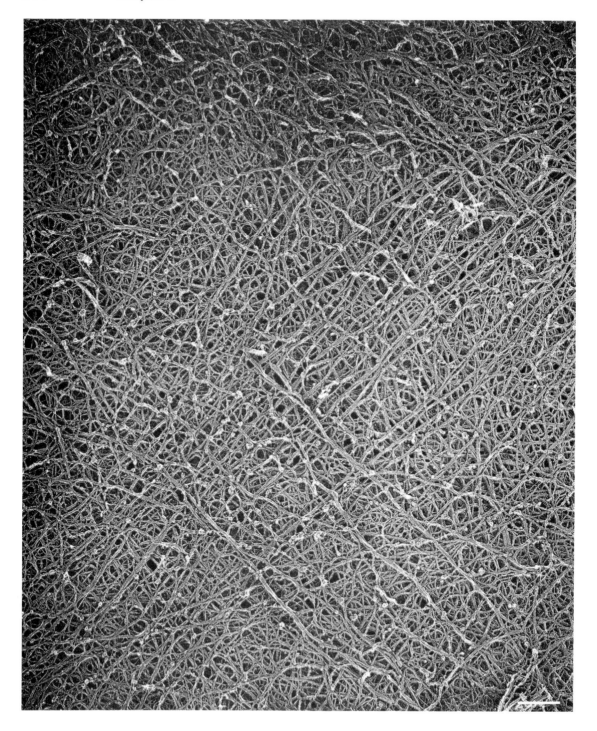

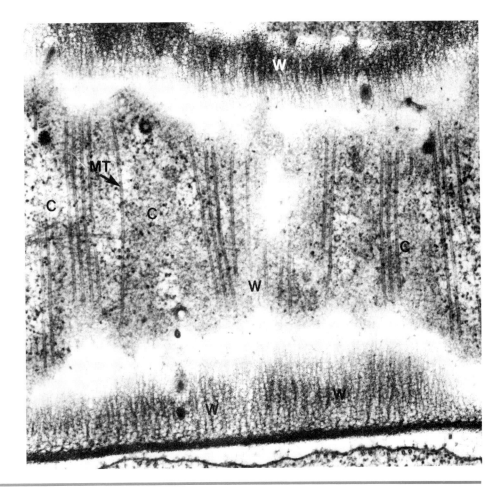

**Figure 5.43 Orientation of cellulose microfibrils in primary cell wall and cortical microtubules**

The cellulose microfibrils and cortical microtubules often have the same orientation. A transmission electron micrograph of a cell that has been sectioned tangentially to reveal the orientation of the fibrous polymers of the cell wall and of the microtubules in the underlying cytoplasm. W, cell wall; MT, microtubule; C, cortical cytoplasm. Reprinted with permission from Gunning and Steer (1975).

**Figure 5.42 Orientation of fibrous polymers in primary cell walls**

The pectins were extracted from the primary cell walls of the onion bulb scale parenchyma cells, before they were rapidly frozen and viewed in the electron microscope by a deep-etch, rotary-shadowed technique. The fibers shown here are the cellulose microfibrils and hemicelluloses, which are interconnected to form a random network. The cellulose microfibrils are bonded to and extensively cross-linked by the hemicelluloses. The cross-links serve not only to connect the microfibrils, but also to keep them apart. Even though the wall fibers are randomly arranged overall, the fibers in any given layer of the wall tend to have the same orientation. Bar = 200 nm. Reprinted with permission from McCann *et al.* (1990).

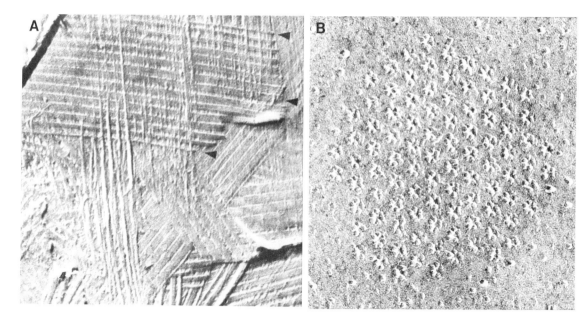

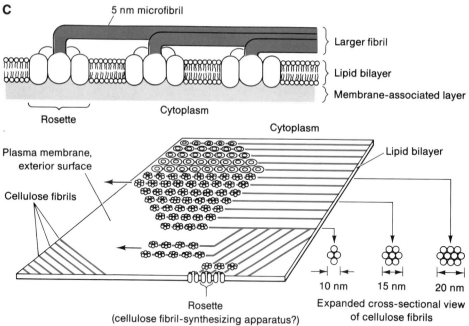

## Figure 5.44 Structure of the secondary wall and plasma membrane of *Micrasterias denticulata*

(A) Bands of parallel cellulose microfibrils crisscross in the secondary wall of this alga. The band of microfibrils marked by arrowheads shows a progression from broad fibrils in the center to narrow fibrils on the sides, with a constant center-to-center spacing.
(B) Freeze-fracture of the cell membrane reveals hexagonal arrays of rosettes, which

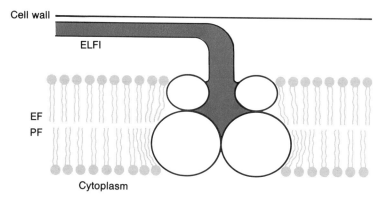

High $Ca^{2+}$

Low $Mg^{2+}$

pH 5.5

UDPG or
nucleotides
absent

Cell wall

ELFI

EF

PF

Cytoplasm

Low $Ca^{2+}$

High $Mg^{2+}$

pH 7.2

UDPG or
nucleotides
present

**Figure 5.45 Schematic diagram showing the conditions that exist during the formation of cellulose by rosette particles (RP) within the cell membrane**
EF, external face; PF, cytoplasmic face; SCS, subunit of cellulose synthase complex; ELFI, elongating cellulose microfibrils. Redrawn with permission from Delmer *et al.* (1985).

are thought to represent cellulose synthetase complexes. (C) Model of cellulose microfibril deposition during secondary wall formation in *Micrasterias*. Each 5-nm cellulose microfibril is thought to be formed by one rosette. Rows of rosettes from several 5-nm microfibrils, which become hydrogen bonded to each other to form larger microfibrils. Reproduced with permission from (A,B) Giddings *et al.* (1980) and the *Journal of Cell Biology* **84,** 327–339, ©The Rockefeller University Press.

in which they were deposited. The continued growth of the cell alters the orientation of older fibrils, pulling them into a random network.

Cortical microtubules can control the orientation of cellulose microfibril deposition    How can we account for the initial transverse orientation of the cellulose microfibrils when they are first deposited in the primary wall? A large body of indirect and correlative evidence suggests that cortical microtubules orient the deposition of the cellulose microfibrils. Both in secondary walls and in elongating cells with primary walls, the orientation of the cortical microtubules is the same as, or very similar to, that of the most recently deposited cellulose microfibrils. Furthermore, agents that destroy cytoplasmic microtubules, such as colchicine, also disrupt the pattern of cellulose microfibrils, but only disrupt the *orientation* of the microfibrils. Although the mechanism by which the cortical microtubules determine the pattern of cellulose microfibril orientation is not known, most likely it is the result of an association of the microtubules with the rosette complexes that synthesize cellulose, either directly or through some guiding mechanism (Fig. 5.46). The link between the membrane or the rosette and the microtubules most likely involves a MAP. Although MAPs have not been isolated and characterized from higher plants, many such proteins are known from animal studies. One well-characterized MAP from vertebrate brain tissue, known as MAP-2, is a long, filamentous protein. One end of this protein contains a specific binding site for the microtubule; the other end interacts with

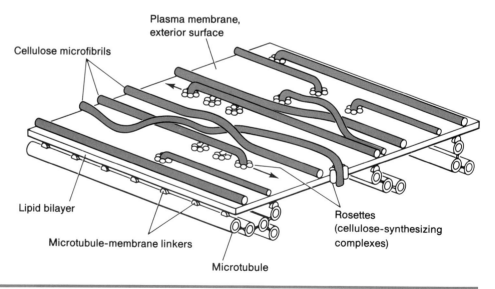

**Figure 5.46 Model illustrating how cortical microtubules might determine the orientation of cellulose microfibril deposition in the cell wall**
Redrawn with permission from Staehelin and Giddings (1982).

a component of membranes. Filamentous cross-bridges have been seen between microtubules and the plasma membrane in electron micrographs of elongating plant cells.

# Cell–Cell Communication and Intercellular Transport in Plants

Water, signaling molecules, and other dissolved substances can move through plant tissues via two different pathways: the symplastic pathway and the apoplastic pathway.

*Symplast.* The interconnected cytosol or cytoplasm of cells in a tissue or organ.

*Apoplast.* The interconnected cell walls of a tissue or organ.

Apoplastic movement of ions and water can occur through the cell wall matrix because primary cell walls are very hydrophilic and the wall matrix is continuous throughout primary tissues. This is particularly important during water uptake from the soil solution by roots. Water and ions can diffuse through the apoplast of the epidermis and cortex until they encounter a barrier such as the Casparian strips on the endodermal cells of roots. Casparian strips are regions of the cell wall that are impregnated with waxy materials, which renders the cell walls hydrophobic, blocking further water movement.

The symplast has been demonstrated to be a pathway for the transport of water and ions. Furthermore, it is clear that symplastic connections between cells could be very important in coordinating the activities of the cells of a particular tissue as they grow and differentiate. Cells that have symplastic connections are also electrically connected. That is, an electrical current, a current that is carried by the flow of ions, will pass between symplastically connected cells. Cells that are electrically coupled have special cytoplasmic structures that pass through their cell walls. These structures are known as **plasmodesmata** (Fig. 5.47). A plasmodesma is an opening through the cell wall, averaging about 50 nm in diameter, that is lined with plasma membrane. Elements of the endoplasmic reticulum also pass through the plasmodesma pore, connecting the ER lumen of one cell with that of its neighbor, as well as the cytosol. Completely free passage of ions and other cytoplasmic components probably does not occur through plasmodesmata. Rather, they are gated openings. The ER and plasma membrane appear to be fused in the neck region of the plasmodesmata (Fig. 5.47). Furthermore, where the ER passes through the pore, it contains some electron-dense material which may fill the cavity. This structure has been designated a **desmotubule,** and in cross section it appears to contain a central rod. A model for the structure of plasmodesmata incorporating these features is shown in Fig. 5.48.

By the use of fluorescence-labeled compounds, it has been shown that the cytoplasmic connections between cells allow the passage of molecules up to about molecular weight 700 to 1000. Calcium ions can close the channels connecting

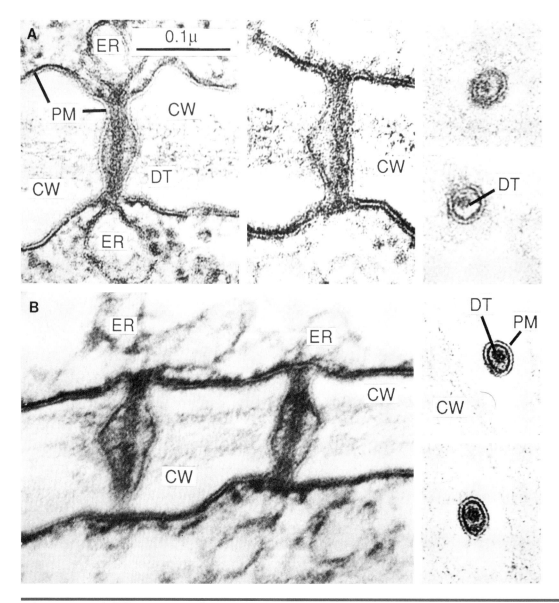

**Figure 5.47 Electron micrographs of plasmodesmata in the cell walls of the fern *Onoclea sensibilis***

Sections were made through cells of the gametophytes of the sensitive fern, *Onoclea sensibilis*, to reveal the structure of plasmodesmata after the tissues were fixed in two different ways: (A) with glutaraldehyde and osmium in a phosphate buffer and (B) with osmium tetroxide. In either case, the photos on the left show the structure of the plasmodesmata as seen in longitudinal section, and the photos on the right are transverse sections through the plasmodesmata in the plane of the cell wall. Endoplasmic reticulum (ER) membranes are seen connecting with the desmotubule (DT) and forming a rodlike structure that passes through the plasmodesmata. CW, cell wall; PM, plasma membrane. All photos are at the same magnification, ×285,000. Reproduced with permission from Tilney *et al.* (1991), *Journal of Cell Biology,* **112,** 739–747, ©The Rockefeller University Press.

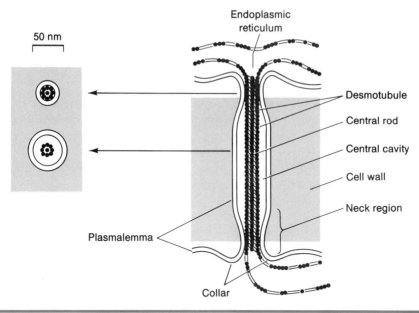

**Figure 5.48 Model for the structure of a plasmodesmata**
Redrawn with permission from Robards and Lucas (1990).

cells. Furthermore, chemical compounds that accelerate tumor growth, known as phorbol esters, also close the connections between cells. This suggests that plasmodesmata behave much as do the gap junctions of animal cells. They are cytoplasmic connections between cells through which information can be exchanged in the form of signaling molecules. These channels could play an important role in development. Virus particles are much larger than the width of a plasmodesma, yet in many cases they can pass readily from cell to cell after the virus initially gains entry into the symplast. Tobacco mosaic virus has been shown to encode a movement protein that may modify the plasmodesmatal opening, allowing the passage of virus particles.

Most plasmodesmata are formed during cytokinesis. As the phragmoplast constructs the new cell wall between daughter cells, remnants of the endoplasmic reticulum persist, entrapped in the developing cell plate, and plasmodesmata form around these. This means that plasmodesmata usually connect cells related to each other. Experimentally, however, it has been shown that plasmodesmata can form between unrelated cells. For example, if two cells are placed in contact in tissue culture and both participate in forming a shoot, plasmodesmata will form between these unrelated cells. Thus, there must be a mechanism that inserts plasmodesmata in walls between cells that are not related by lineage.

## Summary

This chapter describes those cellular structures that are particularly important in determining specific aspects of plant development. The endomembrane system, consisting of the endoplasmic reticulum and Golgi, is involved in the synthesis and processing of proteins targeted for insertion into membranes or for excretion from the cell. These proteins are targeted for synthesis by ribosomes on the rough endoplasmic reticulum, glycosylated in the lumen of the endoplasmic reticulum, and transported to the Golgi apparatus, where they are further modified and packaged in vesicles for excretion or for deposition in the vacuole. The Golgi apparatus (dictyosomes) also is involved in the synthesis of noncellulosic polysaccharides, which are packaged in vesicles for export and insertion into the cell wall.

Plant vacuoles are highly diverse and have many different functional roles, depending on the cell type in which they occur. In small, meristematic cells, vacuoles are also small and have many of the properties of animal cell lysosomes. That is, they are acidic, membrane-bound compartments containing hydrolytic enzymes with acidic pH optimums. These lysosome-like vacuoles may be involved in autophagy, the digestion of cellular structures and organelles found in the cells that contain them. Vacuoles in elongating cells serve as osmotic compartments. They contain dissolved solutes that are responsible for osmotic water uptake and the establishment of the turgor pressure that is the driving force for cell elongation. Vacuoles also may serve as storage compartments for secondary products and for the storage of protein and oil food reserves.

The cytoskeleton is an interconnected network of fibrous proteins that is important in cell structure and for intracellular transport. The cytoskeleton of plant cells is composed principally of microtubules and microfilaments. Both of these cytoskeletal structures are constructed by a self-assembly process. The dimeric protein tubulin self-assembles, forming microtubules, whereas microfilaments are assembled from actin monomers. The monomeric subunits for both polymers are proteins that bind nucleotides and nucleotide binding is necessary for assembly. Actin binds ATP; tubulin binds GTP. Although assembly of the monomers into the polymer requires nucleotide binding and the nucleotide is hydrolyzed after assembly, hydrolysis is not required for assembly. Rather, hydrolysis appears to be necessary for polymer disassembly. Assembly of the polymer occurs as subunits are added to the free ends. Both microtubules and microfilaments are highly polar structures, and assembly is much more rapid at one end than

somes?
occur in

8.  V
age org
oping p

9.  I
What rc
play in 1
the pola
mined e

10.
self-asse
sembly
these fac
tubule s

11.
be prese
What n
were to
havior c

12.
ter (MT
microtu
MTOC

13.
blastine

14.
motor"?
work? V
transpon
be neces
ity in ax

15.
mation.
assembl

16.
why do
the bin
tell us a

17.
actin—n
and my
streamii

18.

at the other. The barbed end of actin filaments is the rapid assembly end; microtubules assemble most rapidly at their plus ends. The assembly of both microtubules and microfilaments is dependent on the soluble monomer concentration. When the monomer concentration is above some critical value, polymers are formed, and the polymer disassembles when the monomer concentration is below the critical concentration. Individual microtubules and microfilaments turn over rapidly in cells, with a half-life on the order of minutes or seconds. Nevertheless, the arrays in which the microtubule and microfilament are found persist much longer. Microtubule arrays are nucleated from microtubule organizing centers, whereas microfilaments may be nucleated by capping proteins that bind one end of the filaments.

Both microtubules and microfilaments are involved in intracellular transport. There are motor proteins specific for both microtubules and microfilaments. These motor proteins are mechanoenzymes that move along microtubule or microfilament tracts, using energy from the hydrolysis of ATP. Myosins are actin-dependent transporters; dynein-like and kinesin-like proteins are microtubule-dependent transporters. The motor proteins coat the outer membranes of vesicles which then move along microtubule or microfilament tracts.

Plant cells have five different cytoskeletal arrays, but these are not present in cells at the same time. Four of these arrays appear at different stages during the cell division cycle. Interphase cells and nondividing cells have cortical microtubules and both cortical and subcortical microfilaments. The microtubules and microfilaments are oriented differently; the microtubules are oriented at right angles to the axis of elongation, whereas microfilaments tend to be oriented parallel to the elongation axis. The cortical microfilaments participate in the orientation of cellulose microfibril deposition, whereas the microfilaments in highly vacuolate cells are responsible for cytoplasmic streaming. Dividing cells exhibit a radial array, a preprophase band, a mitotic spindle, and a phragmoplast. The radial array consists of microtubules and microfilaments which radiate from the nuclear envelope into the surrounding cytoplasm. It is initiated during the G2 phase of the cell cycle and it appears to function in the positioning of the nucleus prior to mitosis. The preprophase band also appears during G2 and is a progressively narrowing band of cortical microtubules that replaces the cortical microtubule array as cells prepare for mitosis. The preprophase band marks the site at which the cell plate will join with the lateral walls to partition the dividing cell into two daughter cells. The mitotic spindle is composed of microtubules that extend from the polar regions of the cell toward the equator. Some of the spindle microtubules are captured by the kinetochores of the chromatids during prometaphase. These microtubules

*continues*

to treat growing roots with trifluralin, what parts of the root would be affected and what would this effect be?

26. What is lignin? How do cell walls become lignified? What types of cell walls become lignified? What is the biological significance of cell wall lignification?

27. What are plasmodesmata? Compare the structure and function of plasmodesmata and gap junctions. Usually, what is the relationship between cells that are connected by plasmodesmata? How are plasmodesmata formed?

## Further Reading

### General References

Alberts, B., Bray, D., Lewis, J., Raff, M., Roberts, K., and Watson, J. D. (1989). *The Molecular Biology of the Cell*, 2nd ed. Garland, New York.

Brett, C., and Waldron, K. (1990). *Physiology and Biochemistry of Plant Cell Walls.* Unwin Hyman, London.

Burgess, J. (1985). *An Introduction to Plant Cell Development.* Cambridge University Press, Cambridge.

Gunning, B. E. S., and Steer, M. W. (1975). *Plant Cell Biology: An Ultrastructural Approach.* Crane, Russak, New York.

Lloyd, C. W. (1991). *The Cytoskeletal Basis of Plant Growth and Form.* Academic Press, London.

### Specific References

Albersheim, P. (1975). The walls of growing plant cells. *Sci. Amer.* **293,** 23–34.

Atkins, D., Hull, R., Wells, B., Roberts, K., Moore, P., and Beachy, R. N. (1991). The tobacco mosaic virus 30K movement protein in transgenic tobacco plants is localized to plasmodesmata. *J. Gen. Virol.* **72,** 209–211.

Averhart-Fullard, V., Datta, K., and Marcus, A. (1988). A hydroxyproline-rich protein in the soybean cell wall. *Proc. Natl. Acad. Sci. USA* **85,** 1082–1085.

Bershadsky, A. D., and Vasiliev, J. M. (1988). *Cytoskeleton* Plenum Press, New York.

Bednarek, S. Y., Wilkins, T. A., Dombrowski, J. E., and Raikhel, N. V. (1990). A carboxyl-terminal propeptide is necessary for proper sorting of barley lectin to vacuoles of tobacco. *Plant Cell* **2,** 1145–1155.

Blanton, R. L., and Northcote, D. H. (1990). A 1,4-$\beta$-D-glucan-synthase system from *Dictyostelium discoideum. Planta* **180,** 324–332.

Darvill, A. G., Albersheim, P., McNeil, M., Lau, J. M., York, W. S., Stevenson, T. T., Thomas, J., Doares, S., Gollin, D. J., Chelf, P., and Davis, K. (1985). Structure and function of plant cell wall polysaccharides. *J. Cell Sci. Suppl.* **2,** 203–217.

Delmer, D. P. (1987). Cellulose biosynthesis. *Annu. Rev. Plant Physiol.* **38,** 259–290.

Delmer, D. P., Cooper, G., Alexander, D., Cooper, T., Hayashi, T., Nitsche, C., and Thelen, M. (1985). New approaches to the study of cellulose biosynthesis. *J. Cell Sci.* **2,** 33–50.

Deus-Neumann, B., and Zenk, M. H. (1984). A highly selective alkaloid uptake system in vacuoles of higher plants. *Planta* **162,** 250–260.

Ding, B., Turgeon, R., and Parthasarathy, M. V. (1991). Microfilaments in preprophase band of freeze substituted tobacco root cells. *Protoplasma* **165,** 209–211.

Fosket, D. E. (1989). Cytoskeletal proteins and their genes in higher plants. In *The Biochemistry of Plants*. Vol. 15. *Molecular Biology* (P. Stumpf and E. E. Conn, Eds.), pp. 392–454. Academic Press, New York.

Giddings, T. H., Jr., Brower, D. L., and Staehelin, L. A. (1980). Visualization of particle complexes in the plasma membrane of *Micrasterias denticulata* associated with the formation of cellulose fibrils in primary and secondary cell walls. *J. Cell Biol.* **84**, 327–339.

Gunning, B. E. S., and Steer, M. (1975). *Ultrastructure and the Biology of Plant Cells*. Edward Arnold, London.

Gunning, B. E. S., and Hardham, A. R. (1982). Microtubules. *Annu. Rev. Plant Physiol.* **33**, 651–698.

Gunning, B. E. S., and Overall, R. L. (1983). Plasmodesmata and cell-to-cell transport in plants. *BioScience* **33**, 260–265.

Gunning, B. E. S., and Sammut, M. (1990). Rearrangements of microtubules involved in establishing cell division planes start immediately after DNA synthesis and are completed just before mitosis. *Plant Cell* **2**, 1273–1282.

Gunning, B. E. S., and Wick, S. M. (1985). Preprophase bands, phragmoplasts and spatial control of cytokinesis. *J. Cell Sci. Suppl.* **2**, 157–179.

Hardham, A. R., and Gunning, B. E. S. (1978). Structure of cortical microtubule arrays in plant cells. *J. Cell Biol.* **77**, 14–33.

Hoffman, L. M., Donaldson, D. D., Bookland, R., Rashka, K., and Herman, E. M. (1987). Synthesis and protein body deposition of maize 15-kd zein in transgenic tobacco seeds. *EMBO J.* **6**, 3213–3221.

Hong, J. C., Nagao, R. T., and Key, J. L. (1987). Characterization and sequence analysis of a developmentally regulated putative cell wall protein gene isolated from soybean. *J. Biol. Chem.* **262**, 8367–8376.

Hood, K. R., Baasiri, R. A., Fritz, S. E., and Hood, E. E. (1991). Biochemical and tissue print analyses of hydroxyproline-rich glycoproteins in cell walls of sporophytic maize tissues. *Plant Physiol.* **96**, 1214–1219.

Keegstra, K., Talmadge, K. W., Bauer, W. D., and Albersheim, P. (1973). The structure of plant cell walls. III. A model of the walls of suspension-cultured sycamore cells based on the interconnection of the macromolecular components. *Plant Physiol.* **51**, 188–197.

Kirschner, M., and Mitchison, T. (1986). Beyond self-assembly: From microtubules to morphogenesis. *Cell* **45**, 329–345.

Knox, J. P., Linstead, P. J., King, J., Cooper, C., and Roberts, K. (1990). Pectin esterification is spatially regulated both within cell walls and between developing tissues of root apices. *Planta* **181**, 512–521.

Kropf, D. L. (1992). Establishment and expression of cellular polarity in fucoid zygotes. *Microbiol. Rev.* **56**, 316–339.

Kutschera, U., Bergfeld, R., and Schopfer, P. (1987). Cooperation of epidermis and inner tissues in auxin-mediated growth of maize coleoptisles. *Planta* **170**, 168–180.

Mandala, S., and Taiz, L. (1985). Proton transport in isolated vacuoles from corn coleoptiles. *Plant Physiol.* **78**, 104–109.

Matile, P. (1978). Biochemistry and function of vacuoles. *Annu. Rev. Plant Physiol.* **29**, 193–213.

Menzel, D., and Elsner-Menzel, C. (1990). The microtubule cytoskeleton in developing cysts of the green alga *Acetabularia:* Involvement in cell wall differentiation. *Protoplasma* **157**, 52–63.

McCurdy, D. W., and Gunning, B. E. S. (1990). Reorganization of cortical actin microfilaments and microtubules at preprophase and mitosis in wheat root-tip cells: A double labeled immunofluorescence study. *Cell Motil. Cytosk.* **15**, 76–87.

McCann, M. C., Wells, B., and Robert, K. (1990). Direct visualization of crosslinks in the primary plant cell wall. *J. Cell Sci.* **96**, 323–334.

Mineyuki, Y., and Palevitz, B. A. (1990) Relationship between preprophase band organization, F-actin and the division site in *Allium*. Fluorescence and morphometric studies on cytochalasin-treated cells. *J. Cell Sci.* **97,** 283–295.

Mineyuki, Y., Mark, J., and Palevitz, B. A. (1991). Relationship between the preprophase band, nucleus and spindle in dividing *Allium* cotyledon cells. *J. Plant Physiol.* **138,** 640–649.

Overall, R. L., and Gunning, B. E. S. (1982) Intercellular communication in *Azolla* roots. II. Electrical coupling. *Protoplasma* **111,** 151–160.

Overall, R. L., Wolfe, J., and Gunning, B. E. S. (1982). Intercellular communication in *Azolla* roots. I. Ultrastructure of plasmodesmata. *Protoplasma* **111,** 134–150.

Parthasarathy, M. V. (1985). F-actin architecture in coleoptile epidermal cells. *Eur. J. Cell Biol.* **39,** 1–12.

Parthasarathy, M. V., Perdue, T. D., Witztum, A., and Alvernaz, J. (1985). Actin network as a normal component of the cytoskeleton in many vascular plant cells. *Am. J. Bot.* **72,** 1318–1323.

Preston, R. D. (1964). Structural and mechanical aspects of plant cell walls with particular reference to synthesis and growth. In *The Formation of Wood in Forest Trees* (M. H. Zimmermann, Ed.), pp. 169–188. Academic Press, New York.

Robards, A. W., and Lucas, W. J. (1990). Plasmodesmata. *Annu. Rev. Plant Physiol. Plant Mol. Biol.* **41,** 369–419.

Roberts, K., Grief, C., Hills, G. J., and Shaw, P. J. (1985). Cell wall glycoproteins: Structure and function. *J. Cell Sci. Suppl.* **2,** 105–127.

Roelofsen, P. A. (1965). Ultrastructure of the wall in growing cells and its relation to the direction of the growth. *Adv. Bot. Res.* **2,** 69–149.

Stafstrom, J. P., and Staehelin, L. A. (1988). Antibody localization of extensin in cell walls of carrot storage roots. *Planta* **174,** 321–332.

Tilney, L. G., Cooke, T. J., Connelly, P. S., and Tilney, M. S. (1991). The structure of plasmodesmata as revealed by plasmolysis, detergent extraction, and protease digestion. *J. Cell Biol.* **112,** 739–747.

Ting, I. P. (1985). Crassulacean acid metabolism. *Annu. Rev. Plant Physiol.* **36,** 595–622.

Tiwari, S. C., Wick, S., Williamson, R. E., and Gunning, B. E. S. (1984). Cytoskeleton and integration of cellular function in cells of higher plants. *J. Cell Biol.* **99,** 63s–69s.

Trinick, J., and Elliott, A. (1979). Electron microscope studies of thick filaments from vertebrate skeletal muscle. *J. Mol. Biol.* **131,** 133–136.

Wyatt, R. A., Nagao, R. T., Key, J. L. (1992). Patterns of soybean proline-rich protein gene expression. *Plant Cell* **4,** 99–110.

# 6

# Light, Hormones, and Cell Signaling Pathways

D evelopment often is considered to be a consequence of the sequential readout of a detailed program encoded in the sequence of bases of the nuclear DNA; however, there may not be a single developmental program for most plants, but rather a series of alternative pathways. The specific developmental pathway that is followed is, to a large measure, determined by environmental factors. Temperature, daylength, light quality and quantity, gravity, water availability, and wind are some of the more significant environmental factors that may initiate or modulate specific developmental pathways. For example, many plants are photoperiodic and will grow vegetatively indefinitely, without initiating reproductive development unless they receive an inductive light–dark diurnal cycle to initiate flowering. In this case, daylength acts as a switch to trigger the reproductive developmental pathway that culminates in flowering and seed formation. Also, seedlings that develop in darkness have a completely different appearance than those that develop in the light. The developmental changes brought about by other environmental factors may be less dramatic, but often they are of considerable ecological significance. Plants that are grown under high humidity may have a thinner cuticle than individuals of the same species growing in a low-humidity environment. Similarly, the wood of a tree branch growing parallel to the ground will develop a different structure than that found in upright branches. Also, the structure of leaves is different depending on whether they develop in direct sunlight or are shaded by other leaves. Light is the single most important environmental factor regulating plant development. As a result, in this chapter we emphasize the various ways in which the plant modifies its development in response to light.

We also examine the mechanisms by which environmental stimuli initiate or modify developmental programs. There are several distinct steps by which environmental factors trigger changes in plant growth and development. First, the stimulus must be perceived. Light, for example, produces a biological effect because it is first absorbed by a pigment. The structures of some pigments active as photoreceptors in plants are shown in Fig. 6.1. Although all pigments absorb some wavelengths of light, not all pigments are photoreceptors. To act as a photoreceptor, a pigment must be coupled to a biochemical mechanism so that absorption of light energy generates a chemical signal that modifies cellular metabolism. Finally, the altered cellular metabolism leads to a change in growth and/or cell differentiation. Although some stimuli initiate whole new developmental pathways, such as the induction of flowering, others simply modify the pattern or direction of growth. Growth is oriented by environmental stimuli so that it occurs toward or away from the stimulus. These growth movements are known as **tropisms.** Tropistic growth movements can occur either toward or away from the stimulus (positive or negative). In many cases, these tropistic responses are mediated by signaling molecules that stimulate or suppress growth and can initiate or alter developmental pathways. These special regulatory molecules are known as **hormones.** In this chapter we use the term *hormone* to mean any naturally occurring molecule that acts as a signal to regulate plant growth and development. There are five generally recognized classes of plant hormones: auxins, cytokinins, gibberellins, ethylene, and abscisic acid. Their activity is not confined to tropistic responses, and some of them play no role in tropistic responses. Hormones regulate many, if not all, aspects of plant growth and development.

Although the mechanisms by which hormones elicit changes in growth and/or development have not been completely worked out, rapid progress is being made to gain such an understanding. Most hormones have a receptor or receptors to which they bind. The binding of a hormone to its receptor initiates a cascade of cellular changes that constitute a signal transduction pathway. In some cases the binding of a hormone to its receptor results in the formation of a second intracellular messenger, such as calcium, and the second messenger then initiates

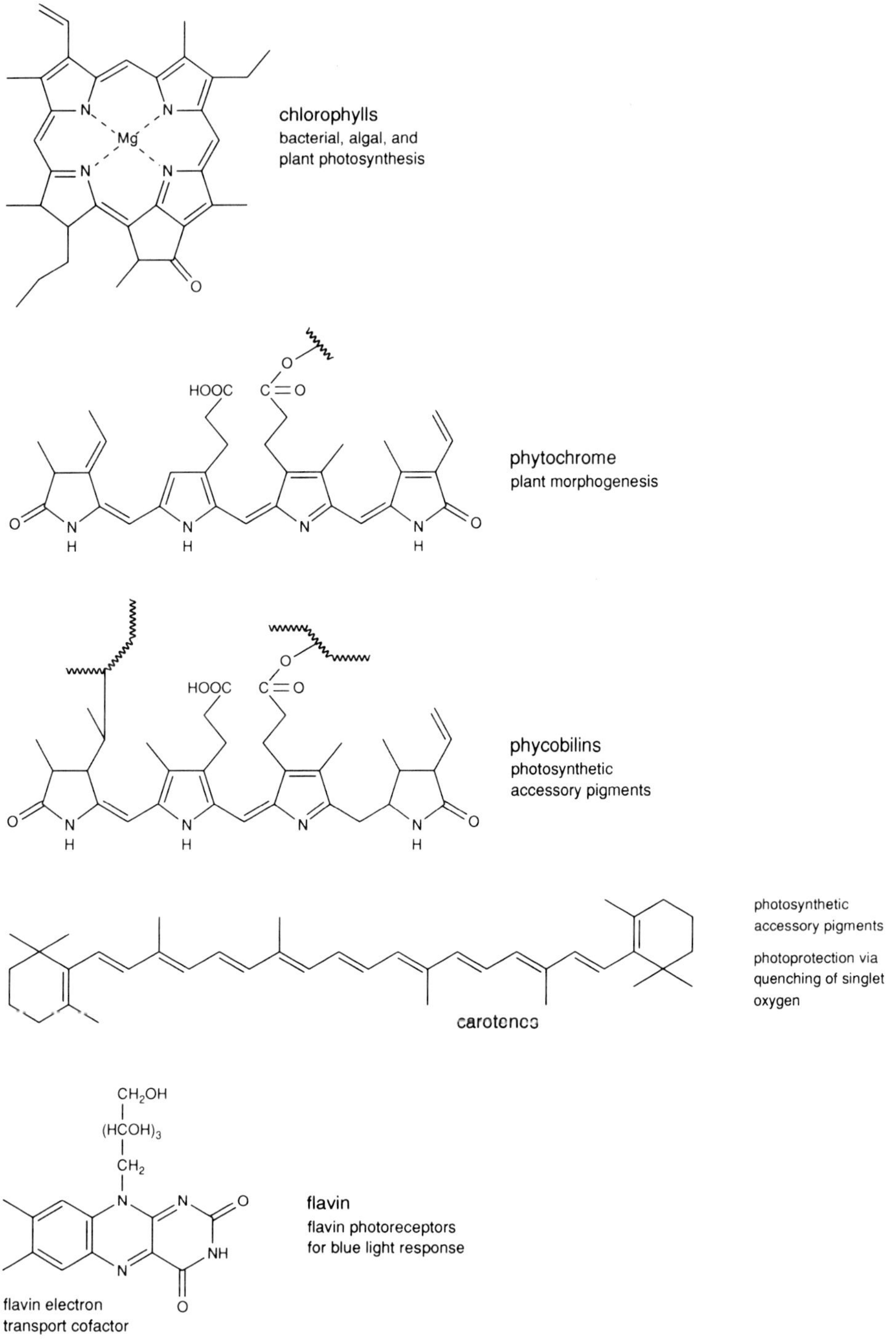

chlorophylls
bacterial, algal, and
plant photosynthesis

phytochrome
plant morphogenesis

phycobilins
photosynthetic
accessory pigments

photosynthetic
accessory pigments

photoprotection via
quenching of singlet
oxygen

carotenes

flavin
flavin photoreceptors
for blue light response

flavin electron
transport cofactor

**Figure 6.1    Structures of some pigments active as photoreceptors**
From Vierstra and Quail (1986).

the regulatory cascade. One common component of this regulatory cascade is the phosphorylation of various cellular proteins, thus altering their biological activity. Protein phosphorylation is known to regulate cellular activities as diverse as the assembly state of microtubules and the activity of transcription factors. Plant hormones have been shown to alter the pattern of gene expression. Some genes have hormone-responsive elements in their promoters, to which transcription factors bind in the presence of the hormone.

# Light

Light regulates many aspects of plant growth and development. Light quality and quantity are perhaps the most significant environmental factors affecting plant development. Of course, high levels of light are essential for photosynthesis. Photosynthesis normally operates in a range of light intensities from 10 to 1000 $W/m^2$. Plants, however, are highly sensitive to light intensities as much as a thousandfold lower and even to infrared and ultraviolet light that is not visible to the human eye, and they respond to these lower levels of light by altering their growth or development. These light-triggered growth and developmental responses have been termed **photomorphogenic responses** and it is these with which we are concerned here.

### Light versus Darkness

The importance of light in plant development may be illustrated most dramatically in the case of the early seedling growth of many dicots. If a pea seed germinates in total darkness, the seedling develops in a very different way than it would if it germinated in the light (Fig. 6.2A). The dark-grown seedling is said to be **etiolated.** The characteristics of an etiolated seedling are as follows:

- There is very rapid and extensive elongation of the internodes.
- Leaf expansion does not occur to any extent.
- Chloroplasts do not develop and the plant is pale yellow.
- The hypocotyl remains "hooked."

The growth of other plant tissues and organs is similarly affected by light. For example, if potatoes sprout in darkness, the shoots exhibit many of the same characteristics as the dark-grown seedling and they too can be said to be etiolated (Fig. 6.2B). In these cases, light appears to overcome an inhibitory factor that is produced in the dark, blocking leaf and chloroplast development.

### Chloroplast Development

Chloroplast development is not an independent phenomenon, but rather one of the components of the light-regulated development of the seedling. Neverthe-

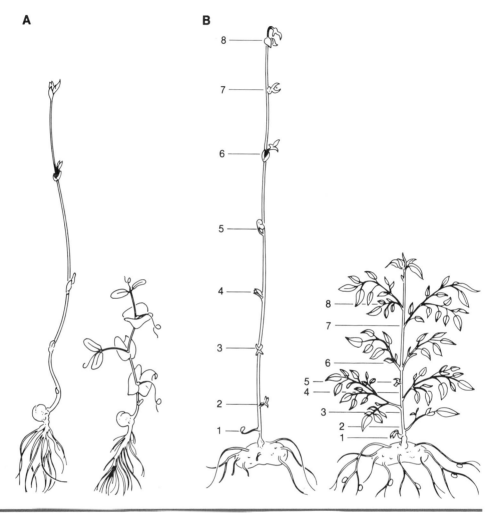

**Figure 6.2  Effect of light on plant development**
(A) Pea seedlings were grown in darkness for 6 days (left) or in light for the same period (right). (B) Potato shoot morphology after sprouting from a tuber bud in darkness (left) or in the light (right). Redrawn from Pfeffer, W. (1904). *The Physiology of Plants*, 2nd ed. (A. J. Ewart, Trans.) Clarendon Press, Oxford.

less, it is the best-studied component of the light-regulated development and the one that we know the most about at the biochemical and molecular levels. The formation of the mature chloroplast in angiosperms requires light. This is not true of all plants. For example, in gymnosperms and algae, chloroplasts will develop in the dark, at least to some extent; however, higher plants have evolved a mechanism that limits chloroplast development to tissues and cells that receive light. Meristematic tissues lack chloroplasts even in the light. Instead, the precursor of the chloroplast is present in the form of small organelles, known as **proplas-**

**tids,** that are surrounded by two concentric membranes, as are all plastids. In contrast to chloroplasts, proplastids contain relatively few internal membranes and they are not photosynthetically competent. When seedlings are germinated in darkness, the proplastids of leaves and stem tissues develop into **etioplasts** (Fig. 6.3). Etioplasts contain an elaborate internal membrane system known as a prolamellar body, but they lack most photosynthetic enzymes and they are not photosynthetically competent. When etiolated seedlings are exposed to light, etioplasts are rapidly converted into chloroplasts (Fig. 6.4). In light-grown seedlings, proplastids develop directly into chloroplasts during the differentiation of the leaf mesophyll cells as the leaf primordium develops.

**Amyloplasts** are starch storage organelles. They are found primarily in storage tissues and organs, such as the potato tuber. Although a mature amyloplast lacks any obvious similarities to a chloroplast, an amyloplast can be converted back into a chloroplast. This occurs to some extent when potato tubers are kept in the light. In fact, functioning chloroplasts may contain a few internal starch granules. The size of these starch granules changes with the diurnal cycle, largely disappearing in the dark and then increasing in the light as the sugars formed during photosynthesis are temporarily stored in the chloroplasts as starch.

**Chromoplasts** synthesize a variety of yellow, red, or orange carotenoid pigments and usually lack any trace of chlorophyll or other components of the photosynthetic machinery. Chloroplasts differentiate into chromoplasts during fruit ripening. Chloroplasts also synthesize $\beta$-carotene, which functions as part of the light-harvesting apparatus. During chromoplast differentiation, however, the chlorophyll and photosynthetic enzymes disappear, and additional and different carotenoid pigments are produced. For example, tomato fruits synthesize a red carotenoid pigment known as lycopene as their chloroplasts differentiate into chromoplasts during ripening. In other cases, such as petal development in some flowers, proplastids differentiate directly into chromoplasts, without ever becoming chloroplasts. Some fruit chromoplasts also can differentiate back into chloroplasts under certain conditions.

**Figure 6.3   Transmission electron micrograph of an etioplast from the leaf of an etiolated maize seedling**
Clearly shown is the paracrystalline prolamellar body, which is composed of membranes. Reprinted with permission from Gunning, B. B. S., and Steer, M. (1975). *Ultrastructure and the Biology of Plant Cells.* ©Edward Arnold Publishers, Ltd.

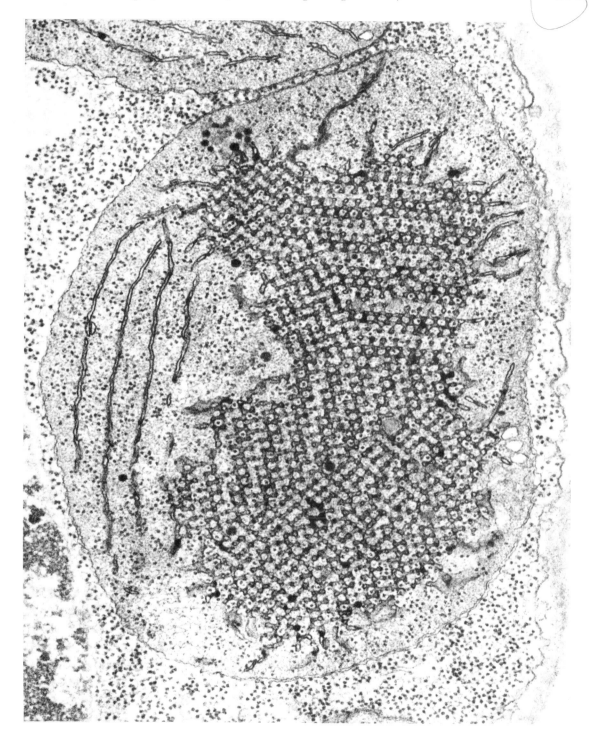

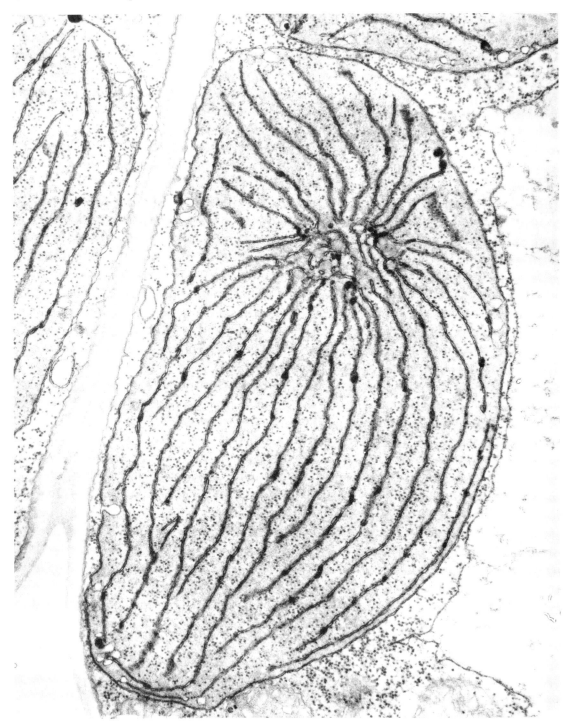

Chloroplasts may be formed via one of several pathways:

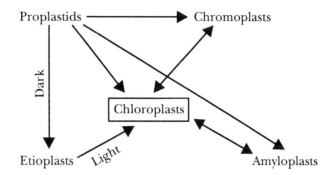

Of these various transitions, the only one that has been studied extensively is the light-regulated transition from etioplast to chloroplast. Morphologically, this involves the growth of the plastid and the development of the prolamellar body into the thylakoid membrane system.

Chloroplasts are self-replicating organelles that contain their own genome which encodes at least some of the chloroplast-specific proteins. To a limited extent, the chloroplast genome participates in its own development (see Chapter 3), however, because the great majority of chloroplast proteins are encoded by nuclear genes and the development of the chloroplast requires the coordinated expression of genes in both compartments.

The formation of the thylakoid membranes involves the synthesis of photosynthetic pigments, chlorophylls, carotenoids, and xanthophylls, as well as the synthesis of the proteins that will form the photosynthetic machinery. Figure 6.5 illustrates the various protein complexes found in the thylakoid membrane: photosystem II (PSII), cytochrome b/f, photosystem I, and the ATP synthase complexes. Each of these complexes is composed of a number of specific proteins. Some of these proteins are encoded by nuclear genes, whereas others are encoded by chloroplast genes. For example, in PSII, the protein designated $Q_B$, a 32-kDa protein also known as the herbicide-binding protein, is encoded by the chloroplast genome, whereas most of the remaining proteins of the complex are encoded in the nuclear genome. Photosystem II is surrounded by light-harvesting complex

**Figure 6.4**   **Transmission electron micrograph of a chloroplast at an early stage in its development from an etioplast**
The prolamellar body has broken down into numerous lamellae. Thylakoids, whose membranes contain the light-harvesting apparatus and photosystems, are initiated at intervals along these lamellae. On further development, numerous thylakoids will be stacked together to form grana. ×40,000. Reprinted with permission from Gunning, B. B. S., and Steer, M. (1975). *Ultrastructure and the Biology of Plant Cells.* ©Edward Arnold Publishers, Ltd.

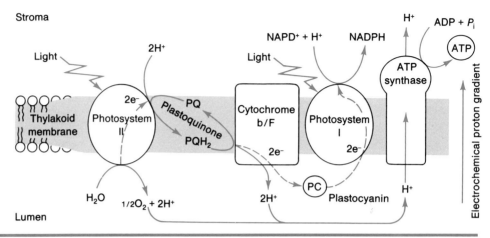

**Figure 6.5   Protein organization scheme of the photosynthetic apparatus within the thylakoid membrane**
Chloroplast electron transfer is brought about by four protein complexes that are a part of the thylakoid membrane: photosystem I, photosystem II, the cytochrome b/f complex, and the large multisubunit enzyme ATP synthase. Redrawn with permission from Taiz and Zieger (1991).

(LHC) II, which is composed of chlorophyll a/b-binding protein to which chlorophyll and the other photosynthetic pigments are bound. The chlorophyll a/b-binding proteins are encoded by the nuclear genome. Photosystem I also is surrounded by LHCI, which is similar to the LHCII, but is composed of a different gene product. Light initiates the expression of the genes encoding these proteins in many angiosperms.

### Ribulose-1,5-bisphosphate Carboxylase–Oxygenase (RUBISCO) Gene Expression

Regulation of RUBISCO gene expression is complex   RUBISCO is a very important photosynthetic enzyme. It is the first enzyme in the Calvin cycle. It is a large multimeric protein composed of two different subunit polypeptides, known simply as the large subunit (LSU) and the small subunits (SSUs). The functional RUBISCO protein consists of eight copies each of these two different polypeptides. The gene encoding the LSU, designated *rbcL,* is in the chloroplast genome, and the genes encoding the small subunits, *rbcS*, are present in the nuclear genomes of angiosperms as small multigene families containing 4 to 13 members.

The synthesis of the functional RUBISCO protein is regulated at several levels. There is a single *rbcL* gene in the chloroplast genome, but each leaf mesophyll cell will have several hundred copies of the *rbcL* gene, as there are several hundred chloroplasts per cell. In contrast, there are up to a dozen *rbcS* genes in the nucleus, but most of the SSUs in a particular tissue will be derived from two or three of these. In *Petunia*, *rbcS* is encoded by six genes, but they do not exhibit coordinate expression. One *rbcS* gene accounts for 47% of all the transcript in the leaf, whereas the other genes are expressed at levels that range from 2 to 23% of the total. Despite these differences, the two subunits are accumulated in approximately equal amounts in the chloroplast. Expression of the *rbcS* genes is regulated by light, tissue and cell type differences, and development. In contrast, the *rbcL* genes probably are expressed constitutively, although the LSU does not accumulate unless the SSU peptide is present so that the complete RUBISCO protein complex is formed. Neither subunit is stable unless it is part of a RUBISCO molecule.

RUBISCO gene expression is regulated by light and development   The RUBISCO protein is the most abundant protein of leaves, where it makes up 50% of the total leaf protein content, but it is a fairly minor component of stems and is almost nonexistent in roots. Not all of the cells of the leaf or stem will express the genes for the RUBISCO polypeptides. Chloroplasts do not develop in the epidermal cells of leaves or stems, other than in the guard cells. As a result, epidermal cells do not express *rbcS* or *rbcL* genes, even when the leaf is exposed to light. The strongest expression of *rbc* genes occurs in mesophyll cells of the leaf, but mesophyll cells do not necessarily express these genes in all plants. The mesophyll cell of plants with C3 photosynthesis synthesize RUBISCO and exhibit the Calvin cycle; however, the mesophyll cells of plants with C4 photosynthesis do not have RUBISCO and do not synthesize LSU and SSU proteins, even though they have chloroplasts with a well-formed light-harvesting apparatus. Only the bundle sheath cells of C4 plants express *rbcL* and *rbcS* genes and accumulate functional RUBISCO protein.

The regulation of *rbcS* gene expression by light has been studied extensively in many different species. The results of studies on tomato *rbcS* gene expression are typical and are presented here as an example of the mechanisms found to regulate the expression of *rbcS* genes in most plants. The tomato genome contains five *rbcS* genes. Three of these—*rbcS3A*, *rbcS3B*, and *rbcS3C*—encode SSU proteins with identical sequences. These three genes are located on the same chromosome in a 10-kb region and they probably arose as a result of recent gene duplication events; however, these three genes and the other two tomato *rbcS* genes, *rbcS1* and *rbcS2*, which are on a different chromosome, are regulated independently and exhibit different patterns of tissue-specific and light-regulated expression. Three of these genes—*rbcS3A*, *rbcS1*, and *rbcS2*—are expressed in the cotyledons of dark-grown seedlings. That is, mRNA produced as a result of the transcription of *rbcS3A*, *rbcS1* and *rbcS2* can be detected when the RNA extracted from these organs is bound to a membrane and hybridized with radioactive probes specific for these genes (Fig. 6.6). The mRNA for the other two genes, *rbcS3B* and *rbcS3C*, can be detected only after the cotyledons have been exposed

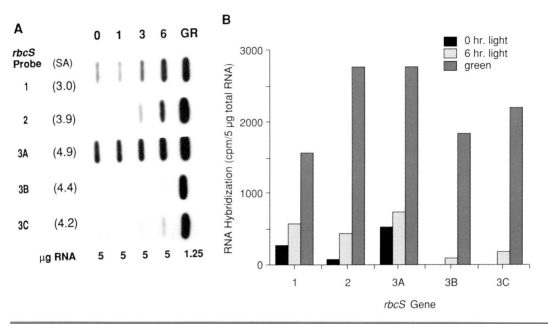

**Figure 6.6  Measurement of mRNA abundance for different tomato *rbcS* genes in etiolated and green seedlings, as well as etiolated seedlings exposed to light**

Total RNA was extracted from tomato seedlings grown for 7 days in complete darkness (lane 0); from 7-day-old dark-grown seedlings exposed to light for 1, 3, or 6 hours (lanes 1, 3, and 6 respectively) immediately before extraction; and from seedlings grown in the light for 7 days (lane GR). Five micrograms of each RNA sample was placed in the appropriate slot of a slot-blotting apparatus where the RNA was adsorbed onto a membrane. Five identical membranes were prepared and then each was hybridized to a radioactive probe whose sequence was specific for one of the tomato *rbcS* genes (1, 2, 3A, 3B, or 3C). (A) Autoradiogram of the blots. (B) Actual amount of radioactivity hybridized in each sample. The amount of radioactivity is proportional to the amount of mRNA for each of the *rbcS* genes. Reprinted (A) and redrawn (B) with permission from Wanner and Gruissem (1991).

to light for several hours. Light strongly enhances the accumulation of mRNA from the *rbcS*3A, *rbcS*1, and *rbcS*2 genes, and it is essential for the appearance of detectable levels of mRNA from the *rbcS*3B and *rbcS*3C genes.

The experiment presented in Fig. 6.6 tells us only that the mRNA for a given gene is or is not present in a given tissue. It does not tell us very much about transcription of the gene itself. A gene could be transcribed, but its mRNA may be rapidly degraded, so the tissue has little mRNA to hybridize to the probe. Alternatively, a gene may be transcribed at a low level, but if its mRNA were unusually stable, it could accumulate to a high level in the tissues. If we want to know whether or not a gene is regulated primarily at the transcriptional level, it is necessary to assay for the transcription of the gene more directly. This can be accomplished by means of a nuclear run-on transcriptional assay in which nuclei are isolated and allowed to complete transcription in a buffer containing all four ribonucleotides triphosphates, one of which is labeled with $^{32}$P, so the primary transcripts synthesized *in vitro* will be radioactive. The transcripts then are hybridized to single-stranded DNA probes that have sequences specific for the different *rbcS* genes. The amount of radioactivity hybridized to the immobilized probe will be proportional to the amount of transcription of each gene.

The results of the nuclear run-on transcription assays demonstrated that *rbcS*3B and *rbcS*3C genes are regulated primarily at the transcriptional level. The other three *rbcS* genes exhibit both transcriptional and posttranscriptional regulation, although transcriptional regulation appears to be the most significant determinant of the abundance of mRNA for all *rbcS* genes in tomato leaves. For example, *rbcS*3A mRNA was the most abundant of the mRNAs of the five *rbcS* genes in both light- and dark-grown seedlings, but the other *rbcS* genes were transcribed at greater rates. This difference in the rates of transcription and mRNA abundance suggests that the mRNAs differ in their stability. Both *rbcS*3A and *rbcS*3C are transcribed weakly in immature green tomato fruits, but this tissue has detectable levels of mRNA from only the *rbcS*1 and *rbcS*2 genes.

Transcription of *rbcS* genes is regulated by a complex interaction of transcription factors with each other and with regulatory elements in the promoters of these genes When the nucleotide sequences of the promoter regions of the five tomato *rbcS* genes are compared, overall they are very different. Each, however, contains several conserved sequences that have been observed in other light-regulated genes. Many of these conserved sequences have been shown to act as regulatory elements. That is, these sequences have the ability to alter the expression of reporter genes in transgenic plants or in transient expression assays when they are inserted into the 5' upstream region containing a minimal promoter. These regulatory elements include sequences known as an L box, an I box, Box II, an AT-rich regulatory element, and a light-regulated element (LRE) (Table 6.1), in addition to the TATA and CAT boxes found in most eukaryotic genes. Protein factors bind to these regulatory elements, several of which have been identified and characterized. These include the GT-1 factor that binds to Box II and Box III and the AT-1 factor that binds to the AT-rich regulatory element (see Table 6.1).

For the most part, the promoters of all five tomato *rbcS* genes contain the same regulatory elements; however, the order and number of these elements, as well as their spatial relationship to each other and to the transcriptional start site, are substantially different among the genes (Fig. 6.7). DNase I footprinting analysis can detect the interaction of protein factors with the regulatory elements in the *rbcS* promoters in different tomato tissues and organs. As described in Chapter 4, the binding of protein factors to a DNA sequence protects it from

**Table 6.1     Conserved sequence motifs in *rbcS* promoters[a]**

| Number symbol | Motif | Sequence | Binding factor | Putative function |
|---|---|---|---|---|
| | L box | AATTAACCAA | | Unknown |
| 1 | | pyrimidine rich | | Unknown |
| | | ⠀⠀⠀⠀⠀⠀⠀⠀⠀⠀C C | | |
| 2 | 1 box | GGATGAGATAAGATTA | GA-1 | Unknown |
| 2 | 1 box | GATAAG | GA-1 | "Enhancer" |
| | | ⠀⠀⠀⠀T | | |
| 2 | GATA | GATGAGATA | ASF-2[b] | Leaf-specific |
| | | ⠀⠀⠀⠀T | | |
| 3 | G box | CACGTGGC | GBF | "Enhancer" |
| 5 | | TTAAATAGAGGGCGTAA | | Unknown |
| | | GA⠀⠀⠀⠀A⠀⠀⠀A | | |
| 8 | Box II | TTGTGPuTAATPuT | GT-1 | Light regulation |
| | | ⠀⠀A⠀⠀⠀⠀C | | |
| 9 | | TTGTAATGTCAA | | Unknown |
| | | ⠀⠀A | | |
| 10 | | GAGCCACA | | Unknown |
| | | ⠀⠀⠀⠀T | | |
| 12 | CAAT box | ATCCAAC | | Unknown |
| 13 | | GGTTAC | | Unknown |
| | | | LRF-1 | |
| 15 | | AGATGAAG | (?) | Light regulation |
| | | ⠀T⠀⠀⠀⠀⠀A⠀⠀A | | |
| 16 | | TTTGTGTCCGTTAGATG | | Unknown |
| 18 | "LRE" | CCTTATCAT | | Unknown |
| | | C⠀⠀⠀⠀T | | |
| 19 | TATA box | TTATATAAA | TFIID[b] | Transcription initiation |
| | A-T rich | (varies) | | Unknown |
| | A-T rich | AATATTTTTATT | AT-1 | Unknown |
| | A-T rich | AAATAGATAAATAAAAACATT | 3AF-1[b] | "Enhancer" |

[a] From Manzara *et al.* (1991).

[b] These genes have been cloned.

digestion with DNase I, whereas DNA that does not have protein bound to it will be digested. Thus, even when the nature of factors interacting with a regulatory element in DNA has not been determined, its presence in a given tissue or organ can be detected by this method. DNase I footprinting has demonstrated that factors are associated with most of the *rbcS* promoter regulatory elements in green leaf tissue where these genes are transcriptionally active. Also, protein factors do not bind to most of the regulatory elements in the *rbcS* genes in roots and other nongreen tissues in which the genes are not transcribed. These results suggest that transcription of the *rbcS* genes requires a complex interaction of protein factors with the regulatory sequences in their promoters. In genes that have been extensively studied, transcription requires a combination of transcription factors that interact with each other through structural domains such as

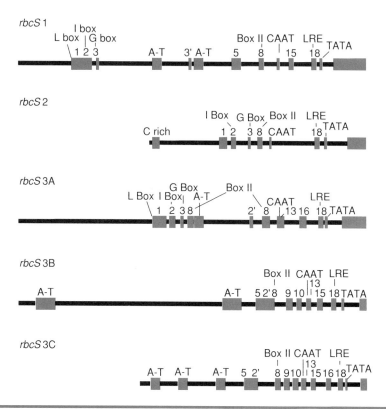

**Figure 6.7** **Regulatory elements in the promoter regions of the five**
*rbcS* **genes of tomato**
The horizontal line represents the region of each *rbcS* gene upstream (in the 5′
direction) from the start of transcription. The sequence of each of the regulatory
elements is shown in Table 6.1. Redrawn with permission from Manzara *et al.* (1991).

leucine zippers. As presented in Chapter 4, transcription factors usually bind
more tightly and specifically to regulatory sequences after they have formed
dimers. Some transcription factors can form homodimers; others dimerize with
other transcription factors.

At present, we cannot explain exactly how light and development regulate
the expression of genes. It seems likely that the signal transduction pathway
initiated by light results in the formation or activation of perhaps several transcription factors. Additional transcription factors are generated by the specific tissues
in which these genes are expressed. These transcription factors then may dimerize
and interact with the promoters of genes containing the appropriate regulatory
elements, leading to gene transcription.

## Mediation of Photomorphogenic Responses by Three Light Receptors

At least three kinds of light receptors are known to mediate these photomorphogenic responses:

*Protochlorophyllide.* Chlorophyll synthesis in angiosperms requires light. In dark-grown plants, the synthesis of chlorophyll is blocked at a step one removed from the final product; however, the biosynthetic pathway for chlorophyll operates in darkness to produce the pigment protochlorophyllide. Protochlorophyllide has a structure similar to chlorophyll a, except that it has a double bond between two carbons in one of the four pyrrole rings. Protochlorophyllide is not green and it cannot absorb light for photosynthesis, but it is a photoreceptor that has an absorption maximum at $\lambda = 650$ nm. Light absorption by protochlorophyllide results in the addition of protons to the double bond in the D ring of the tetrapyrrole, converting the protochlorophyllide to chlorophyll a. Protochlorophyllide is attached to a protein, and the protochlorophyllide–apoprotein complex can act as the enzyme protochlorophyllide reductase; however, the enzyme is active only after its chromophore, protochlorophyllide, has absorbed light. Each molecule of protochlorophyllide must be autocatalytically reduced before dark-grown seedlings can form photosynthetically competent chloroplasts. As a result, the greening reaction of dark-grown seedlings requires continuous exposure to light.

*Blue light receptors.* A number of plant developmental responses are triggered by light of approximately 450 nm. These are known as blue light responses. Usually these blue light responses have a second absorption maximum between $\lambda = 370$ nm and $\lambda = 380$ nm and also are triggered by light in the near ultraviolet (UVA). The best-studied blue light response is phototropism, the growth of plant organs toward or away from a source of unidirectional light (Fig. 6.8). Blue light responses in higher plants include inhibition of internode elongation, induction of stomatal opening, initiation of chloroplast development, and chloroplast orientation within the cell. In addition, there are numerous blue light responses in algae and fungi, and in ferns the orientation of cell division in the gametophyte is controlled by blue light. The blue light receptor has never been identified with certainty in any organism. Two candidates for the blue light receptor are $\beta$-carotene and riboflavin (Fig. 6.9). Both absorb blue light and UVA, and there is circumstantial evidence, but no convincing proof, that each could act as the receptor for blue light-mediated responses.

*Phytochrome.* Phytochrome is the photoreceptor for numerous light-regulated developmental responses. The following are some of the developmental responses regulated by phytochrome:

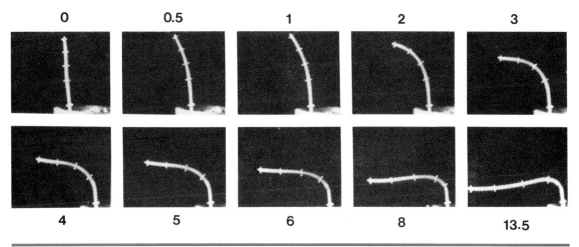

**Figure 6.8    Phototropism in the coleoptile of *Avena***
The coleoptile was exposed to a unidirectional source of blue light which was to the
left of the seedling. It was photographed at the intervals shown (hours). Surface
markers are chromatography beads placed to indicate the different regions of the
coleoptile. Phototropic curvature in the apical region is nearly complete after 2 hours,
after which the apical tissues begin to straighten while bending continues in the more
basal portions of the coleoptile. The straightening response is known as autotropism.
Reprinted with permission from Firn (1986).

- Regulation of internode elongation
- Initiation of chloroplast development
- Hypocotyl hook unfolding
- Initiation of leaf abscission
- Induction of seed germination
- Initiation of bud dormancy
- Initiation of reproductive development

The unique characteristics of phytochrome-regulated responses are that they
are initiated by red light and, in most cases, can be reversed or blocked by far-
red light. For example, the germination of many seeds requires, or is stimulated
by, light. Only about 20% of the seeds of the Grand Rapids cultivar of lettuce
will germinate in complete darkness; however, if the seeds are exposed to 3
minutes or red light after they have imbibed water, nearly 100% germinate.
When the red light treatment is followed immediately by 3 minutes of far-red
light, the red light stimulation of germination is blocked (Fig. 6.10A). Examination
of the effectiveness of various wavelengths of light on lettuce seed germination
demonstrates that the light near 660 nm gives optimal germination, whereas
light near 730 nm actually blocks germination (Fig. 6.10B).

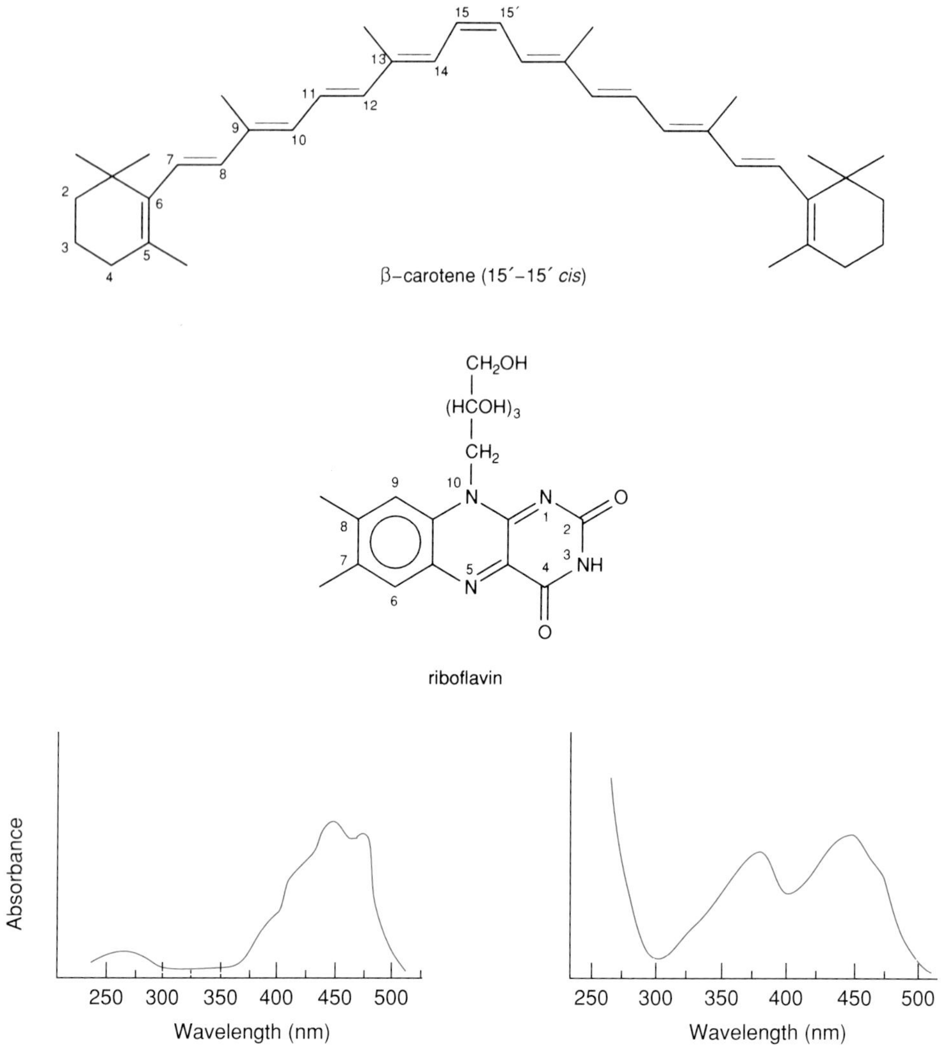

β–carotene (15′–15′ cis)

CH₂OH
(HCOH)₃
CH₂

riboflavin

**Figure 6.9** **Two candidates for the blue-light photoreceptor**
The light absorption spectra of β-carotene and riboflavin are shown along with their structural formulas. Adapted with permission from Hart (1988).

Phytochrome exhibits two interconvertible light-absorbing forms

The unique characteristic of lettuce seed germination and other phytochrome-regulated responses is their photoreversibility. Photoreversibility is possible because the phytochrome exists in two photointerconvertible forms; one absorbs red light (660 nm) and the other absorbs far-red light (730 nm) (Fig. 6.11).

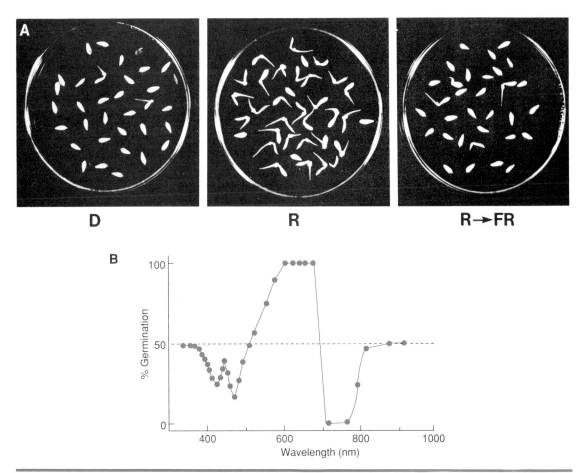

**Figure 6.10 Stimulation of lettuce seed germination by red light and its reversal by far red light**

(A) Lettuce seeds (c.v. Grand Rapids) were allowed to imbibe water and either germinated for 36 hours in complete darkness, **D**, or given 3 minutes of red light, **R** or 3 minutes of red light followed immediately by 3 minutes of far red light, **R-FR**, and then returned to the dark for 36 hours. (B) Action spectrum for the effect of light on lettuce seed germination. The seeds were soaked in water and given sufficient red light to induce 50% germination. The seeds were then exposed to light of the indicated wavelength and returned to the dark for 36 hours, after which the percentage germination was determined. (A) From Kone and Kendrick (1986). (B) Redrawn with permission from Hart (1988).

$$P_r \underset{730\,nm}{\overset{660\,nm}{\rightleftharpoons}} P_{fr}$$

$P_{fr}$ is the biologically active form of the pigment. Not only can it be converted to the $P_r$ form by absorbing far-red light, but the form of $P_{fr}$ present in etiolated

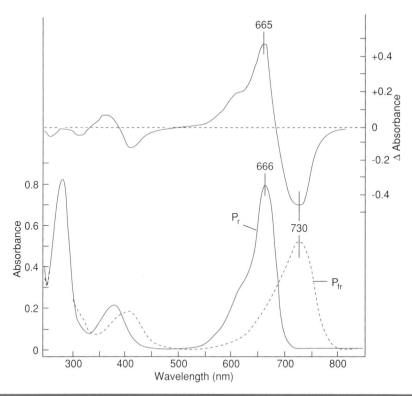

**Figure 6.11 Absorption spectra and difference spectrum for the $P_r$ and $P_{fr}$ forms of type I phytochrome of dark-grown *Avena* seedlings**

The difference spectrum was determined by subtracting the $P_r$ spectrum from the $P_{fr}$ spectrum. Redrawn with permission from Vierstra and Quail (1986).

plants also is destroyed in the dark. Even in light-grown plants, the proportion of phytochrome in the $P_r$ and $P_{fr}$ forms will vary with the time of day and with the quality of light the plant receives. As a result, the pigment continues to play an important role in determining the course of plant development throughout the life of the plant. The ratio of red (R) to far-red (FR) light differs substantially at different times of day. At noon on a sunny day the R:FR ratio is approximately 1.19, whereas at twilight the R:FR ratio is 0.7 to 0.9. Similarly, the R:FR ratio under a forest canopy is substantially different from that of unfiltered daylight and is strongly shifted toward the far red. These different ratios of R:FR light result in differing ratios of $P_r$:$P_{fr}$ in the plant, with important consequences for plant development. This can be demonstrated experimentally by growing plants under light regimes with different R:FR ratios. Plants receiving the same total level of illumination, but a greater proportion of far-red light relative to red light, exhibit much greater internode elongation (Fig. 6.12).

Phytochrome contains a single 124-kDa polypeptide and a covalently attached tetrapyrrole group that acts as the chromophore (Fig. 6.13). The protein folds loosely into two domains, with the chromophore attached to a hydrophobic region in the amino-terminal domain. Phytochrome aggregates in solution. $P_r$ aggregates to form a dimer and this interaction occurs through the carboxy-terminal domain of the protein. The $P_r$ form of the protein changes its conformation when it absorbs red light. One of these conformational changes occurs in the amino-terminal domain and involves a change from a random coil configuration in $P_r$ to an amphipathic $\alpha$ helix in $P_r$. Formation of the amphipathic $\alpha$ helix could create a site for the interaction of $P_{fr}$ forms of phytochrome with other proteins through hydrophobic bonding.

The cellular localization of a receptor can give us important clues as to how it functions. As a result, considerable effort has been devoted to the identification of the cellular localization of $P_r$ and $P_{fr}$ forms of phytochrome. Although there is general agreement that $P_r$ is a soluble, cytoplasmic protein and is not associated with membranes or other cellular organelles, there has been substantial disagreement about the cellular localization of $P_{fr}$. There are some reports that $P_{fr}$ is associated with the nucleus and other reports that it is associated with the plasma membrane; however, it now seems unlikely that neither of these is likely to be true. In one approach to the identification of the cellular localization of a protein, cells containing the protein are homogenized and fractionated, and then the organelle or structure with which the protein is associated is determined. The $P_{fr}$ form of phytochrome A forms large aggregates in cells and these aggregates can be pelleted by centrifugation, giving rise to much confusion as to their true cellular localization. The large aggregates do not, however, appear to be associated with the nucleus, organelle, or cell membrane fraction of intact cells. This ability to form large cytoplasmic aggregates may be an important clue to the function of $P_{fr}$, although at present we do not know what this means.

**Most plants probably have several different phytochromes** Southern analysis of *Arabidopsis* genomic DNA demonstrates that it has five genes encoding phytochrome, and cDNAs for three of these have been isolated and sequenced. The predicted amino acid sequences of these three phytochromes are only about 50% identical. This raises the possibility that phytochrome is involved in several different regulatory pathways, each with somewhat unique characteristics. Alternatively, the specific amino acid sequence of phytochrome may not be important for its function as a photoreceptor, so the amino acid sequence changes rapidly and randomly over evolutionary time. This question can best be answered by changing the coding sequence of a phytochrome gene to alter the amino acid sequence of the protein in a systematic way and then examining the effects of these mutations on the ability of the protein to function when the gene is intro-

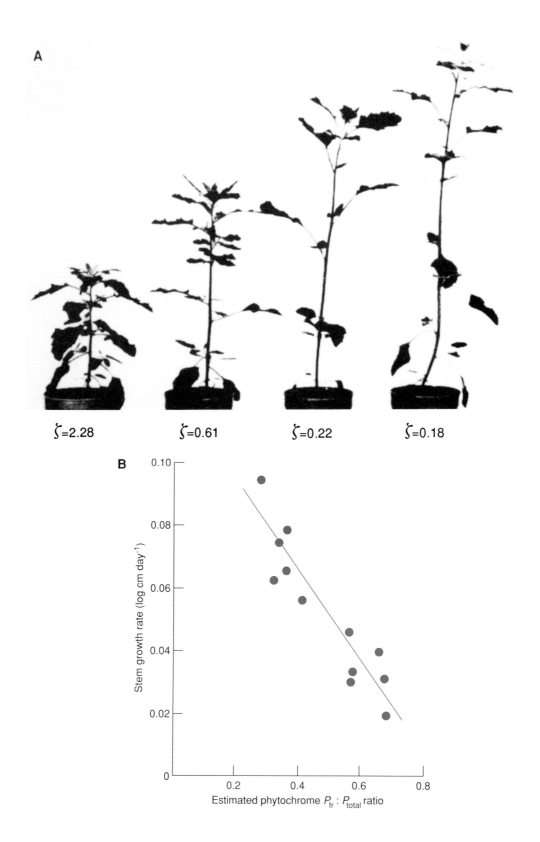

A

$\zeta$=2.28    $\zeta$=0.61    $\zeta$=0.22    $\zeta$=0.18

B

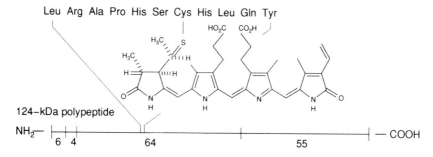

Leu Arg Ala Pro His Ser Cys His Leu Gln Tyr

124–kDa polypeptide

NH₂— |—|—|——————————————————————————|— COOH
     6  4              64                         55

**Figure 6.13** **Structure of the 124-kDa *Avena* phytochrome molecule and the position of attachment of its chromophore**
The horizontal line represents the length of the phytochrome protein from its amino terminus at the left to the carboxyl terminus. The amino acid sequence of the chromophore attachment site is shown, as are the major proteolytic cleavage sites, indicated by vertical lines. Reprinted with permission from Vierstra and Quail (1986).

duced back into phytochrome-deficient plants. This has not been done at the present time; however, there is good evidence that at least two functionally different phytochrome receptors exist, and it is most likely that this difference is the result of differences in the amino acid sequence of the phytochrome protein.

Physiologically, plant tissues have two different phytochromes, which have been designated type I and type II. These two phytochrome types differ in their behavior after they are exposed to red light. Both are converted to the $P_{fr}$ form by absorbing red light. The $P_{fr}$ form of type II phytochrome is stable. In contrast, type I $P_{fr}$ is unstable and turns over very rapidly with a half-life of approximately 60 minutes. Etiolated tissues contain comparatively high levels of phytochrome and nearly all of this is type I. It is synthesized in the dark in the $P_r$ form and is present as a soluble dimer that is distributed evenly throughout the cytoplasm. Light-grown green tissues contain much lower levels of phytochrome than etiolated tissues, but nearly all of the phytochrome found in green tissues is type II phytochrome. Recently, it was demonstrated that type I phytochrome in *Arabidopsis* is the product of a single phytochrome gene, *phyA*, so this phytochrome is now designated phytochrome A. The products of phytochrome genes *phyB* and *phyC* both behave as type II phytochromes; that is, their $P_{fr}$ forms are stable. The

**Figure 6.12** **Effects of light quality on internode elongation in *Chenopodium album***
(A) Seedlings were grown for 14 days under photoperiods in which they received light of differing Red: Far Red ratios (ζ), as indicated below each plant. (B) The rate of internode elongation is plotted as a function of the ratio of $P_{fr}$ to $P_{total}$. Redrawn with permission from Hart (1988).

mechanism responsible for the rapid destruction of phytochrome A is not known, but some evidence suggests that it occurs through a process known as ubiquitination. Ubiquitin is a small polypeptide found in nearly all eukaryotes whose amino acid sequence is highly conserved in evolution. It is attached to proteins that are then marked for destruction by proteases. Ubiquitin has been found to be attached to some of the $P_{fr}$ forms of phytochrome A after etiolated tissues have been exposed to red light.

Mutations affect phytochrome responses    Two different classes of mutations have been identified that affect phytochrome function. The first are mutations that reduce the phytochrome content of plant tissues; the second affect the phytochrome signal transduction pathway. In both cases the mutations are generated by screening for plants that develop abnormally in the light or dark. Three characteristics of etiolated plants, as presented earlier, are that the leaves do not expand in the dark, chloroplast development is arrested, and the hypocotyls are longer than those of seedlings germinated in the light. Light arrests hypocotyl elongation and initiates chloroplast development and leaf expansion in wild-type seedlings, whereas light does not arrest hypocotyl elongation in mutants. These mutants include the *aurea* (*au*) mutant of tomato, the long hypocotyl (*lh*) mutant of cucumber, and six different *long hypocotyl* mutants of *Arabidopsis,* designated *hy*1, *hy*2, *hy*3, *hy*4, *hy*5, and *hy*6. Each of these represents a different gene or group of genes that must function in this photoresponse. The *hy*1, *hy*2, and *hy*6 mutations are defective in chromophore biosynthesis, so they lack functional phytochrome. The phytochrome levels of *hy*4 and *hy*5 are normal, indicating that these mutations most likely affect components of the signal transduction pathway or another light-regulated mechanism regulating hypocotyl elongation. There is evidence that the *hy*4 mutation is defective in the blue light receptor transduction pathway, but the defect in *hy*5 is not known. In contrast, *hy*3 mutants show a dramatically reduced level of type II phytochrome B, but normal levels of phytochrome A and C. Phytochrome C is encoded by the *phyC* gene and represents a small fraction of the total type II phytochrome in wild-type plants. This demonstrates that phytochrome B plays an important role in the regulation of hypocotyl elongation by light. This conclusion is strengthened by the fact that overexpression of the *phyB* gene in transgenic *Arabidopsis* plants results in plants with markedly shortened hypocotyls (Fig. 6.14). Both the *au* mutant of tomato and the *lh* mutant of cucumber result from markedly reduced levels of type I phytochrome. Overexpression of the *phyA* gene in *Arabidopsis* results in a dwarf phenotype with markedly shortened hypocotyls. The *hy*1, *hy*2, and *hy*6 mutants of *Arabidopsis* completely lack type I or type II phytochromes, yet these mutations are not lethal. Although leaf development and greening are slower, these processes are initiated when these mutants are grown in white light. As these mutants completely lack functional phytochromes, photomorphogenesis, including leaf and chloroplast development, must be initiated by some other

**Figure 6.14 Effects of overexpression of an *Arabidopsis* gene**
Overexpression of the *Arabidopsis* gene encoding phytochrome B (*phyB*) causes a short hypocotyl phenotype in seedlings of transgenic *Arabidopsis*. Shown are light-grown seedlings of an *Arabidopsis* plant transformed with a chimeric *Arabidopsis phyB* gene whose expression was controlled by a cauliflower mosaic virus (CaMV) 35S promoter. The strong CaMV 35S promoter overexpressed the *phyB* gene, resulting in a short hypocotyl phenotype in approximately three-fourths of the light-grown seedlings. Seedlings with longer hypocotyls did not inherit the chimeric CaMV 35S *phyB* gene. Overexpression of *phyB* does not affect hypocotyl elongation in darkgrown seedlings. The rate of hypocotyl elongation is markedly reduced by exposure to light in wild-type *Arabidopsis*, and overexpression of *phyB* reduced this elongation even further to give the short hypocotyl phenotype. The seedlings with longer hypocotyls lacked the chimeric *phyB* gene and had hypocotyl lengths similar to those of untransformed control plants. Reprinted with permission from Wagner *et al.* (1991).

photoregulated mechanism, perhaps the blue light receptor. This fact, plus the likelihood that the *hy*4 mutants are in the blue light receptor pathway, demonstrates that there are several regulatory pathways by which light controls hypocotyl elongation.

*Arabidopsis* mutants have been isolated that exhibit the light-grown phenotype when the seedlings are grown in darkness. These have been designated *deetiolated* (*det*) and *constitutive photomorphogenic* (*cop*) mutants. In both cases these mutants exhibit expanded cotyledons, the initiation of primary leaves, the inhibition of hypocotyl elongation, the accumulation of anthocyanin pigments, and the expression of many nuclear genes encoding photosynthetic proteins when the seedlings are grown in complete darkness (Fig. 6.15). The various *cop* and *det* mutations differ somewhat from each other. The *det*1 mutation affects light regulation of photomorphogenesis only during the seedling stage of development. Adult *det*1 plants respond normally when they are placed in darkness; stem elongation increases, newly formed leaves do not expand, and chloroplast development is blocked. The *det*1 mutants also do not require light for seed germination, as is normally the case in the wild-type *Arabidopsis*. In contrast, *cop*1 and *det*2 mutations affect light-regulated development in both seedlings and adult plants, but these

*det1*                                                    wild type

*det1*                                                    wild type

mutations do not affect seed germination. Chloroplast development is initiated in the dark in the *det*1 and *cop*1 mutants, but not in the *det*2 mutants. All three of these mutations are recessive. This means that the protein products of these genes in wild-type plants repress the developmental program induced by light. In normal plants, the signal generated by light must overcome the repression of light-induced development by DET1, DET2, and COP1 proteins.

*Phytochrome regulates the expression of nuclear genes* Phytochrome regulates the expression of many plant genes, and it may control plant development primarily through its role in regulating gene expression. As presented earlier, light regulates the synthesis of RUBISCO, the chlorophyll a/b-binding protein, and many other chloroplast proteins. Phytochrome plays a central role in this regulation, although other light receptors also are involved in regulating the expression of genes encoding chloroplast proteins as well as other genes involved in photomorphogenesis. Nevertheless, it is important to separate the individual regulatory pathways and to understand how these contribute to the overall developmental program. As a result, it is important to determine the specific mechanism by which phytochrome regulates gene expression. Phytochrome exerts both positive and negative control of gene expression. Although many genes encoding chloroplast proteins are positively regulated by phytochrome, the expression of others is inhibited. The most notable example of the latter is the gene for phytochrome A, *phyA*, whose transcription is blocked when etiolated seedlings are exposed to red light. Also, when light-grown plants are placed in darkness, transcription of the *phyA* gene is reinitiated.

One of the most interesting investigations to determine the mechanisms by which phytochrome regulates gene transcription is being conducted in the small water plant *Lemna gibba* by Elaine Tobin, at the University of California, Los Angeles. The *Lemna* genome contains 12 to 14 *rbcS* genes, at least 6 of which have been shown to be regulated by phytochrome. *Lemna* can be grown heterotrophically in the dark through many generations by culturing the plant aseptically on a medium containing sucrose. Under these conditions, the plant grows, but slowly, and is pale yellow. No chloroplast development occurs in the dark-grown plants and most of the nuclear genes encoding chloroplast proteins are not transcribed. If the plants are given 10 minutes of red light every 8 hours, there is an enormous increase in the growth rate, and transcription of the *rbcS* genes

**Figure 6.15** **Growth and development of the *det*1 mutant and the wild-type *Arabidopsis thaliana* in the dark (A) and in the light (B)**
The *det*1 mutants develop in darkness as though they had been exposed to light and lack the etiolated characteristics of the dark-grown wild-type seedlings; however, they are dwarfs when grown in the light. Reprinted with permission from Chory *et al.* (1989). ©Cell Press.

is initiated. The cells now accumulate *rbcS* mRNA, although the RUBISCO protein is not synthesized and the plastids remain etioplasts. Completion of chloroplast development requires a higher level of light, but the transcription of the *rbcS* genes is initiated by a brief exposure to red light. Furthermore, this mRNA accumulation is far-red light reversible. Tobin's group has identified a DNA-binding factor, which they have designated LRF-1 (light-regulated nuclear factor) that binds to the sequence GATAAG which is present in the promoter region of many *rbcS* genes and other light-regulated genes.

## Hormones

Plant hormones are naturally occurring substances, effective in very small amounts, that act as signals to stimulate or inhibit growth or regulate some developmental program. They are all small molecules, with molecular weights under 1000. In many cases, plant hormones are active in specific target tissues, which are different from the tissues in which the hormone is produced. They are chemical signals that facilitate intercellular communication. In these cases, the hormone is either actively transported or passively moved to the target tissues from its site of synthesis, and it may have little or no effect on the tissues through which it moves or on the tissue that produced it. In other cases, the effects of the hormone are more generalized, and it may act on the tissue that produced the signal as well as other tissues. Typically, there is a specific, regulated mechanism for the biosynthesis of the hormone, as well as a regulated mechanism for destroying the hormone. As a result, the levels of the hormone in the plant are modulated. Environmental factors such as light, photoperiod, and gravity can affect the biosynthesis, destruction and distribution of plant hormones and, in turn, the hormones modify the developmental program or the growth response.

The following molecules or groups of molecules have been shown to exert a powerful effect on some aspects of plant growth and development: auxins, gibberellins, cytokinins, abscisic acid, ethylene, brassinolide, thiamine, the peptide systemin, various oligosaccharides, jasmonic acid, methyl jasmonate, and certain phenolic compounds. Not all of these are hormones. Thiamine is included in this list mostly to illustrate the differences between plant hormones and other molecules that might stimulate growth under some conditions. As you know, thiamine is a vitamin. This means that it is an essential nutrient for humans. Humans cannot synthesize it and must obtain it from their diet. It is not a vitamin as far as the intact plant is concerned as plants synthesize all the thiamine they require. Thiamine is only synthesized by the shoot of the plant in the light, however. It is not synthesized by roots and roots depend on thiamine that is made in the shoot for their growth. Isolated roots can be cultured aseptically in a nutrient solution that contains essential minerals, a metabolizable carbon source (usually sucrose), and thiamine. They will not grow without thiamine. Because thiamine also is required in very small amounts, it has been argued that it is a plant hormone. There is, however, something fundamentally different about thiamine as compared with a plant hormone such as auxin. Once thiamine enters the cell, it is phosphorylated and then it serves as a cofactor for carboxylase enzymes. In other words, it participates in the respiratory metabolism that is essential to provide the energy for growth and development. Plant hormones are not merely permissive factors. They play a significant regulatory role in determining when or where growth occurs, and/or they initiate or inhibit developmental processes. Most plant hormones have been shown to initiate gene expression, and a number of genes have been shown to contain specific base sequences in their promoters that act as hormone response elements. In the presence of the hormone, a transcription factor will bind to these hormone response elements to permit the initiation of transcription.

Of the molecules listed earlier, five are generally recognized to be plant hormones: **auxins, gibberellins, cytokinins, abscisic acid,** and **ethylene** (Fig. 6.16). The recently discovered 18-amino-acid polypeptide systemin also is a plant hormone and must be added to this list. **Systemin** is produced in response to wounding or insect foraging and is rapidly transported through the plant where it triggers the synthesis of proteinase inhibitors. In addition to these, specific cell wall fragments, sometimes called **oligosaccharins,** also initiate part of the plant's response to insect or pathogen attack and clearly are important signaling molecules. Oligosaccharins derived from the plant cell wall may act as signals that regulate other aspects of plant development as well, but this is less firmly established. Recently, certain bacterially synthesized **lipooligosaccharides** were discovered to be important signaling molecules in the induction of root nodule formation during the establishment of the nitrogen-fixing symbiosis between *Rhizobium* bacteria and legumes. Finally, plant pathogens produce a variety of molecules that act as elicitors of the plants' defense against disease. An **elicitor** is a signaling molecule that conveys information about the presence of the pathogen to the plant under attack, and in response to this information, the plant initiates defensive measures that may stop the attack. In the following we present some of the most important things we have learned about the five main plant hormones in an effort to illustrate some of the features of the hormonal regulation of plant growth and development.

## Auxin

Auxin was the first plant hormone to be identified  The word *auxin* is derived from a Greek word that means "to increase." The first auxin was discovered by biologists who were looking for the factor(s) that regulated the

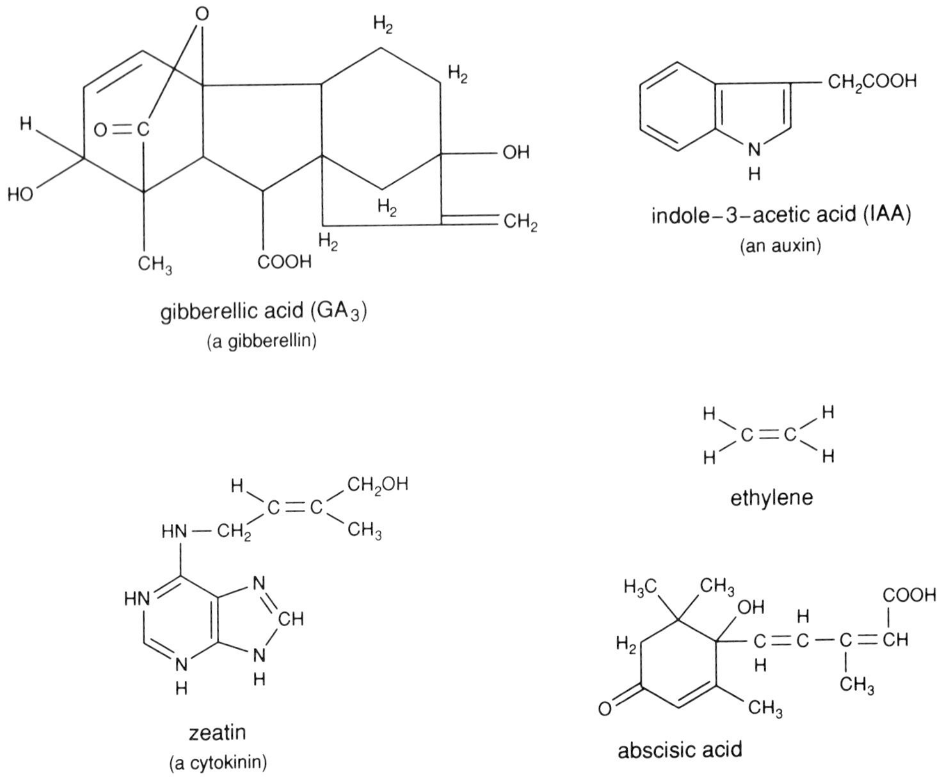

**Figure 6.16 Structures of representatives of each of the five main classes of plant hormones**

growth rate of stem tissue. Many noted biologists contributed to the discovery of auxin, including Charles Darwin. Darwin initiated the research that led to the discovery of auxin with his investigations on the power of movement in plants. Darwin noted that the coleoptile, an ensheathing organ of the grass seedling, would bend toward a light source, and that this bending was the result of uneven growth. The coleoptile is a hollow, capped tube found only in the seedling stage of the grass plant, where it covers and protects the first leaves as they emerge from the soil. Darwin noted that coleoptile growth stopped if the tip was cut off and growth would resume if the tip was replaced. Nearly 40 years later, Fritz Went demonstrated that growth of the coleoptile was dependent on a diffusible substance that was transported from the coleoptile tip into the subjacent tissues. He named the substance auxin and showed that it regulated the rate of cell elongation in the tissues below the tip in which it is produced. Light apparently interferes with the movement of this substance, leading to asymmetrical growth and bending of the coleoptile. Kenneth Thimann identified a naturally occurring, highly potent growth-regulating molecule, indole-3-acetic acid, which subsequently was shown to be the diffusible substance regulating cell elongation in this system and in many other plant tissues and organs. Indole-3-acetic acid (IAA) is the principal naturally occurring auxin in most plants (Fig. 6.17A). Since the discovery of native auxins, chemists have synthesized a large number of molecules that have been found to be very potent auxins. These include the potent weed killers 2,4-dichlorophenoxyacetic acid (2,4-D) (Fig. 6.18B) and 2,4,5-trichlorophenoxyacetic acid (2,4,5-T).

*The auxin biosynthetic pathway is not known with certainty*
Auxin (IAA) is an indole derivative, and several other indole derivatives are found in plants, including the amino acid tryptophan. In fact, tryptophan can act as intermediary in the biosynthesis of auxin. Several different enzyme-catalyzed reactions lead to the synthesis of IAA from tryptophan, and it is possible that plants may use some or all of these different mechanisms to produce the hormone. It has been shown that tryptophan can be enzymatically deaminated to form indolepyruvic acid, which in turn can be decarboxylated to form in-

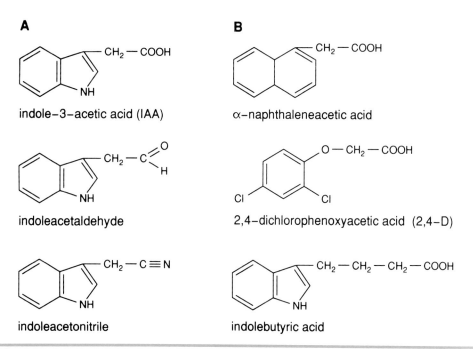

**A**

indole–3–acetic acid (IAA)

indoleacetaldehyde

indoleacetonitrile

**B**

α–naphthaleneacetic acid

2,4–dichlorophenoxyacetic acid  (2,4–D)

indolebutyric acid

**Figure 6.17 Structures of some molecules that act as auxins**
(A) Naturally occurring auxins, all of which are indole derivatives. (B) Some synthetic auxins.

doleacetaldehyde. Indoleacetaldehyde then can be oxidized to form IAA (Fig. 6.18). Both indolepyruvic acid and indoleacetaldehyde have been shown to be active as auxins in some tissues, probably as a result of their conversion to IAA. The native auxin biosynthetic pathway is not, however, known with certainty. The T-DNA of *Agrobacterium* Ti plasmids contains two genes, *iaaH* and *iaaM*, which participate in the synthesis of IAA from tryptophan. The *iaaM* and *iaaH* genes, along with a third gene encoding an enzyme involved in cytokinin biosynthesis, are inserted into plant cells as part of the Ti-derived T-DNA which transforms them during *Agrobacterium* infection. The crown gall tumors that result from this transformation synthesize auxin as well as cytokinin, and it is this hormone synthesis that is responsible for the tumor phenotype. The synthesis of auxin is very carefully regulated during plant development and this regulation is an important component of growth control. In the case of crown gall tumor

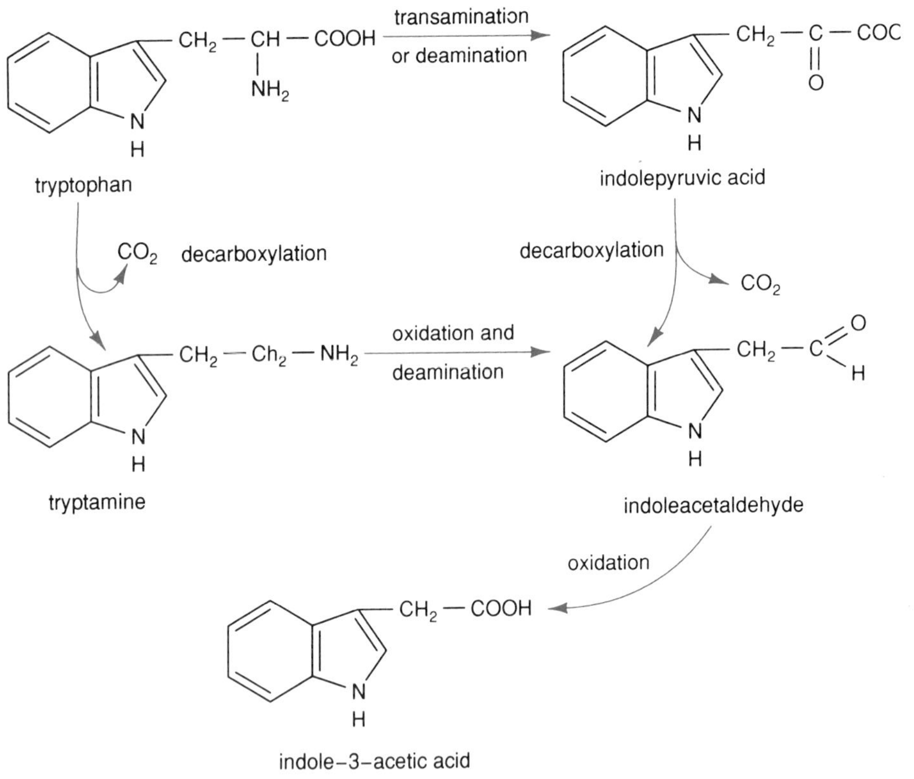

**Figure 6.18 Some chemical reactions that could lead to the synthesis of indole-3-acetic acid**
Redrawn with permission from Noggle, G. R., and Fritz, G. J. (1976). *Introductory Plant Physiology*, Prentice-Hall, Englewood Cliffs, New Jersey.

tissues, however, hormone synthesis is not regulated by the transformed cells, which may be part of the reason the cells proliferate uncontrollably, forming a tumor.

Auxin probably is not synthesized by this mechanism, except in plants that have been transformed with wild-type *Agrobacterium* and thus contain the *iaaH* and *iaaM* genes. Mutants of *Arabidopsis* that are deficient in tryptophan biosynthesis actually have higher levels of IAA than wild-type plants. These mutants also have higher levels of indole, a precursor for tryptophan biosynthesis, probably because tryptophan biosynthesis is blocked. This suggests that IAA may be synthesized from indole and that indole is the direct precursor of IAA. Now that mutants are beginning to be generated in this pathway, we may soon learn how this important hormone is synthesized.

**Auxin regulates many aspects of plant growth and development**   It is possible that auxin is essential for plant life. In addition to the regulation of cell elongation, auxin has been implicated in the control of the following phenomena:

- Phototropism
- Inhibition of abscission zone formation
- Inhibition of lateral bud development
- Vascular tissue differentiation
- Vascular cambium cell division   *secondary growth.*
- Maintenance of tissue polarity
- Leaf blade expansion

Auxin is synthesized in the apical buds and young leaves of a typical plant and it is transported down the stem by a specific mechanism known as the auxin transport system. Auxin is well established as the signaling molecule in phototropism. It regulates the elongation of young internodes and the enlargement of the young leaf blades. It also initiates cell differentiation in both leaves and internodes as it stimulates their enlargement or elongation. While it is promoting the growth of these tissues and organs, it inhibits the development of lateral buds. This leads to a condition known as apical dominance. Auxin inhibits the formation of the abscission zone at the base of the leaves, fruits, and other organs that allows them to drop from the plant. Auxin also regulates cell division in the vascular cambium, which of course is involved in secondary growth, and it has been implicated in the maintenance of tissue polarity.

Auxin can have either a positive or a negative effect on any of these phenomena, depending on its concentration. It must be above some threshold concentration to be effective, and above this concentration, the amount of growth increases as a linear function of the increasing concentration of the hormone. At some point, however, further increases in the auxin concentration cease to promote growth and in fact become inhibitory. Different tissues vary widely in the effective auxin concentration for promoting their growth and/or development. These facts are clearly illustrated by the data presented in Fig. 6.19. Auxin stimulates the curvature of decapitated oat coleoptiles over the concentration range 0.01 to 0.2 mg/liter, whereas concentrations above 0.2 mg/liter are supraoptimal and actually inhibit further elongation (Fig. 6.19A). In contrast, the growth of pea epicotyl segments is stimulated over a fivefold greater concentration range, although concentrations of auxin above the optimum also are clearly inhibitory to further growth (Fig. 6.19B).

These conclusions were reached largely by examining the effects of applied auxin. Typically, an attempt is made to eliminate the endogenous sources of auxin, such as actively growing buds, and then the effect of exogenously applied hormone is examined. The problem with this approach is that it can be difficult to determine how much hormone actually reaches the tissues. Recently, Henry Klee and his co-workers have taken a different approach by transforming petunia or tobacco plants with the *iaaM* gene from *Agrobacterium*. Transgenic plants expressing the *iaaM* gene have up to a 10-fold higher level of auxin. Although these plants grow, flower, and set seed, they have a number of abnormalities, including much stronger apical dominance and more vascular tissue. Additionally, Klee *et al.* have transformed plants with a bacterial gene, *iaaL*, encoding an IAA–lysine synthase that conjugates IAA with the amino acid lysine. As IAA–lysine conjugates are not active as an auxin, the auxin level of the plant is reduced. Transgenic plants expressing the *iaaL* gene not only have up to a 20-fold lower level of auxin, but they also exhibit markedly reduced apical dominance and less vascular tissue. Thus, the molecular–genetic approach confirms the earlier conclusion that auxin plays an important role in apical dominance and vascular differentiation.

With the discovery of gibberellin-deficient mutants of pea and maize, which have markedly reduced internode elongation as a result of the deficiency of this hormone (see next section), some individuals concluded that gibberellin was the major hormone regulating internode elongation in light-grown plants and that auxin does not participate in stem elongation. It now seems likely that both auxin and gibberellin regulate stem elongation, although they are likely to do so by different mechanisms. Mutants of *Arabidopsis* and tobacco have been isolated that are unusually resistant to applied auxin. These auxin-resistant mutants are likely to be defects in the auxin receptor or signal transduction pathway, as they have normal amounts of the hormone. All of these mutants have unusually short internodes, in addition to other defects characteristic of auxin-regulated developmental processes, such as reduced vascular differentiation, diminished apical dominance, and lack of tropistic response. The mechanism by which auxin regulates stem elongation is discussed in Chapter 7.

### Gibberellins

The gibberellins were first discovered by Kurosawa in Japan in the course of his studies on a fungal disease of rice. Specifically, he was investigating the mechanism

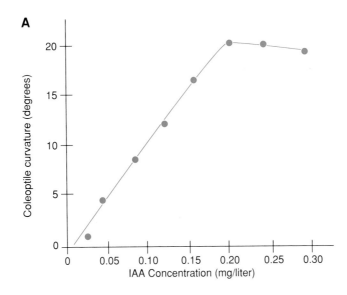

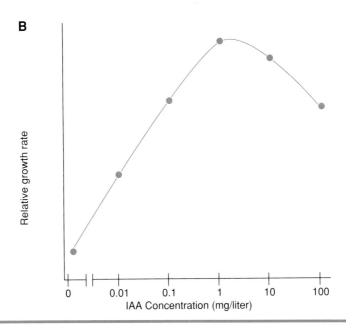

Figure 6.19 **Dependence of growth on auxin concentration**
(A) Auxin can either inhibit or promote coleoptile curvature, depending on the
concentration. Auxin at the indicated concentration was applied to one side of
coleoptiles and the amount of curvature was measured several hours later. (B) The
amount of growth of pea stem segments is dependent on auxin concentration. Pea
stem segments were placed in Petri dishes containing a sugar, salts, and indole-3-acetic
acid at the indicated concentration. The length of the segments was measured 24
hours later and the growth rate determined. (A) Reprinted with permission from
Went, F. W., and Thimann, K. V. (1937), *Phytohormones,* MacMillan, New York. (B)
Reprinted with permission from Galston, A. W., and Hand, M. E. (1949). *Am. J. Bot.*
**36,** 85–94.

by which the fungal pathogen of rice, *Gibberella fujikuroi,* brought about the abnormally elongated stems that characterize seedlings infected with this pathogen. Kurosawa showed that this abnormal elongation was due to a water-soluble compound produced by the fungus which he called gibberellin A. Later, gibberellin A was shown to be a mixture of six different gibberellins. More importantly, gibberellins were shown to be synthesized by higher plants as well as some fungi. It is fairly common for microorganisms, particularly those that grow as pathogens of higher plants, to produce plant hormones. The secretion of hormones into plant tissues can have a profound effect on the growth and development of the plant and these changes often are beneficial to the microorganism, although harmful to the plant.

**Gibberellins have an unusual chemical structure that is not closely related to any primary metabolite** There are more than 80 known gibberellins, all of which have the same 20-carbon *ent*-gibberellane skeleton, although some have lost one carbon. In addition to the number of carbons in the skeleton, the gibberellins differ in the number and position of hydroxyl groups, the oxidation state of $C_{20}$, and the presence or absence of a lactone bridge between $C_{10}$ and $C_{19}$ (Fig. 6.20). Most of these different gibberellins are precursors or metabolites of $GA_1$, which is generally considered to be the biologically active hormone, at least for stem elongation (GA = gibberellin).

Gibberellins are 20-carbon terpenoids that are synthesized by the condensation of four isoprenoid subunits. Gibberellin biosynthesis occurs via the mevalonic acid pathway. Mevalonic acid is phosphorylated and decarboxylated to form isopentenyl pyrophosphate (IPP). These IPP subunits then are condensed sequentially, to form first the $C_{10}$ geranyl pyrophosphate, then the $C_{15}$ farnesyl pyrophosphate, and finally the $C_{20}$ geranylgeranyl pyrophosphate (Fig. 6.21). This biosynthetic pathway is common to many biological molecules, including carotenoids and steroids. The next step, the cyclization of the geranylgeranyl pyrophosphate to form the *ent*-kaurene, is the first reaction that is unique to the gibberellin biosynthetic pathway. The first gibberellin is $GA_{12}$-aldehyde, which is formed as a result of the oxidation of $C_{19}$ and a contraction of the B ring of *ent*-kaurene from 6 to 5 carbons. A number of plant growth retardants are known that act to block the synthesis of gibberellins. For example, AMO-1618 inhibits *ent*-kaurene synthesis, and paclobutrazol blocks for formation of $GA_{12}$-aldehyde (see Fig. 6.21).

**Gibberellins regulate stem elongation** As gibberellins were discovered because of their ability to stimulate stem elongation, we should not be surprised that the ability to stimulate stem elongation is a common property of these hormones. Why, you may wonder, does this not make them simply another auxin? Part of the reason is that auxins and gibberellins affect different tissues. For example, although auxins regulate the elongation of coleoptile tissues in

plants that exhibit a
elongate. Application
the same height as th
control one of the tra
of the laws of hered
individuals homozygo
contents of elongatin
tall plants have more
to the application of
application of $GA_{20}$,
As a result, it seems
$GA_{20}$ to $GA_1$. $GA_1$
elongating internod

The *le* gene res
completely block $GA$
the elongating tissu
of pea, known as *na*
aldehyde from *ent*
not synthesize any
internodes. The ro
also has been estab
versity of Californ
unable to synthesi
resulted in defecti
more, these gibbe
with gibberellin (
a central role in
supported by the
chemicals that, v
to grow as dwarf
synthesis.

Studies of the
suggest that the
plant developme
involved in the

- Reversio
  phases
- Control
- Inductic
- Breakin
- Sexual
- Mobiliz

---

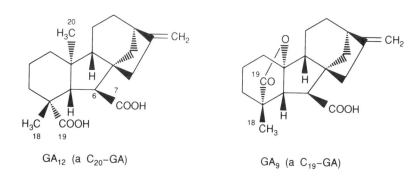

**Figure 6.20 General structure for the gibberellins (*ent*-gibberellane) and structures of two different gibberellins ($GA_{12}$ and $GA_9$)**

grass seedlings, gibberellins do not occur in these organs in measurable amounts, and application of gibberellins to decapitated coleoptiles does not stimulate their growth. Gibberellins are involved in regulating internode elongation in intact plants. This is most dramatic in biennial plants, which require a cold period and long day for the induction of flowering. These cold-requiring plants include cabbage, cauliflower, and other members of the broccoli group. They grow vegetatively, forming a basal rosette of leaves, separated by very short internodes, often with the leaves tightly packed into a head, before they receive the stimulus that induces reproductive development. A period of cold temperatures followed by long days induces them to flower, and flowering is accompanied by an enormous stimulation of internode elongation that is known as bolting. Gibberellins will substitute for the cold period in inducing the stem elongation that normally accompanies flowering in these plants. Auxins have no effect in this system.

Gibberellins also have a profound effect on the growth of some genetic dwarf plants. For example, there are single gene mutants of pea, maize, and other

acetoacetyl-Co

+

$CH_3-C-S-$

acetyl-CoA

geranylgeranyl p

copalyl pyroph

ent-

$H_3C$ $COOH$

$GA_1$

Figure 6.21 The m
of gibb

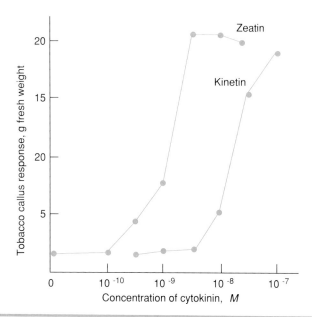

Figure 6.25 **Dependence of cultured tobacco tissue on cytokinin for their growth**

Cultured tobacco callus was transferred to fresh medium that contained an auxin and either zeatin or kinetin at the indicated concentrations. The tissues were weighed after growing for 1 month. The extent of cell proliferation in these tissues is directly proportional to their fresh weight. Redrawn with permission from Leonard *et al.* (1968). ©Proceedings of the National Academy of Sciences.

the ability to stimulate the proliferation of a wide spectrum of cell types. Cytokinin may be the primary plant cell division hormone.

Many tissues, nevertheless, do not require cytokinin to proliferate in culture. Furthermore, some tissues initially require cytokinin, but after prolonged culture involving many subculturings, they lose their cytokinin requirement and will continue to grow without the addition of this hormone to the culture medium. This phenomenon is known as habituation. Most likely a plant tissue becomes habituated for cytokinin because it has acquired the capacity to synthesize the hormone. Certainly, habituated tissue has been shown to contain cytokinin. This tells us that cytokinin synthesis is likely to be regulated in the plant and that this regulatory mechanism can be bypassed in some cases, leading to unscheduled cytokinin synthesis, giving a tissue the potential for further cell proliferation if its nutritional requirements are met.

Shortly after the discovery of kinetin, cultured tobacco pith segments or tobacco callus tissues were observed to produce roots or shoots, depending on the ratio of auxin to cytokinin in the culture medium. High levels of auxin relative to kinetin stimulated the formation of roots, whereas high levels of cytokinin relative to auxin led to the formation of shoots. At intermediate levels the tissue grew as an undifferentiated callus. Thus, the auxin/cytokinin ratio regulates growth and morphogenesis in these cultured tissues. This suggests that auxin and cytokinin could exert a major influence on the type of development that plant tissues exhibit.

Plant hormones do not work alone. Even in those cases in which a response can be evoked by application of a single hormone, the tissue may contain additional endogenous hormones which could contribute to the response. In some cases we know that a response is evoked by two or more hormones; however, until we have a way of blocking the action of the endogenous hormones, we cannot be certain that the actual regulation of the phenomenon being studied is not more complex than it appears to be. Nevertheless, applied cytokinins can evoke a variety of physiological, metabolic, biochemical, and developmental processes when they are applied to higher plants, and it is probable that they play an important role in the regulation of these events in the intact plant. Cytokinins have been shown to participate in the regulation of numerous additional aspects of plant development, including the following:

- Initiation of chloroplast development
- Delay of leaf senescence
- Enhancement of cotyledon cell expansion
- Mobilization of nutrients
- Initiation of buds

## Abscisic Acid

Abscisic acid was discovered in a search for factors that regulated the formation of the abscission zone in fruits and leaves and as a result of investigations on the regulatory signals controlling the onset of dormancy. It was named **abscisic acid** because initially it was believed that the hormone was the primary signal regulating leaf abscission; however, although abscisic acid will promote leaf abscission in cotton and a few other species under experimental conditions, we now know that in most plants, leaf and fruit abscission layer formation is initiated by ethylene, not abscisic acid. Abscisic acid nevertheless regulates several important aspects of plant growth and development. These include the initiation of seed and possibly bud dormancy, the control of stomatal closing under water stress, and the initiation of senescence. Abscisic acid is sometimes considered to be a plant stress hormone because it is synthesized in response to many kinds of stress, including cold, salt, and water stress. Abscisic acid can inhibit or reverse the actions of other growth-promoting hormones. Gibberellins initiate growth from dormant seeds and shoots, whereas abscisic acid inhibits both gibberellin- and auxin-induced growth and promotes the onset of dormancy.

Abscisic acid synthesis may be regulated by stress and photoperiod   The pathway for the biosynthesis of abscisic acid is not known with certainty. There is evidence that either of two different routes may be used. In both cases abscisic acid is synthesized via the mevalonate pathway, as are all terpenoids. In one case, however, the hormone is synthesized directly from the 15-carbon precursor, farnesyl pyrophosphate, whereas in the other case, it is synthesized indirectly by the cleavage of a 40-carbon carotenoid precursor known as violaxanthin (Fig. 6.26). Carotenoids, including violaxanthin, also are synthe-

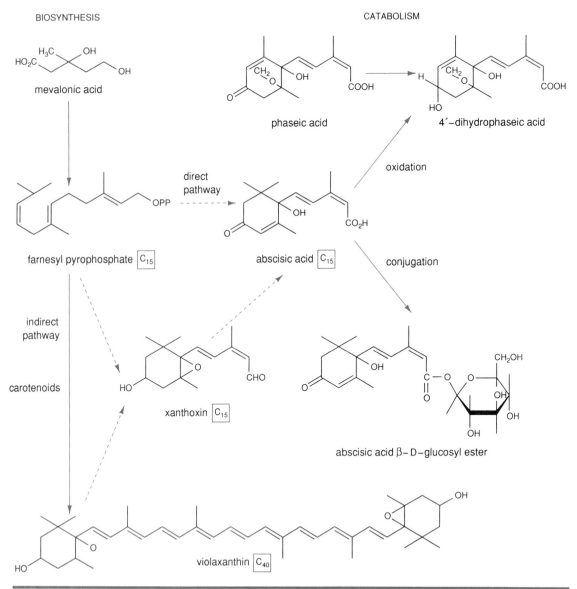

**Figure 6.26** **Biosynthesis and metabolism of abscisic acid**

sized via the mevalonate pathway. Both of these pathways may be used for the synthesis of abscisic acid, but in different tissues or under different conditions. Abscisic acid is synthesized in mature leaves and transported both to the root and throughout the shoot system, primarily via the phloem. Roots also can synthesize abscisic acid and transport the hormone into the shoot in response to water stress. Water stress results in a large increase in abscisic acid synthesis, as do short days in some photoperiodically sensitive plants.

*Abscisic acid induces stomatal closure*    Abscisic acid can induce the closure of stomates within a few minutes. Water stress in the roots can lead to as much as a 40-fold increase in the abscisic acid levels of the plant. Guard cells use turgor pressure to control the stomatal aperture. An increase in ion uptake triggered by light or changes in carbon dioxide concentration results in the osmotic uptake of water by the guard cells. The resultant increase in turgor pressure deforms the guard cell, opening the stomate. Stomates close as a result of the loss of ions, leading to a loss of water and a reduction in turgor pressure. Guard cells have an abscisic acid receptor on the outer surface of their plasma membranes. Binding of abscisic acid to its receptor in the guard cell membrane opens ion channels and activates proton pump.

*Gene expression is regulated by abscisic acid*    Mutants of *Arabidopsis* and maize have been selected that either are unable to synthesize abscisic acid or are unable to respond to the hormone. Analysis of three abscisic acid-insensitive mutants of *Arabidopsis*—*abi*1, *abi*2, and *abi*3—indicates that there are two different abscisic acid regulatory mechanisms in these plants. The product of the *abi*3 gene, ABI3, acts primarily on seed development. The embryos of seeds bearing this mutation do not go dormant in the later stages of seed development. In contrast, ABI1 and ABI2 act not only during seed development, but also during vegetative growth. There is also a viviparous (*vp*1) mutant of maize in which the embryos do not become dormant. This mutation reduces the sensitivity of the embryo to abscisic acid, without affecting abscisic acid synthesis. The wild-type VP1 gene product probably is involved in either the receptor mechanism for abscisic acid or the signal transduction pathway. The *vp*1 mutation also has pleiotropic effects on anthocyanin synthesis during seed development. In the wild type, anthocyanin pigments are synthesized and accumulate in the aleurone layer of the seed, but in the *vp*1 mutant this synthesis is blocked because the VP1 gene product is required for expression of the *c*1 gene. The *c*1 gene encodes a transcription factor that regulates genes encoding enzymes required for anthocyanin synthesis. The *vp*1 gene has been cloned by transposon tagging and sequenced. The VP1 gene product most likely is a transcription factor involved in regulating the expression of other genes. The VP1 protein appears to activate the transcription of the *Em* gene, a gene involved in maize seed maturation. Transcription of the *Em* gene is promoted by abscisic acid, but direct binding of VP1 to regulatory elements in the promoters of either the *Em* or *c*1 gene has not yet been observed.

### Ethylene

Ethylene is a very simple regulatory molecule. It is a gas with the formula

$$H_2C{=}CH_2$$

It has been known since the 1900s that ethylene gas can modify the growth of plants. Many physiological and developmental changes occur when plants are exposed to smoke, various gases, and air pollutants. Puerto Rican pineapple growers and Philippine mango growers start bonfires near their crops in an effort to initiate flowering or to ripen the fruits. In 1864, Girardin noted that trees near leaking gas mains in Paris had shed their leaves. At least some of these effects are known to be caused by ethylene. This was initially demonstrated in Russia by Neljabou, who in 1901 found that pea plants grown in his laboratory, which was illuminated with gas lights, showed what has come to be called the triple response: (1) the roots were no longer positively geotropic, (2) the stems swelled, and (3) the petioles of the leaves were epinastic (they curved down toward the stem). He analyzed the components of the coal gas used for the gas lights and tested the components for their effects on pea seedlings. He showed that ethylene was present in the coal gas at a concentration of approximately 10 parts per million, and that at this concentration, ethylene could produce all the effects on pea seedling growth he observed. Acetylene would produce similar effects, but only at much higher concentrations. It was not until 1934, however, that Gane demonstrated that ethylene gas was produced by ripening apples. We now know that ethylene is produced by fruits during their ripening and that many of the events that occur during fruit ripening, such as softening of the flesh, changes in color, and conversion of starch to sugar, are triggered by ethylene. As a result of this work, ethylene became known as a fruit-ripening hormone, and it was considered to have a rather limited role in regulating plant development. A true assessment of the significance of ethylene in the regulation of plant processes awaited the development of sensitive methods for its detection, principally gas chromatography. With a flame ionization detector, gas chromatography is able to detect concentrations of ethylene as low as 1 part per billion in air. As a result of these improved methods for its detection, we now know that ethylene is produced by all parts of the plant, but that certain parts of the plant, notably ripening fruits, senescing tissues, and meristematic regions, produce much larger quantities. Ethylene also is produced in many other parts of the plant in response to stress, such as wounding and invasion by pathogenic organisms.

**Ethylene is derived from the amino acid methionine**    A key enzyme in the synthesis of ethylene, 1-aminocyclopropane-1-carboxylic acid synthase (ACC synthase), produces the intermediate, 1-aminocyclopropane-1-carboxylic acid, from S-adenosylmethionine. This is the rate-limiting step in ethylene biosynthesis. ACC is converted to ethylene. Methionine then is regenerated in a series of reactions known as the Yang cycle (Fig. 6.27). This enables ethylene to be

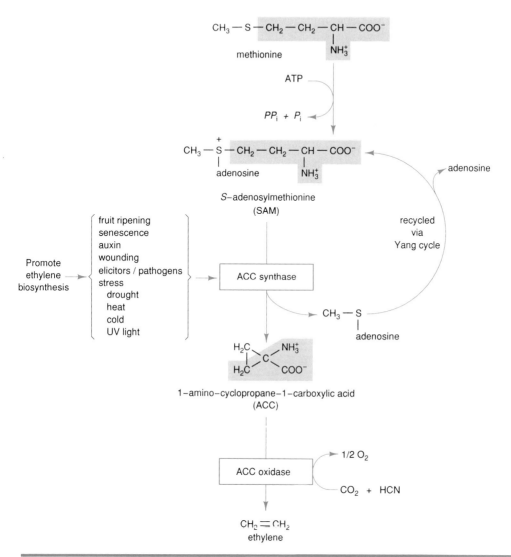

**Figure 6.27 Ethylene biosynthesis from methionine**

Ethylene ultimately is derived from the third and fourth carbons of the amino acid methionine; however, the immediate precursor is 1-aminocyclopropane-1carboxylic acid (ACC). The rate-limiting step in ethylene biosynthesis is the formation of ACC from S-adenosylmethionine (SAM) by the activity of the enzyme ACC synthase. ACC synthase activity is regulated by many internal and environmental factors, some of which are listed. Methionine is regenerated by a series of reactions known as the Yang cycle in which the $CH_3$–S– group is reused. Redrawn with permissions from Yang and Hoffman, 1984.

synthesized without depleting the cellular methionine pools. Ethylene synthesis can be triggered by many stimuli, including auxins, stress, wounding, and fruit ripening.

Ethylene production varies in different plant tissues    Basal levels of ethylene production range from 0.5 to 500 nl/g of tissue per hour, depending on the tissue. These levels increase to approximately 100 nl/g/h in ripening fruits, whereas senescing flowers may produce about half this amount. Ethylene levels are initially high during the early stages of fruit development, when growth is accompanied by rapid cell proliferation. Ethylene production then declines and is low until the final phase of growth and ripening, when the level again rises. Ripening can involve numerous steps, depending on the fruit. These include the differentiation of chromoplasts from chloroplasts with the subsequent loss of photosynthetic capacity, the synthesis of special carotenoid pigments, and the synthesis of wall-degrading enzymes, leading to the softening of the fruit. Ripening is triggered by ethylene in many fruits. Modern fruit storage methods use this fact to delay ripening and increase the storage life of fruits, both by reducing the concentration of ethylene around the fruits and by increasing the concentration of carbon dioxide gas, a natural antagonist of ethylene action. Recently, a biotechnological solution to the problem of fruit storage has been found. Tomato plants have been transformed with an artificial gene construct that expresses an antisense mRNA for the ACC synthase gene. An antisense RNA has a nucleotide sequence complementary to a specific mRNA. As a result, it hybridizes with the mRNA and prevents its translation into a functional protein product. In many cases the antisense RNA:mRNA hybrid is rapidly degraded in the cell. Tomato plants expressing the antisense ACC synthase gene do not produce ethylene because they are unable to synthesize the key enzyme for ethylene biosynthesis. The fruits of these transformed tomatoes remain green until they are treated with ethylene. This result demonstrates the importance of ACC synthase in ethylene biosynthesis and the key role of ethylene in controlling fruit ripening, and such treatment may become the preferred method for controlling the ripening of tomato fruits. In the near future most tomatoes sold in supermarkets may contain this synthetic gene expressing an antisense ACC synthase RNA. The fruits will be stored for many months as mature green fruit. Shortly before they are to be sold, they will be exposed to ethylene to initiate the final stages of ripening.

Ethylene also is important for its role in controlling abscission layer formation. The abscission zone is a specific layer that forms in the petiole at the base of a leaf and represents a region of structural weakness that allows the leaf to become detached from the plant. Its formation is inhibited by auxins, but promoted by ethylene.

Ethylene synthesis can be induced    Ethylene biosynthesis can be induced or enhanced both by stress, such as wounding and drought, and by auxin treatments. Auxin applications can bring about up to 100-fold increase in ethylene production, and many of the inhibitory actions of auxin have been shown to be caused by auxin-induced ethylene. In these cases ethylene appears to be a kind of second messenger. In fact, many of the effects that were first attributed to auxin have been shown to be mediated by auxin-induced ethylene; however, ethylene cannot substitute for auxin in promoting stem elongation in most plants or in the promotion of cell division in cultured plant tissues.

# Initiation of Regulatory Cascades by Hormones

At present, we do not have complete knowledge about how any plant hormone carries out its regulatory role. It is, however, well established that these regulatory molecules act via signal transduction pathways that are initiated as a result of the binding of the hormone to its **receptor.** As we saw in Chapter 5, steroid hormone receptors are DNA-binding proteins that are able to initiate the transcription of specific genes after they have bound to the hormone. Animal hormones also include water-soluble molecules, many of which are proteins or peptides, that do not have to enter the cell to initiate a response. In these cases, the hormone receptor is on the surface of the cell, anchored to the plasma membrane (Fig. 6.28). Receptors have been identified for auxin, gibberellin, and abscisic acid. Of these, the gibberellin and abscisic acid receptors are on the cell surface, and the auxin receptors are found both on the cell surface and in the cytoplasm.

Hormone binding to its cell surface receptor initiates a chain of events that leads to the generation of a short-lived **second messenger.** Several different second messengers may be used, but a specific hormone acting on a particular cell type will trigger the appearance of a specific second messenger. These second messengers include cyclic adenosine monophosphate (cAMP), cyclic guanosine monophosphate (cGMP), 1,2-diacylglycerol, inositol 1,4,5-triphosphate, jasmonic acid, and calcium ions. Of these, jasmonic acid, inositol phosphates, diacylglycerol, and calcium ions act as second messengers in plant hormone responses, as discussed later in this chapter. Almost all organisms from bacteria to humans use cAMP as a second messenger. Angiosperms may be an exception. There is no conclusive evidence that cAMP has a regulatory role in higher plants. In contrast, there is strong evidence that 1,2-diacylglycerol, inositol 1,4,5-triphosphate, jasmonic acid, and calcium ions are part of both the signal transduction pathway for hormones and the transduction pathway for environmental factors such as light, gravity, and touch.

## Signal Transduction in Response to Insect Attack

Potato and tomato leaves synthesize proteinase inhibitors I and II in response to insect attack, a response illustrating the unique aspects of some plant signal transduction pathways. Important components of the signal transduction path-

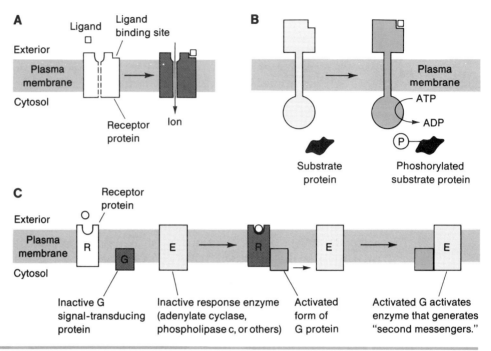

**Figure 6.28 Examples of some cell surface receptors in animal systems**
(A) Ligand-regulated ion channel. Binding of the ligand to the receptor opens a specific ion channel. (B) Binding of ligand to its cell surface receptor activates a protein kinase activity in the receptor's cytoplasmic domain. (C) Binding of the ligand to its receptor activates a G protein required for activating an enzyme that will generate a second messenger. Redrawn with permission from Darnell *et al.* (1990).

way controlling the synthesis of these proteinase inhibitors have been identified through the efforts of Clarence Ryan at Washington State University. Many other plants also synthesize proteinase inhibitors in response to insect foraging and the presence of these inhibitors in leaves can result in the death of the insects that feed on these leaves as a result of protein malnutrition. Proteinase inhibitor I is a powerful inhibitor of the digestive enzyme chymotrypsin, and proteinase inhibitor II blocks the degradation of proteins by trypsin and chymotrypsin. It is the insect's chewing on the leaves that induces proteinase inhibitor synthesis, so any mechanical wounding will serve to induce the synthesis of these proteins. The mRNAs for the proteinase inhibitors appear within 3 to 4 hours of wounding, and the proteins are secreted into the vacuole where they persist for several weeks. Mechanical wounding initiates the production of signaling molecules that trigger the transcription of the genes encoding the proteinase inhibitors. Several different signaling molecules will induce this response, including the plant hormone abscisic acid, but the most effective inducers are **oligosaccharins** and a small peptide known as **systemin.**

The effective oligosaccharins are **oligogalacturonides,** and they are cell wall fragments derived from the pectic fraction of plant cell walls. Several enzymes are known to degrade pectin to form oligouronides that are effective in inducing the response, and the activity of these enzymes appears to be increased by wounding. The oligouronides are produced locally at the site of wounding and induce a localized response. That is, the cells around the wound respond and synthesize proteinase inhibitors, whereas the rest of the plant does not respond. The response is localized because the oligouronides are not readily transported in plant tissues. A plasma membrane receptor for oligouronides has been identified, and the local response appears to be initiated by the binding of the oligouronide to the receptor. Systemin is an 18-amino-acid peptide that also is produced in response to wounding. It is highly effective in inducing proteinase inhibitor synthesis, but, unlike the oligouronides, it induces a systemic response. Systemin is rapidly transported through the plant and induces proteinase inhibitor synthesis throughout the shoot. A systemin receptor resides on the cell surface, and the response is initiated by the binding of systemin to its receptor. The systemin receptor and the oligouronide receptor are different, but they appear to activate proteinase inhibitor synthesis through a common mechanism.

Although the details of the signal transduction pathway are not completely known, the lipid jasmonic acid is a powerful inducer of proteinase inhibitor synthesis. Jasmonic acid is synthesized from linolenic acid, a fatty acid commonly esterified to glycerophosphate in the cell membrane, and synthesis occurs via the pathway shown in Fig. 6.29. The binding of an oligouronide or systemin to its respective membrane receptor could activate a lipase, which would hydrolyze the ester bond, freeing linolenic acid and initiating the chain of reactions leading to jasmonic acid formation. Methyl jasmonate, a derivative of jasmonic acid, also is a potent inducer of proteinase inhibitor synthesis. Methyl jasmonate is a volatile liquid, whereas jasmonic acid is not volatile. Methyl jasmonate is converted to jasmonic acid after it enters plant cells, and this conversion is responsible for its activity. Methyl jasmonate also may be produced by plants after they are attacked by pathogens or insects, and airborne methyl jasmonate is an effective inducer of proteinase inhibitor synthesis. Methyl jasmonate can transmit a signal from a plant under attack to neighboring plants and even a different species, inducing them to synthesize proteinase inhibitors.

The genes encoding proteinase inhibitors I and II both contain jasmonate response elements in their promoters. These are sequences that, when introduced into the promoters of other genes, cause them to be methyl jasmonate or wound induceable. It is unlikely that jasmonic acid itself binds to these regulatory elements. Rather, a transcription factor may acquire the ability to bind to the response element after it first binds jasmonic acid. A model for the signaling pathway leading to proteinase inhibitor synthesis that incorporates many of these elements is shown in Figure 6.30.

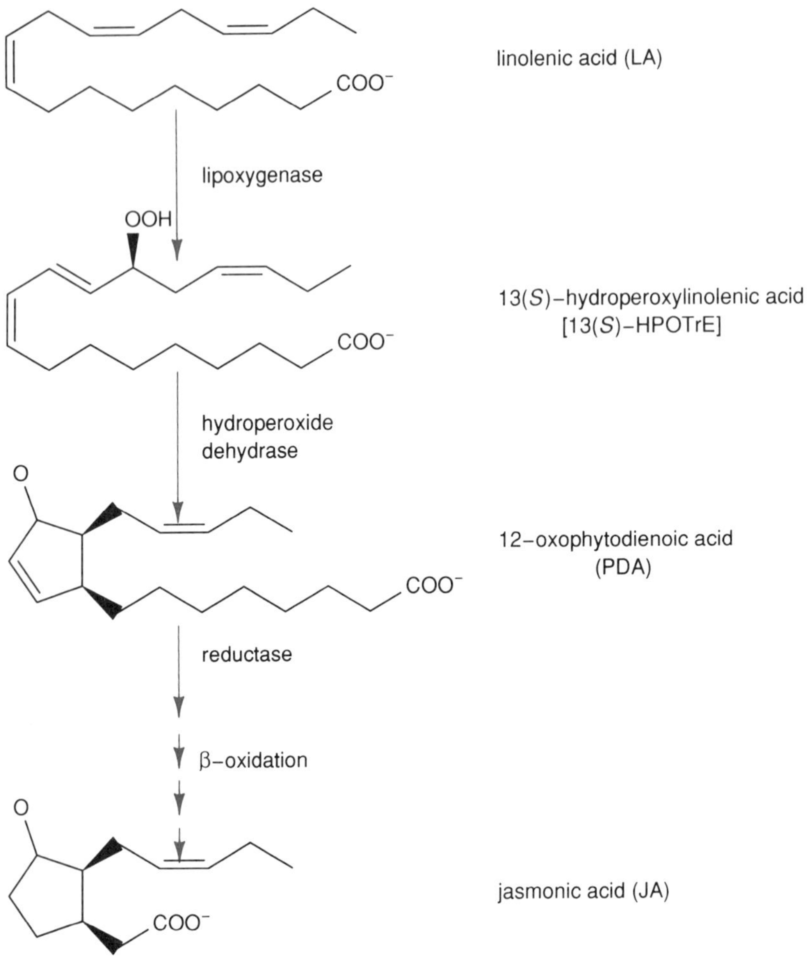

COO⁻
linolenic acid (LA)
lipoxygenase
OOH
COO⁻
13(S)-hydroperoxylinolenic acid
[13(S)-HPOTrE]
hydroperoxide
dehydrase
O
COO⁻
12-oxophytodienoic acid
(PDA)
reductase
β-oxidation
O
COO⁻
jasmonic acid (JA)


**Figure 6.29 Proposed biosynthetic pathway for jasmonic acid**
From Farmer and Ryan (1992).

## Calcium Ion Regulation of the Activity of Some Protein Kinases

Calcium ion concentration can have a powerful effect on cell behavior. Both plant and animal cells have very low cytosolic levels of calcium ions. Frequently the cytosolic $[Ca^{2+}]_i$ is less than micromolar. The extracellular fluid, as well as some intracellular membrane-bound compartments, often contains comparatively high levels of calcium ions, on the order of millimolar concentrations. Under these conditions, calcium ions would tend to cross the cell membrane and

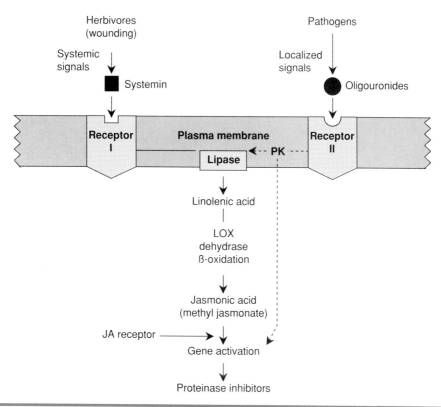

**Figure 6.30 Model for the signaling pathway leading to the induction of proteinase inhibitor genes**
PK, protein kinase; LOX, lipoxygenase; JA, jasmonic acid. Redrawn with permission from Farmer and Ryan (1992).

diffuse into the cytoplasm; however, the steep concentration gradient across the plasma and ER membranes is maintained by means of ATPase calcium ion pumps, which use energy obtained from the hydrolysis of ATP to actively pump calcium out of the cytoplasm (Fig. 6.31).

Many plant hormones stimulate an increase in cytoplasmic calcium levels, and this in turn may trigger a variety of cellular responses (Fig. 6.32). The elevated calcium level is transient, as the cytosolic calcium is rapidly pumped back out of the cell or into one of the calcium stores. For example, gibberellin stimulates the synthesis and secretion of $\alpha$-amylase by barley aleurone cells in the presence of millimolar levels of external calcium. The gibberellin triggers the transcription of genes encoding $\alpha$-amylase as well as the synthesis and secretion of the enzyme. This response is preceded by a threefold gibberellin-induced increase in $[Ca^{2+}]_i$. Abscisic acid blocks gibberellin-induced $\alpha$-amylase synthesis, as well

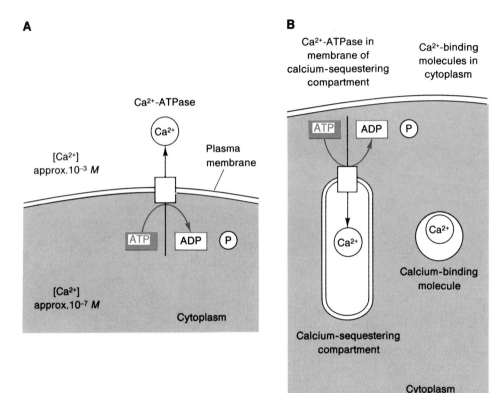

**A**

Ca²⁺-ATPase

Ca²⁺

[Ca²⁺]
approx.10⁻³ $M$

Plasma
membrane

ATP    ADP    P

[Ca²⁺]
approx.10⁻⁷ $M$

Cytoplasm

**B**

Ca²⁺-ATPase in
membrane of
calcium-sequestering
compartment

Ca²⁺-binding
molecules in
cytoplasm

ATP    ADP    P

Ca²⁺

Ca²⁺

Calcium-binding
molecule

Calcium-sequestering
compartment

Cytoplasm

**Figure 6.31 Mechanisms for reducing cytosolic calcium concentration**
The cytoplasmic calcium concentration is reduced by calcium-specific ion pumps
localized in the plasma membrane (A) and by pumps in calcium-sequestering
compartments or by reaction with molecules that bind calcium, such as calsequestrin
(B). Redrawn with permission from Alberts *et al.* (1989).

as the increase in $[Ca^{2+}]_i$ that precedes it. The plasma membrane of the aleurone
cells contains a receptor for gibberellin, so it is likely that the increase in cytosolic
calcium is triggered by the binding of the hormone to its receptor. Although it
is not known how this gibberellin-induced $[Ca^{2+}]_i$ increase triggers the expression
of the α-amylase genes, one of the most common responses to elevated levels of
calcium is the phosphorylation of other cellular proteins by protein kinases. The
activity of many enzymes is regulated by phosphorylation of serine, tyrosine, or
threonine residues. The ability of some transcription factors to bind to regulatory
elements in the promoters of genes is determined by phosphorylation state.
     Cytoplasmic calcium levels can be measured by injecting $Ca^{2+}$ indicator dyes
into the cytoplasm. These dyes give off fluorescent light when they are complexed
with calcium and the amount of fluorescent light given off is proportional to the

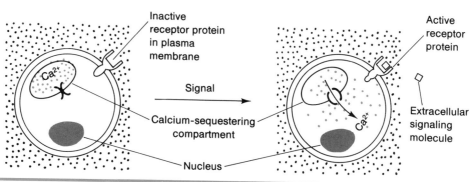

**Figure 6.32** **Increases in cytosolic calcium**
Cytosolic calcium increases in response to signals that open calcium ion channels in the plasma membrane (A) or in the membrane of calcium-sequestering compartments (B). Redrawn with permission from Alberts *et al.* (1989).

$[Ca^{2+}]_i$. Calcium also can be monitored with the calcium-sensitive luminescent protein aequorin. Tobacco plants have been transformed with the gene encoding the aequorin apoprotein which has been placed under the control of the cauliflower mosaic virus 35S promoter. Expression of the aequorin protein does not interfere with normal growth and development of the transgenic plants. As the cytosolic level of calcium can be determined in any cell simply by measuring the light given off by that cell with a luminometer, it is relatively easy to determine the effect of environmental and hormonal stimuli on $[Ca^{2+}]_i$ in these transgenic plants. So far it has been shown that cold temperatures, fungal elicitors, and touch induce a transient increase in $[Ca^{2+}]_i$ using this approach.

The importance of the plant touch response has been fully appreciated only recently. Some specialized plant organs, such as the tendrils of vines, the pulvinus of *Mimosa pudica* leaves, and the leaves of some insectivorous plants, are sensitive to touch and respond rapidly when so stimulated. These are not growth responses, but rather are brought about by rapid changes in turgor pressure. It was recently demonstrated that touching *Arabidopsis* plants brings about a dramatic decrease in their growth rate and induces a transient increase in the protein calmodulin (Fig. 6.33). Calmodulin frequently is involved in calcium-regulated processes.

## Calmodulin Requirement for Many Calcium Ion-Induced Responses

All eukaryotes, including higher plants, contain a small calcium-binding protein known as calmodulin. Calmodulin consists of only 148 amino acids and it has four binding sites for calcium ions, two in the carboxyl-terminal domain and two in the amino-terminal domain. After calmodulin binds calcium ions, it is able to activate other cellular proteins. The calcium–calmodulin complex has no enzyme

**Figure 6.33 Effect of touch on plant growth**
Rain, wind, and touch all reduce plant growth, as illustrated by this experiment in
which *Arabidopsis* plants were either sprayed periodically with water (left) or not
sprayed (right). Spraying induced a transient increase in calmodulin, a calcium-binding
protein that acts as a regulatory subunit for other proteins, including some protein
kinases. Reprinted with permission from Braam and Davis (1990). ©Cell Press.

activity itself; however, it often acts as a regulatory subunit for other enzyme
complexes, such as calcium–calmodulin-dependent protein kinases (Ca-kinases).
These protein kinases are inactive unless they have bound calcium–calmodulin
complex and must be activated by binding calmodulin–calcium before they are
able to add phosphate groups to serine and/or threonine hydroxyl groups on
other proteins (Fig. 6.34).

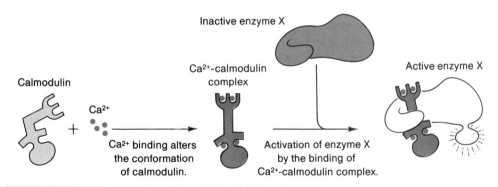

**Figure 6.34 Activation of certain enzymes by an increase in cytosolic calcium**
The protein calmodulin is an essential regulatory subunit for some protein kinases; however, calmodulin cannot bind to and activate the enzyme unless it first binds four calcium ions. Redrawn with permission from Alberts *et al.* (1989).

### Inositol Phosphates and 1,2-Diacylglycerol as Important Second Messengers

The binding of a hormone to its receptor frequently triggers the breakdown of phosphatidylinositol (PI), one of the less common phospholipids in the cytoplasmic half of the plasma membrane. Three different phosphatidylinositides are found in the plasma membrane and they differ in their degree of phosphorylation (Fig. 6.35). The hydrolysis of one of these, PI 4,5-bisphosphate (PIP$_2$), by the enzyme phospholipase C yields two products: 1,2-diacylglycerol and inositol 1,4,5-trisphosphate (InsP$_3$) (Fig. 6.36). InsP$_3$ is water soluble and rapidly diffuses through the cytosol, whereas 1,2-diacylglycerol is lipid soluble and remains in the membrane. Both compounds are important second messengers. InsP$_3$ triggers a transient increase in cytosolic calcium by opening calcium channels in the plasma membrane or in a calcium-sequestering compartment. Diacylglycerol activates a specific calcium-requiring protein kinase known as protein kinase C (Fig. 6.37).

### Key Role of Membrane-Associated G Proteins in Signal Transduction Pathways

G proteins are composed of several subunits, one of which binds GTP tightly. Another G protein subunit can interact with the cytoplasmic domain of a specific hormone receptor, but only after it has bound its hormone. The binding of the G protein to the receptor activates the G protein so that it can in turn activate phospholipase C (see Fig. 6.37).

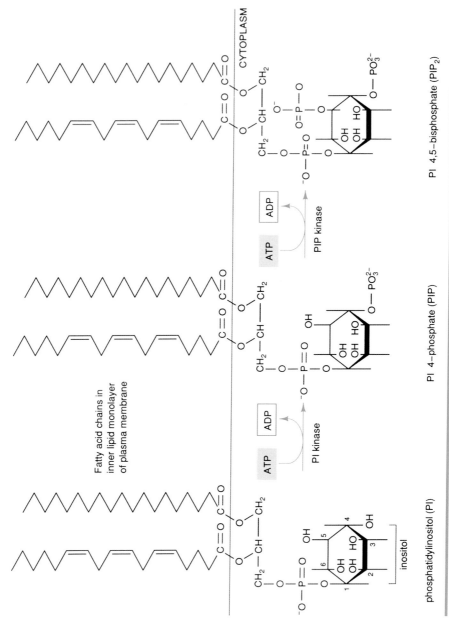

**Figure 6.35 The inositol phospholipids of cell membranes**

332

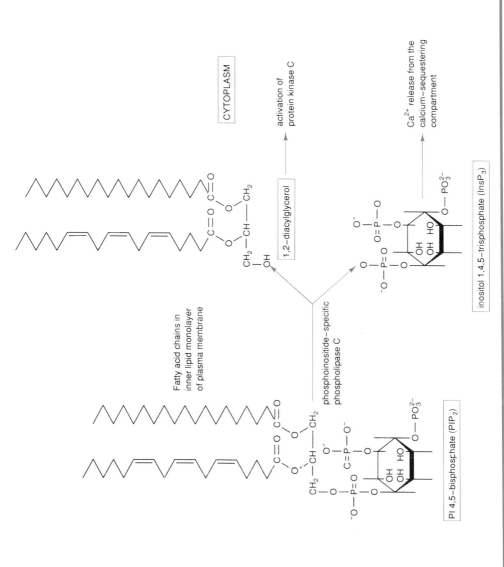

**Figure 6.36 Generation of the second messengers 1,2-diacylglycerol and inositol triphosphate by a phosphoinositide-specific phospholipase C**

333

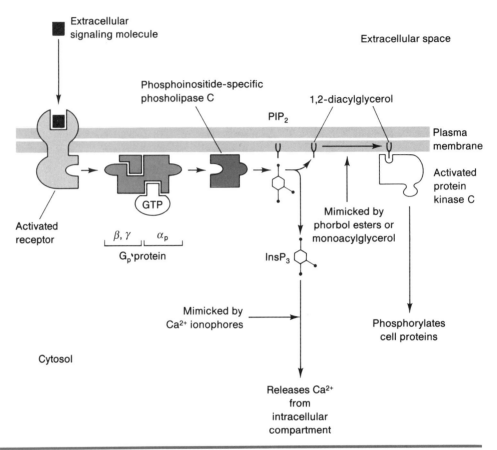

**Figure 6.37 Generation of second messengers by ligand binding to its receptor and two possible consequences of the appearance of these messengers**

The activation of a phosphoinositide-specific phospholipase C by the binding of a ligand to its receptor is mediated by a G protein. Redrawn with permission from Alberts *et al.* (1989).

## Summary

Angiosperms develop very differently in the light and in darkness. Seedlings germinated in complete darkness are said to be etiolated. Skotomorphic development is characterized by a high rate of internode elongation, little leaf blade enlargement, failure of chloroplasts to develop, and a hooked hypocotyl. Small quantities of light trigger photomorphic development in which the rate of internode elongation is reduced, leaf blades expand, the photosynthetic apparatus is formed within the chloroplasts, and the hypocotyl straightens. Light overcomes an inhibitory factor that blocks photomorphic development. Chloroplast development requires the synthesis of numerous protein components of the photosynthetic apparatus, many of which are encoded in the nuclear genome. In many cases, light induces the specific expression of the genes encoding these photosynthetic proteins. RUBISCO, the first enzyme of the Calvin cycle, is a soluble protein of the chloroplast stroma. The functional RUBISCO protein is composed of two different subunit polypeptides, the large subunit (LSU) and small subunits (SSUs). The SSUs are encoded by a small multigene family of nuclear genes, whereas the LSU is encoded in the chloroplast genome. Transcription of the *rbcS* genes encoding the SSUs is regulated not only by light but also by cell-, tissue-, and organ-specific factors. Several regulatory elements that bind transcription factors present in the nuclei of light-grown leaves have been identified in the promoters of *rbcS* and other light-regulated genes. Transcription of the *rbcL* genes encoding the LSU, in contrast, is constitutive and not affected by light, although the mRNA for this protein is stable only in the light-grown leaves, so the synthesis of the LSU protein requires light.

To induce a photomorphogenic response, light must first be adsorbed by a pigment that is coupled to a signal transduction pathway. Two of the best-characterized photomorphogenic systems are the blue light receptor system and phytochrome. The blue light receptor itself has not been identified with certainty, but could be either carotene or riboflavin. Both have absorption maxima near 450 nm, the wavelength most effective in inducing such blue light phenomena as phototropism. Phytochrome, the best-characterized photomorphogenic pigment, is a protein with a tetrapyrrole chromophore. It exhibits two photointerconvertible forms, a red-light-absorbing form ($P_r$) with an absorption maximum near 660 nm, and a far-red-absorbing form ($P_{fr}$) with an absorption maximum near 730 nm. $P_r$ is abundant in etiolated plants. It is converted to $P_{fr}$ when exposed to red light which then can trigger numerous photomorphogenic responses. A number of

Chapter6

mutants have been isolated that give either the etiolated phenotype in light-grown seedlings or the light-phenotype in dark-grown seedlings. *Arabidopsis hy1, hy2, hy3, hy4, hy5,* and *hy6* are independent mutations that result in long hypocotyls in light-grown seedlings that are homozygous for these mutations. The *hy3* mutants have markedly reduced levels of one form of phytochrome; *hy4* and *hy5* apparently affect some component of the signal transduction pathway as plants carrying one of these mutations in the homozygous state have normal levels of functional phytochrome. The mutations *deetiolated* (*det*) and *constitutive photomorphogenic* (*cop*) have expanded leaves and the expression of many light-regulated genes when seed carrying these mutations are germinated in darkness.

Hormones are chemical signals that facilitate intercellular communication. Angiosperms have several different classes of hormonal signaling molecules. In addition to the five generally recognized classes of plant hormones, auxins, cytokinins, gibberellins, abscisic acid, and ethylene, it seems likely that oligosaccharins and systemin also are plant hormones. Also, several molecules or groups of molecules act as signals between microorganisms and plants. These include the lipo-oligosaccharides produced by symbiotic nitrogen-fixing bacteria that induce root nodule formation in legumes, as well as fungal elicitors, a heterogeneous class of chemical compounds that trigger the plant's defense against disease organisms. The native auxin is indole-3-acetic acid. Auxin is the signaling molecule responsible for phototropism and it is an important factor regulating vascular tissue differentiation, leaf blade expansion, maintenance of tissue polarity, and the inhibition of lateral bud development. Gibberellins regulate stem elongation and some mutations in pea and maize resulting in dwarf stature have been shown to occur in genes encoding enzymes involved in the biosynthesis of gibberellins. Gibberellins are involved in the regulation of numerous additional aspects of plant development. Cytokinins are adenine derivatives that are capable of stimulating cell proliferation in cultured tissues in the presence of an auxin. Cytokinins also delay leaf senescence, promote chloroplast development and are involved in the initiation of buds. Abscisic acid is a signal mediating many stress responses. It regulates the onset of seed and bud dormancy and stomatal closing. Ethylene initially was shown to be the signal triggering fruit ripening. Additionally, ethylene synthesis is induced by stress and by high levels of auxin.

Hormones act by binding to a receptor either on the cell surface or localized in the cytosol of those cells that are targets of the hormone. The binding of the hormone to its receptor then initiates a regulatory cascade that may include the release of second intracellular messengers, such as calcium ions, inositol phosphate, and jasmonic acid. Calcium ion levels are very low in the cytosol but high in some membrane-bound compartments and in the extracellular environment. Some hormones cause cytosolic calcium levels to increase transiently. As a consequence, some proteins may be phosphorylated by calcium-dependent protein kinases. Some transcription factors must be phosphorylated to bind to regulatory elements in genes, so the calcium-dependent protein kinase activity could lead to changes in gene expression.

# Questions for Study and Review

1. What is the difference between a plant hormone and a growth-regulating substance? Why is phytochrome not considered a plant hormone?

2. How can a hormone initiate changes in cellular activities without entering the cell? What molecules have been shown to act as second messengers?

3. Compare the calcium concentration in an extracellular environment such as the cell wall with that of the cytosol. How is the calcium ion concentration in the cytosol regulated? What effect would a metabolic poison such as dinitrophenol have on the cytosolic calcium concentration? Why?

4. There is a transient increase in the cellular levels of the protein calmodulin when *Arabidopsis* plants are touched (see Fig. 6.34). What is the effect of this increase on plant growth? How could this effect be brought about? Could inositol trisphosphate play a role in this response and, if so, what role might it play?

5. What are G proteins? What role do G proteins play in signal transduction pathways?

6. How might you explain the phototropic and geotropic behavior of plants?

7. The plant tumor known as crown gall results from a genetic change in plant cells brought about by the bacterium *Agrobacterium tumefaciens*. Crown gall tumors have high levels of both auxin and cytokinin, which is the reason for their abnormal proliferation. How could a genetic change in the plant cells result in high levels of plant hormones? We discuss this question in greater detail later in this book, but for now simply describe ways in which this change could theoretically come about.

8. By what mechanism does auxin move through the plant? What are the developmental consequences?

9. Describe five processes that usually are regulated by the plant hormone auxin.

10. Both auxin and gibberellin induce elongation in some systems. Why then are they considered to be different classes of hormones? How do auxin and gibberellin differ in the types of regulatory roles they perform?

11. The chemical compound known as AMO-1618 is an inhibitor of GA biosynthesis. If you applied it to tall pea plants, what effect would it have on their growth and development? What effect would it have on dwarf pea plants?

12. Some bacteria and fungi synthesize and secrete plant hormones. How does the secretion of cytokinin by *Corynebacterium fascians* cause

"witch's broom"? What does this suggest about the role of cytokinins in the regulation of plant development?

13. Compare the synthesis of free cytokinin with the synthesis of the cytokinins in transfer RNA.

14. How do cytokinins move through the plant? How is this different from auxin movement?

15. What is meant by the term *habituation* with reference to the growth of cultured tissues?

16. What two plant hormones are synthe-

sized from an amino acid and which amino acid is involved in each case? What chemical transformations of the amino acid are required to produce the plant hormones?

17. What plant hormones are synthesized, in part, via the mevalonic acid pathway?

18. What changes occur during the ripening of a fruit such as a peach? How does ethylene regulate fruit ripening?

19. What factors can lead to an increase in ethylene production? Could ethylene be considered a "second messenger"?

## Further Reading

### General References

Boss, W. F., Morre, D. J., and Allen, N. S. (Eds.) (1989). *Second Messengers in Plant Growth and Development.* Alan R. Liss, New York.

Cohen, P. (1989). The structure and regulation of protein phosphatases. *Annu. Rev. Biochem.* **58,** 453–508.

Davies, P. J. (Ed.) (1987). *Plant Hormones and Their Role in Plant Growth and Development.* Martinus Nijhoff, Dordrecht.

Galston, A. W., Davies, P. J., and Satter, R. L. (1980). *The Life of the Green Plant.* Prentice-Hall, Englewood Cliffs, NJ.

Hart, J. M. (1988). *Light and Plant Growth.* Unwin Hyman, London.

Kendrick, R. E., and Kronenberg, G. H. M. (Eds.) (1986). *Photomorphogenesis in Plants,* Martinus Nijhoff, Dordrecht.

Roberts, J., Kirk, C., and Venis, M. (Eds.) (1989). *Hormone Perception and Signal Transduction in Animals and Plants.* The Company of Biologists Limited, Cambridge.

Taiz, L., and Zeiger, E. (1991). *Plant Physiology.* Benjamin/Cummings, Redwood City, CA.

Went, F. W., and Thimann, K. V. (1937). *Phytohormones.* Macmillan, New York.

### Specific References

Allende, J. E. (1988). GTP-mediated macromolecular interactions: The common features of different systems. *FASEB J.* **2,** 2356–2367.

Bell, R. M. (1986). Protein kinase C activation by diacylglycerol second messengers. *Cell* **45,** 631–632.

Berridge, M. J., and Irvine, R. F. (1989). Inositol phosphates and cell signalling. *Nature* **341,** 197–205.

Bleecker, A. B., Estelle, M. A., Somerville, C., and Kende, H. (1988). Insensitivity to ethylene conferred by a dominant mutation in *Arabidopsis thaliana. Science* **241,** 1086–1089.

Braam, J., and Davis, R. W. (1990). Rain-, wind-, and touch-induced expression of calmodulin and calmodulin-related genes in *Arabidopsis. Cell* **60,** 357–364.

Buzby, J. S., Yamada, T., and Tobin, E. M. (1990). A light-regulated DNA-binding activity interacts with a conserved region of a *Lemna gibba rbcS* promoter. *Plant Cell* **2,** 805–814.

Estelle, M. A., and Somerville, C. (1987). Auxin-resistant mutants of *Arabidopsis thaliana* with an altered morphology. *Mol. Gen. Genet.* **206,** 200–206.

Casey, P. J., and Gilman, A. G. (1988). G protein involvement in receptor–effector coupling. *J. Biol. Chem.* **263,** 2577–2580.

Chory, J. (1991). Light signals in leaf and chloroplast development: Photoreceptors and downstream responses in search of a transduction pathway. *New Biol.* **3,** 538–548.

Chory, J., Peto, C., Feinbaum, R., Pratt, L., and Ausubel, F. (1989). *Arabidopsis thaliana* mutant that develops as a light-grown plant in the absence of light. *Cell* **58,** 991–999.

Cone, J. W., and Kendrick, R. E. (1986). Photocontrol of seed germination. In *Photomorphogenesis in Plants* (R. E. Kendrick and G. H. M. Kronenberg, Eds.), pp. 443–465. Martinus Nijhoff, Dordrecht.

Drøbak, B. (1993). Plant phosphoinositides and intracellular signaling. *Plant Physiol.* **102,** 705–709.

Farmer, E. E., and Ryan, C. A. (1992). Octadecanoid precursors of jasmonic acid activate the synthesis of wound-inducible proteinase inhibitors. *Plant Cell* **4,** 129–134.

Firn, R. D. (1986). Photoperiodism. In *Photomorphogenesis in Plants* (R. E. Kendrick and G. H. M. Kronenberg, Eds.), pp. 367–389. Martinus Nijhoff, Dordrecht.

Firtel, R. A., Van Haastert, P. J. M., Kimmel, A. R., and Devreotes, P. N. (1989). G protein-linked signal transduction pathways in development: *Dictyostelium* as an experimental system. *Cell* **58,** 235–239.

Gilmartin, P. M., Sarokin, L., Memelink, J., and Chua, N.-H. (1990). Molecular light switches for plant genes. *Plant Cell* **2,** 369–378.

Guiltinan, M. J., Marcotte, W. R., Jr., and Quatrano, R. S. (1990). A plant leucine zipper protein that recognizes an abscisic acid response element. *Science* **250,** 267–271.

Hunter, T. (1987). A thousand and one protein kinases. *Cell* **50,** 823–829.

Jacobs, M., and Gilbert, S. F. (1983). Basal localization of the presumptive auxin transport carrier in pea stem cells. *Science* **220,** 1297–1300.

Kende, H. (1989). Enzymes of ethylene biosynthesis. *Plant Physiol.* **91,** 1–4.

Kikkawa, U., Kishmoto, A., and Nishizuka, Y. (1989). The protein kinase C family: Heterogeneity and its implications. *Annu. Rev. Biochem.* **58,** 31–44.

Knight, M. R., Campbell, A. K., Smith, S. S., and Trewavas, A. J. (1991). Transgenic plant aequorin reports the effect of touch and cold-shock and elicitors on cytoplasmic calcium. *Nature* **352,** 524–526.

Leonard, N., Hecht, S. M., Skoog, F. S., and Schnitz, R. Y. (1968). Cytokinins: Synthesis of 6-(3-methyl-3-butenylamino)-9-8-D-ribofuranosyl purine and the effect of side-chain unsaturation on the biological activity of isopentylaminopurines and their ribosides. *Proc. Natl. Acad. Sci. USA* **59,** 15–21.

Letham, D. S., and Palni, L. M. S. (1983). The biosynthesis and metabolism of cytokinins. *Annu. Rev. Plant Physiol.* **34,** 163–197.

Li, Y., Hagen, G., and Guilfoyle, T. J. (1991). An auxin-responsive promoter is differentially induced by auxin gradient during tropisms. *Plant Cell* **3,** 1167–1175.

Majerus, P. W., Connolly, T. M., Deckmyn, H., Ross, T. S., Bross, T. E., Ishii, H., Bansal, V. S., and Wilson, D. B. (1986). The metabolism of phosphoinositide-derived messenger molecules. *Science* **234,** 1519–1526.

Manzara, T., Carrasco, P., and Gruissem, W. (1991). Developmental and organ-specific changes in promoter DNA–protein interactions in the tomato *rbcS* gene family. *Plant Cell* **3,** 1305–1316.

Palme, K. (1992). Molecular analysis of plant signaling elements: Relevance of eukaryotic signal transduction models. *Int. Rev. Cytol.* **132,** 223–282.

Parks, B. M., and Quail, P. H. (1991). Phytochrome-deficient *hy*1 and *hy*2 long hypocotyl mutants of *Arabidopsis* are defective in phytochrome chromophore biosynthesis. *Plant Cell* **3,** 1177–1186.

Pearce, G., Strydom, D., Johnson, S., and Ryan, C. A. (1991). A polypeptide from tomato leaves induces wound-inducible proteinase inhibitor proteins. *Science* **253,** 895–898.

Phinney, B. O., and Spray, C. (1982). Chemical genetics and the gibberellin pathway in *Zea mays* L. In *Plant Growth Substances* (P. F. Wareing, Ed.), Academic Press, New York.

Rasmussen, H. (1989). The cycling of calcium as an intracellular messenger. *Sci. Am.* **261,** 66–73.

Rood, S. B., Pearce, D., Williams, P. H., and Pharis, R. P. (1989). A gibberellin-deficient *Brassica* mutant-rosette. *Plant Physiol.* **89,** 482–487.

Ryan, C. A. (1987). Oligosaccharide signalling in plants. *Annu. Rev. Cell Biol.* **3,** 295–317.

Schulman, H., and Lou, L. L. (1989). Multifunctional $Ca^{2+}$/calmodulin-dependent protein kinase: Domain structure and regulation. *Trends Biochem. Sci.* **14,** 62–66.

Theologis, A. (1992). One rotten apple spoils the whole bushel: The role of ethylene in fruit ripening. *Cell* **70,** 181–184.

Vierstra, R. D., and Quail, P. H. (1986). Phytochrome: The Protein. In *Photomorphogenesis in Plants* (R. E. Kendrick and G. H. M. Kronenberg, Eds.), pp. 35–60. Martinus Nijhoff, Dordrecht.

Wagner, D., Tepperman, J. M., and Quail, P. H. (1991). Overexpression of phytochrome B induces a short hypocotyl phenotype in transgenic *Arabidopsis. Plant Cell* **3,** 1275–1288.

Wanner, L. A., and Gruissem, W. (1991). Expression dynamics of the tomato *rbcS* gene family during development. *Plant Cell* **3,** 1289–1303.

Yang, S. F., and Hoffman, N. E. (1984). Ethylene biosynthesis and its regulation in higher plants. *Annu. Rev. Plant Physiol.* **35,** 155–189.

# 7

# Cell Division, Polarity, and Growth in Plant Development

Plant biologists define growth as an irreversible increase in volume. This means that most plant growth is brought about by cell enlargement. In some cases cell enlargement is accompanied by cell division and in other cases it is not. Cell division is not in itself a growth mechanism. There are developing plant tissues and organs in which cell division simply decreases the average cell size without increasing the overall volume of the structure. The most notable example of this is in early embryogenesis. Nevertheless, cell division plays a vital, although indirect, role in plant growth, as well as a key role in plant development. Cell division increases the growth potential of a tissue or organ. This is because there is a limit on the final size that can be attained by a cell through cell enlargement. The limit may be imposed by the ability of a single nucleus to provide sufficient mRNA to maintain the protein required for a large cell to function. As a result, cell division increases the number of cells that can grow. Tissues with more cells have a greater potential for growth than those with fewer cells. Cell division plays several additional roles in plant development. Some critical decisions that lead to cell differentiation are made during cell division. Also, the plane in which a cell divides is critical in determining what functional role its progeny will play in the plant body. For these reasons it is important to understand how cell division is regulated in plant development.

Plant cell enlargement takes place by a turgor pressure-driven expansion of the cell wall, accompanied by water uptake and a large increase in the size of the vacuole. Cell expansion in growing plant tissues, however, often is highly polar. The growing cells do not simply enlarge; they elongate. This means that at some point before the enlargement began, the growing cells became polarized. This chapter is concerned with the cell cycle and its regulation, the role of cell division in plant development, the mechanism of cell polarization, and the mechanism and control of cell enlargement.

# The Cell Division Cycle and Its Control

### The Two Control Points in the Cell Cycle

Cell proliferation in eukaryotes is a complicated process that involves a number of distinct biochemical and cytological events, all of which are integrated in some manner. The major events of the cell cycle are the replication of the chromosomes, which requires DNA synthesis, the separation of daughter chromosomes into different nuclei by the process of mitosis, and the partitioning of the daughter nuclei and the cytoplasm into two cells by the process of cytokinesis. When cells divide repeatedly, there is an interval of time, known as interphase, between each mitotic event. DNA synthesis occurs in interphase, but typically it does not occupy the entire interphase period. There is a pre-DNA synthetic interphase, which is called **G1** (the "G" stands for "gap"), and a post-DNA synthetic interphase, which is designated **G2.** The DNA synthetic period is known as **S phase** (the "S" is for DNA synthesis). The different parts of interphase usually are linked so that one follows the other in an apparently deterministic fashion, with mitosis following G2. This sequence is called the **cell cycle** (Fig. 7.1). In a dividing cell population, such as an actively growing meristem, the *average* cell size remains constant. This means that, after it is formed, each cell will approximately double its cytoplasmic mass, including all of its organelles, before it divides again. Most of the growth in cytoplasmic mass occurs in the G1 phase of the cell cycle, whereas the onset of the S phase sees the replication of not only the DNA but other components of the nucleus as well. Although there is great variation in the times required for different kinds of cells to complete their passage through their cycle, it is not uncommon for the duration of the cell cycle to be approximately 24 hours. Mitosis is relatively short, lasting 1 to 2 hours, whereas the bulk of the cell cycle is concerned with the events of G1, S, and G2.

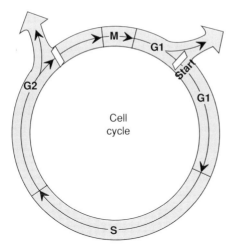

**Figure 7.1   Phases of the cell cycle**
The cell cycle is represented as the face of a clock, with the relative duration of each phase indicated by the space it occupies on the clock face. Cells come into existence as a result of mitosis (M) and progress into interphase, which is composed of three distinct phases: G1, which precedes DNA replication; the S phase, in which DNA is replicated; and a post-DNA synthetic G2 phase. The entire complement of nuclear DNA is replicated once during S phase. G2 phase ceases when the cells enter mitosis, during which the chromosomes condense, the nuclear envelope breaks down, and a mitotic spindle forms that separates the two chromatids of each chromosome and transports them to opposite ends of the cell. Mitosis ends with the re-formation of the nuclear envelope around each daughter nucleus, while cytokinesis physically divides the cell.

Proteins known as cyclins trigger the start of the cell cycle and initiate mitosis   A group of proteins known as cyclins are synthesized during the cell cycle. They play important roles in triggering the start of the cell cycle as well as mitosis. The levels of cyclin in cells change dramatically as cells progress through their division cycle. Cyclin levels are near zero in the cytoplasm in early G1. Cyclins accumulate as the result of the synthesis of the protein at a constant rate, with cyclin protein reaching a peak as mitosis is initiated. Then, at some point in mitosis, the cyclin level of the cell rapidly drops back to zero and each daughter cell begins the next cell cycle with essentially no cyclin. Cyclin genes have been cloned from a variety of organisms and the cyclin proteins they encode are highly conserved. When mRNA from a cloned cyclin gene is injected into *Xenopus* oocytes, which are arrested in G2, it induces them to enter mitosis. Apparently the entry of G2 cells into mitosis requires some critical concentration of the cyclin protein. The mechanism responsible for cyclin destruction during mitosis has not been identified.

The G1 regulatory point known as Start controls entry into the S phase of the cycle   **Start** is the name appropriately given for a key control point for the cell cycle. Start is the point in G1 at which the cell becomes committed to enter the cell cycle and replicate its nuclear DNA. Prior to Start, the cell was uncommitted. It could either leave the cell cycle and differentiate or enter another cell cycle. After Start, it is committed to the cell cycle. In vertebrate cells, all the events of the chromosome cycle (DNA replication and mitosis) follow in a deterministic fashion after Start is activated. In plants, mitosis need not follow from the replication of the chromosomes, although it usually does so in meristematic cells. In both cases, once S is initiated, DNA synthesis continues until the entire genome is replicated. Different parts of the nuclear DNA are replicated at different times during S, but all of the nuclear genome is replicated and it is only replicated once. There is some mechanism, yet to be identified, that prevents the DNA from undergoing a second round of replication during any given S phase. In addition to the nuclear DNA, chromosomal proteins, such as the histones, also are synthesized during S and assembled onto the newly replicated DNA so that the entire chromosome is reproduced.

After DNA replication is completed, the cell enters the G2 phase, where it may remain until a second signal is received to trigger its entry into mitosis. The cell may initiate another complete cycle (G1 through mitosis) or it may leave the cell cycle and differentiate after it has completed mitosis. The choice to leave the cell cycle is made at either of the critical control points, Start in G1 or the G2/M transition. Cells that leave the cell cycle in G2 are, of course, tetraploid as they have replicated their DNA. Differentiated plant cells may be either diploid or tetraploid, depending on where they left the cell cycle. Polyploid cells and tissues are common in mature plant organs. Leaf mesophyll cells, for example, frequently are tetraploid. Even higher levels of ploidy are found in some specialized plant cells such as differentiating vessel elements and the suspensor cells that attach the embryo to the placenta. Polyploid cells occur in an otherwise diploid plant because a cell may go through successive cycles (G1–S–G2) without entering mitosis.

**Yeasts have provided important insights into how a cell's progression through its cell cycle is controlled**   Two different species of yeasts have been used to investigate the molecular–genetic mechanisms of cell cycle progression: *Saccharomyces cerevisiae* is the common yeast used to make bread and beer and is known as a budding yeast because of the way the cells divide, and *Schizosaccharomyces pombe* is a fission yeast in which the cells divide by splitting in two. Yeasts are fungi, belonging to a group known as the Ascomycetes. They are only distantly related to plants, but they are eukaryotic organisms with small genomes (one-fifth the size of *Arabidopsis*) and have been very useful for the molecular analysis of the cell cycle. These two yeasts are as distantly related to each other as either of them is to angiosperms. As a result, the discovery that the cell cycle is controlled in similar ways in these yeasts suggests that there may be a common mechanism regulating the cell cycle in all eukaryotic organisms. Indeed, key components of the regulatory mechanism have been found in dividing cells of higher plants as well as mammals.

The cell cycles of these two yeasts differ in a number of ways. *S. cerevisiae* lacks a well-defined G2 period and the cells pass from S phase directly into mitosis. As they enter the S phase of the cell cycle a small bud forms as an outpocketing of the cell wall. Toward the end of S, the nucleus migrates into the neck of the bud and initiates mitosis (Fig. 7.2). All fungi including yeasts have a spindle pole body which directs the formation of the intranuclear spindle apparatus. Mitosis occurs without the disintegration of the nuclear envelope. The cell cycle of *S. pombe* is more similar to that of most other eukaryotes in that it has distinct G1, S, G2, and M phases. *S. pombe* is a cylindrical cell with a central nucleus. The cell elongates continuously through the cell cycle. Mitosis, which occurs at the end of G2, is followed by the formation of a cell plate between the nuclei which divides the parental cell into two daughters. Subsequently, the daughters split apart (Fig. 7.2). Both yeasts can grow as either haploids or diploids. The haploids can fuse (conjugate) with each other if they are of the opposite mating type to form a diploid cell. When nutrients are provided, the diploid cell will divide mitotically to form a population of cells. If nutrients are scarce the diploid zygote will undergo meiosis immediately after fusion and form four haploid cells which then differentiate into resistant spores. The spores will germinate again when nutrients are available and divide repeatedly by mitosis to form a population of haploids cells (Fig. 7.3). Both yeasts use sugars for their metabolism, and nutrients are the principal external factors determining whether or not these yeasts divide. When nutrients become scarce, they leave the cell cycle if they cannot mate. Both yeasts leave the cell cycle at a specific point: *S. pombe* leaves in G2, and *S. cerevisiae* leaves the cell cycle in G1. If the cells are at some other point in their cell cycle when nutrients become scarce, they will complete the cycle up to the exit point in G1 or G2 and then stop.

Conditional mutations have identified specific genes whose products are required for progression through the cell cycle   It has been possible to identify genes that are essential for specific steps in the cell cycle by isolating mutants that prevent cells from completing their cell cycle. As a mutation in a gene essential for cell division would never be recovered if it was completely nonfunctional, investigators select for conditional mutations to identify cell cycle genes. Usually these conditional mutations are temperature sensitive; the gene product can function when the cells are grown at one temperature (the permissive temperature) but not at another (the restrictive temperature). It is important to realize that the unique characteristic of these cell division cycle (*cdc*) mutants is that they block cell division at a specific place in the cell cycle (Fig. 7.4). The product of the gene is needed at that point if the cell cycle is to continue. At the present time approximately 50 *cdc* genes have been identified by isolating conditional mutations in these two yeast species. In some cases, mutations in *cdc* genes cause cells to arrest in G1 at the restrictive temperature; in other cases arrest occurs in S, G2, or M. Analysis of the effects of these different mutations demonstrates that the cell cycle is actually composed of three loosely connected cycles: a cytoplasmic cycle that leads to bud emergence and nuclear migration,

A

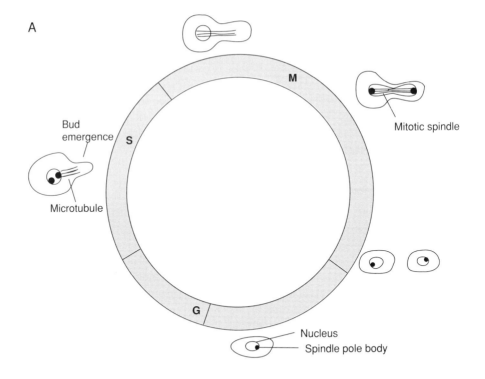

B

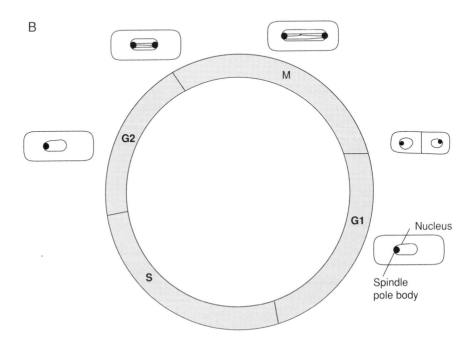

a chromosome cycle that is concerned with the replication of the DNA and nuclear division, and a centrosome cycle that leads to spindle pole body duplication and migration (Fig. 7.5). A mutation in a *cdc* gene that must be expressed in the chromosome cycle will not necessarily block either the cytoplasmic or the centrosome cycle, although it will block the chromosome cycle.

The *cdc*2 gene product is necessary for Start and for entry into mitosis   We know the function of the proteins encoded by a number of the *cdc* genes. Of these, the *cdc*2 gene of *S. pombe* and the homologous *cdc*28 gene of *S. cerevisiae* have proven to be very interesting and important. Both the *cdc*28 and *cdc*2 genes encode a 34,000-Da protein (p34$^{cdc2}$) that is an essential component of both Start and the mechanism controlling entry into mitosis. Genes encoding similar proteins are present in many different organisms and, perhaps, all eukaryotes including higher plants. In fact, the p34$^{cdc2}$ gene from *Arabidopsis* will allow yeast cells with mutations in their *cdc*28 genes to divide when the mutant yeast cells are transformed with the *Arabidopsis* gene. The p34$^{cdc2}$ protein is a kinase that phosphorylates serine and threonine residues in proteins. The kinase activity of p34$^{cdc2}$ is controlled by its interaction with cyclins. That is, cyclins activate the p34$^{cdc2}$ protein kinase. The yeast p34$^{cdc2}$ protein associates with different cyclins in different phases of the cell cycle. As many as 12 different cyclins have been identified at this point, but we do not know the function of most of these. The trigger leading to the initiation of mitosis is the activation of p34$^{cdc2}$ by cyclin B.

Among the proteins that are phosphorylated at the initiation of mitosis are the lamins. Lamins are fibrous proteins, closely related to the intermediate filament proteins of the cytoskeleton, that form a network attached to the inner surface of the nuclear envelope. The lamin fibrous network is responsible for the integrity of the nuclear envelope. Phosphorylation causes the lamin fibers to disaggregate. The nuclear envelope fragments as the lamin network depolymerizes. The nuclear envelope re-forms in late telophase when the lamin proteins are dephosphorylated by a phosphatase, and these proteins reassemble.

**Figure 7.2   The cell cycles of two yeasts, *Saccharomyces cerevisiae*, a budding yeast, and *Schizosaccharomyces pombe*, a fission yeast**

(A) The budding yeast (*S. cerevisiae*) cell cycle is unusual in that it lacks a G2 phase and the cells progress directly from S into mitosis, which is prolonged. An intranuclear mitotic spindle and the bud begin to form during S. Toward the end of S the nucleus migrates near the bud and the spindle elongates without chromosome condensation or the breakdown of the nuclear envelope. (B) The cell cycle of the fission yeast (*S. pombe*) is more similar to that found in most other eukaryotes in that there is a distinct G2 phase, as well as S and G1. Also, the chromosomes condense during mitosis, as is the case in nearly all other eukaryotes; however, the nuclear envelope does not break down at any stage of mitosis and an intranuclear mitotic spindle is organized by the spindle pole body, as in the budding yeast and other fungi.

A

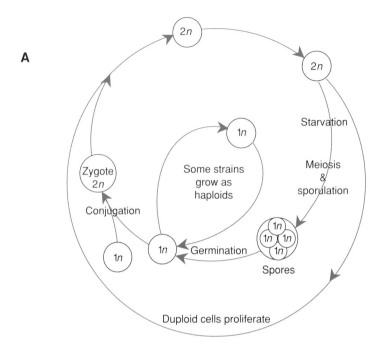

B

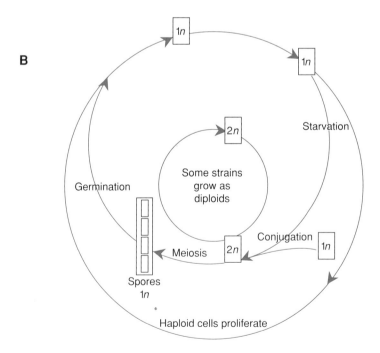

The interphase array of microtubules also is depolymerized with the onset of mitosis. In many plant cells, the cortical array of microtubules becomes tightly focused into the preprophase band, and the position of the latter array determines the plane of cytokinesis. The preprophase band begins to appear after completion of the S phase of the cell cycle, reaching a peak as a dense band of crosslinked microtubules in preprophase. The preprophase band disappears as the mitotic spindle is formed. p34$^{cdc2}$ has been shown to be associated with the preprophase band, although its function in this array has not been determined. In other organisms, phosphorylation of proteins by p34$^{cdc2}$ kinase at the onset of mitosis depolymerizes the interphase microtubule array.

Although the precise mechanism by which the events of the cell cycle are initiated is still being worked out, it is clear that there is a regulatory network in which phosphorylation and dephosphorylation turn on and off the activity of key proteins to initiate the major steps of the cycle. It is possible that other factors, such as plant hormones, may intervene in this regulatory network to initiate or reinitiate progress through the cell cycle. In developing wheat leaves and carrot cotyledons, the level of p34$^{cdc2}$ is high during active periods of cell division and falls dramatically as cells leave the cell cycle and differentiate. Furthermore, when cell division was reinitiated in mature carrot cotyledons by auxin treatment, the level of p34$^{cdc2}$ protein increased.

Cell division is regulated during plant development    Cell division normally occurs in only a limited number of plant tissues and structures. It is confined to meristematic tissues, such as the shoot and root apical meristems and in the cambium, and to the early phase of organ development. Different regions of the meristem move through the cell cycle at different rates, but mature plant cells

---

Figure 7.3    **The life cycles of two yeasts, *Saccharomyces cerevisiae* and *Schizosaccharomyces pombe***

The two species differ in that the budding yeast (*S. cerevisiae*) usually grows as a diploid unicellular organism, whereas the fission yeast *S. pombe* usually grows vegetatively as haploid cells. In both yeasts, the vegetative cells actively divide when they have adequate nutrients, with the cells completing their cell cycle in approximately 2 hours. When nutrients become limiting for growth they enter a sexual cycle. In *S. pombe,* haploid cells of opposite mating type conjugate to form a diploid zygote that will immediately undergo meiosis, forming four haploid spores. When nutrients are again available, the spores germinate, again initiating the vegetative growth. In the budding yeast, starvation induces the vegetative cells to undergo meiosis, leading to the formation of four haploid spores. The spores germinate when fed, and cells of opposite mating type undergo conjugation to form vegetative diploid cells, which then proliferate as long as the nutrients last. In either case, laboratory strains have been isolated that will grow vegetatively as haploid or diploid cells, regardless of the wild-type behavior of the species.

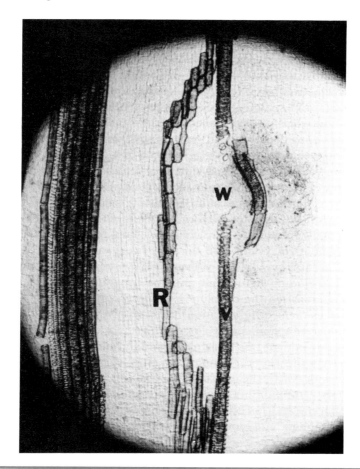

**Figure 7.6   Formation of tracheary elements in response to wounding**
A wound was made by inserting a thin scalpel into the fifth internode of a *Coleus* stem to sever two of the smaller vascular bundles, which appear as dark vertical stripes in this light micrograph. The differentiation of vascular elements, including both the tracheary elements of the xylem and the sieve elements of the phloem, restored the vascular connection that was broken when the vascular bundles were ruptured by wounding. R, regenerated xylem and phloem cells; W, wound; V, vascular bundle. Reprinted with permission from Roberts, L. W., and Fosket, D. E. (1962). *Bot. Gaz.* **123,** 247–254. ©The University of Chicago.

but there is experimental evidence against a necessary relationship between the division and subsequent differentiation. Lateral roots are formed, not in the root meristem, but in more mature regions of the root, where they are initiated by cell divisions in a highly differentiated tissue known as the pericycle (Fig. 7.8). The derivatives of the dividing pericyclic cells become an organized meristem

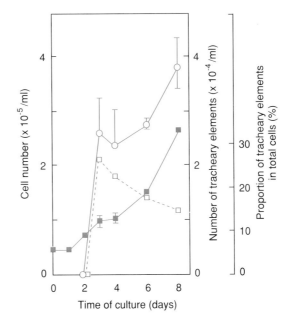

**Figure 7.7** **Time course of cell division and tracheary element differentiation in cultured *Zinnia* leaf mesophyll cells**
The mesophyll cells of young *Zinnia* leaves are isolated mechanically and placed in culture as a cell suspension in a medium containing an auxin and a cytokinin. Over the subsequent 14 days a very large percentage of the mesophyll cells differentiate as tracheary elements. Although some of the mesophyll cells divide, most of them differentiate directly without dividing. Redrawn with permission from Fukuda and Komamine (1980).

which produces the branch root. Foard and his colleagues showed that the pericycle cells would undergo enlargement and their growth axis would change to form a "pseudoprimordium," even when they blocked cell division with the drug colchicine. When the colchicine was removed, the pseudoprimordia developed as lateral roots. Although this result argues that cell division per se is not necessary for the change in pericycle cell commitment, it is possible that these decisions are made in some other part of a "quantal" cell cycle, which of course would not necessarily be blocked by the colchicine. In summary, there is no compelling experimental evidence that major changes in gene expression, cell commitment, or cytodifferentiation must be initiated during special formative or quantal divisions in higher plants. Cell division nevertheless usually precedes a change in the differentiation state of cells in growing, developing plant tissues and organs. This relationship is particularly apparent where unequal cell divisions precede differentiation.

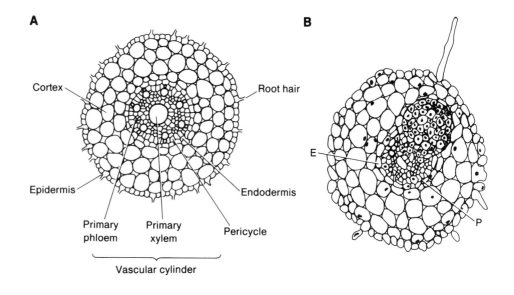

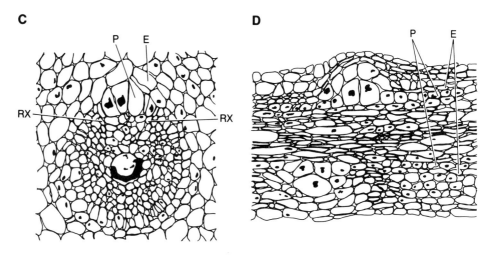

**Figure 7.8** **Initiation of lateral roots from the pericycle**
(A) The pericycle (P) is a one-cell-thick tissue layer immediately inward from the endodermis (E). Lateral roots are initiated with periclinal cell divisions in pericycle cells that lie over the xylem (RX). The lateral root meristem becomes organized as additional cell proliferation occurs from the derivatives of the pericycle cells.
(B) Subsequent growth of the primordium compresses the cortical cells of the main root, ultimately rupturing the epidermis as it emerges. (C) If the division of pericycle cells is blocked with colchicine, tumor-like outgrowths occur in the region that would give rise to lateral roots resulting from the expansion of the pericycle cells. The repolarization of the pericycle cells occurs without cell division, with the establishment of a pseudolateral root promeristem. (D) A longitudinal section through the tumorlike outgrowths in the colchicine-treated roots demonstrates their similarity to lateral root promeristems. Redrawn with permission from Foard *et al.* (1965).

## Unequal Cell Divisions

Formation of the stomatal guard cells occurs late in leaf development and is initiated by cell divisions that occur in epidermal cells after almost all other mitotic activity in the leaf has ceased. By the time these late cell divisions occur, most of the epidermal cells have left the cell cycle and enlarged. The division preceding the formation of the stomatal complex is asymmetrical. These formative divisions are said to be unequal because the mitotic spindle is not centered in the cell and because the daughter cells are very different in size (Fig. 7.9). More importantly, these cell divisions are unequal because the two daughter cells have very different fates: the smaller, less vacuolate daughter goes on to form the guard cells; the other daughter retains the characteristics of epidermal parenchyma. Guard cells are very different from epidermal parenchyma. In addition to forming a unique differentially thickened cell wall that permits stomatal opening and closing, guard cells have chloroplasts, which are absent in epidermal parenchyma. After the initial, unequal division, the smaller daughter undergoes at least one or more divisions to form the guard cells (see Fig. 7.9).

Root hairs are vital for the absorption of water and nutrients from the soil. They are formed from the epidermal cells in a limited region of the root near the tip (Fig. 7.10). The hair is an outgrowth of a single epidermal cell and only a fraction of the epidermal cells form root hairs. The cells that will produce hairs are formed as a result of late cell divisions within the epidermis, after most of the cells have left the cell cycle. These late divisions also are often unequal so that one of the daughter cells receives most of the cell volume, including the vacuole. It is the smaller derivative that subsequently differentiates as a root hair cell. Even in cases in which the division giving rise to the root hair initial is not asymmetrical as far as the cytoplasmic partitioning is concerned, the divisions are unequal also in the sense that the fates of the two daughter cells are different; only one daughter goes on to form a root hair.

## Perpetuation and Loss of Determined States

The process of determination is a change in the developmental potential of a tissue, an organ, or even a whole plant. For example, environmental signals such as the length of the day sometimes initiate flowering. This process, which will be discussed in greater detail in Chapter 9, may convert a vegetative meristem into a floral meristem that will form floral organs instead of leaves. When very small floral meristems are placed in culture, they grow and develop to produce anatomically normal flowers. This demonstrates that a determined state (flowering) can be perpetuated through proliferative divisions. Determination remains a somewhat nebulous concept. Although it can be defined operationally, the molecular–genetic mechanism responsible for determination has not been identified. Furthermore, the fact that differentiated cell characteristics can be maintained through cell division and distributed to the daughter cells suggests that some mechanism can stabilize the determined or differentiated state.

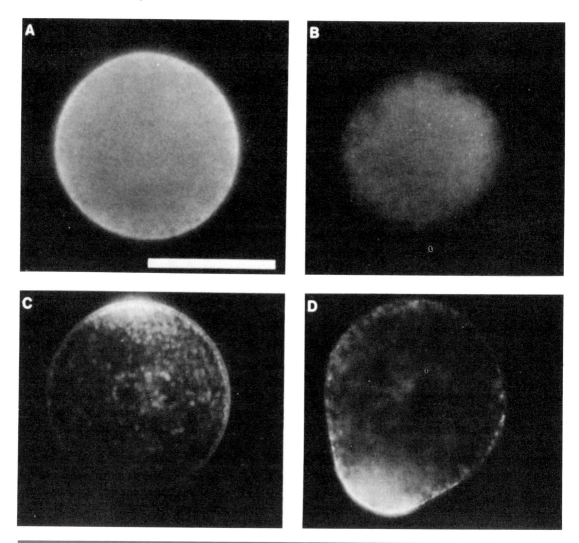

**Figure 7.14** *Pelvetia* **zygotes before and after the establishment of a polar axis**

(A) The zygote 3 hours after fertilization is completely apolar. (B) By 12 hours after fertilization, the rhizoid begins to extend from the side away from light. (C) The first division of the zygote has begun by 16 hours after fertilization, while the rhizoid has extended further. (D) Somewhat later, cytokinesis divides the zygote into a larger thallus cell (upper) and a rhizoid cell (lower). The arrows indicate the division plane. Reprinted with permission from Kropf (1992).

Two events appear to be correlated with the establishment of the growth axis: the establishment of an ionic current through the embryo and the secretion of an amorphous polysaccharide jelly outside the cell wall in the growing rhizoid tip. The ionic current consists of a positive charge that flows into the zygote at the rhizoid pole and out through the thallus pole. Calcium ions appear to be responsible for this inward flowing current and it may establish a calcium gradient through the zygote (Fig. 7.15A). As calcium ions are an important second messenger, it is likely that this calcium gradient polarizes the cellular architecture. Calcium can stimulate protein kinases, whereas protein phosphorylation can determine when, where, and in what orientation cytoskeletal elements will form. The cytoskeleton, in turn, may redistribute cellular organelles, thus polarizing the zygote (Fig. 7.15B).

The cytoskeleton plays an important role in the establishment and maintenance of cell polarity in angiosperms. This has been most clearly demonstrated where cell polarity changes in response to wounding and when a new growth axis is established. In some succulent plants, such as *Graptopetalum paraguayense*, a new shoot apical meristem is formed at the base of mature leaves after they are detached from the plant. Epidermal cells, as well as the underlying cells, become activated and organize a shoot apical meristem which will initiate growth at right angles to the original growth axis of the leaf. The epidermal cells participating in meristem formation change their polarity as the meristem is initiated so they are oriented at right angles to the longitudinal axis of the leaf. As this reorientation occurs, the cortical microtubules become realigned to conform with the new growth axis, even before the meristem initiates active growth. The most recently formed cellulose microfibrils in the cell wall also are reoriented so they are parallel with the cortical microtubules and at right angles to the new growth axis. This altered pattern of cellulose microfibril reinforcement is believed to be ultimately responsible for the change in the growth axis.

Cell polarity also can be altered by wounding. When highly polarized, mitotically quiescent parenchyma cells, such as stem or root cortical tissue, are wounded, some of the cells in the vicinity of the wound divide, enlarge, and differentiate. Many of the cells stimulated by wounding develop a new axis of polarity at right angles to their original polar axis. The first indication of this altered polar axis is a reorientation of the cortical microtubules and microfilaments which begins shortly after wounding, before the cells begin to divide.

As we have seen in Chapter 5, the orientation of the cellulose microfibrils in the most recently deposited layers of the cell wall is the principal determinant of the growth axis of enlarging plant cells. If the cells are to become polarized and elongate, there must be a reinforcement of the lateral cell walls at right angles to the axis of cell elongation, and the cellulose microfibrils serve this function. Cortical microtubules play a major role in determining the orientation in which cellulose microfibrils are deposited; however, it is likely that the cortical microtubule orientation is a manifestation of a more fundamental axial polarization of the cell. This is not to say that microtubule guidance is not important for

**A**

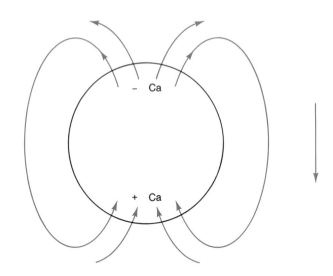

**B**

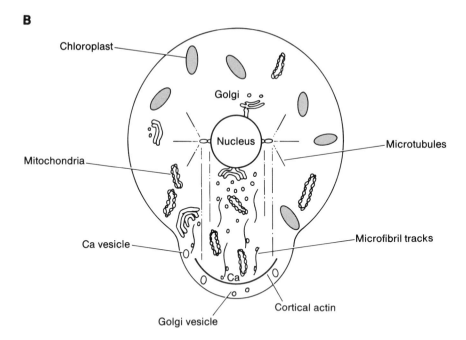

cellulose microfibril orientation, but only that we still cannot explain how the microtubules come to be oriented.

Cells that have developed axial polarity also may have an asymmetrical distribution of cytoplasmic organelles and/or other cellular components. The cellular distribution of organelles is not random and one of the functions of the cytoskeleton is to localize or distribute cytoplasmic structures. This has been demonstrated most clearly in the gravity-perceiving cells of the root cap known as **statocytes.** Statocytes are highly polarized cells in which the endoplasmic reticulum (ER) is asymmetrically distributed, with a complex of ER cisternae flattened against the plasma membrane farthest from the apical meristem. Amyloplasts sediment in a gravitational field and are thought to generate a signal by pressing against this distal ER complex. This signal generated may include hormones that change the pattern of growth within the root so that it maintains a particular orientation within a gravitational field. A network of cortical microtubules lies between these ER cisternae and the plasma membrane. These cortical microtubules are cross-bridged to the plasma membrane and to the distal ER complex. Drugs that destroy the microtubules also prevent the asymmetrical distribution of the ER complex. Thus, these cortical microtubules appear to be responsible for the asymmetrical distribution of the ER cisternae in these cells.

The position of the nucleus also is determined by the cytoskeleton, particularly as cells are preparing for division. This is most apparent in cells with a large central vacuole where the nucleus migrates from the cell periphery into the center of the cell prior to their division. A system of transvacuolar strands will develop in these cells during the G2 phase of their cell cycle. These transvacuolar cytoplasmic strands contain a system of microtubules and microfilaments that is responsible for the movement of the nucleus from the cell periphery into the center of the cell. The nucleus is held in this position by these cytoskeletal elements through the remainder of the cell cycle, until the completion of mitosis.

Nuclear and organelle positioning is particularly important during unequal cell divisions. Unequal cell divisions also may separate cytoplasmic determinants

**Figure 7.15 Establishment of a transcellular ionic current within the zygote and the subsequent asymmetrical distribution of cytoplasmic organelles**

(A) A positive inflowing ionic current has been detected from the presumptive rhizoid site in *Pelvetia* zygotes after they are subjected to unilateral light. (B) After establishment of the polar axis, cytoplasmic organelles become distributed in a highly structured way. Chloroplasts tend to be confined to the half of the cell that will become the thallus, whereas strands of microtubules and microfilaments extend from the rhizoid toward the nucleus. Microfilaments transport Golgi vesicles to the rhizoid. Calcium vesicles and Golgi vesicles accumulate preferentially in the region of the rhizoid. Reprinted with permission from Kropf (1992).

of cell differentiation, such as mRNAs. Before unequal cell divisions can achieve the differential partitioning of cellular components, there must be a sorting and positioning of cell components. Most likely the sorting and separating of cell components are brought about by the cytoskeleton. Generative cells of most angiosperms lack plastids so plastid inheritance is strictly maternal. The first division of the male microspore is unequal. The nucleus migrates to one end of the cell, and the plastids are excluded from the daughter cell that becomes the generative cell. Both nuclear migration and plastid distribution within the dividing microspore are dependent on a system of microtubules that radiate from the nucleus.

## Cell Enlargement and Its Hormonal Regulation

Differential cell enlargement plays a key role in plant development and, from what you have already learned, you know that cell enlargement usually is hormonally regulated. In the following we consider the process by which plant cells enlarge by elongation, and we examine the mechanism by which the hormone auxin regulates this process. Plant growth usually is defined as an irreversible increase in volume. Plant cell growth occurs by cell enlargement and may or may not be accompanied by cell division.

Cell division, when considered in the context of the cell cycle, leads to a doubling of the cell number and usually also to a doubling of all components of the cell, such as the amount of nuclear DNA, the number of cellular organelles, and the total protein and RNA content. The volume of the cell, however, might not increase at all as a given plant cell goes through its cell division cycle. The two daughter cells may not occupy any more volume than the parent cell did at the start of the cycle. As a result, by the definition of plant growth as an irreversible increase in volume, cell division may not lead to growth. In contrast, plant cell elongation leads to a large increase in cell volume, with no necessary net increase in the number of cell organelles, nuclear DNA, or cytoplasmic protein. During elongation, however, the cell wall does increase in area and usually also in the total mass of cell wall material. Cell elongation is a special case of cell enlargement in which growth occurs more strongly in one direction, usually along the axis of the plant, than another. The following discussion is concerned with cell elongation and its regulation.

At one time it was believed that plant cell elongation was absolutely dependent on auxin; that is, without auxin no elongation would occur. Auxin was seen as the regulatory compound that would determine where in the plant and when elongation would take place. We now know that things are not quite this simple. Other hormones, particularly gibberellins, profoundly influence elongation in some plant tissues and under some conditions. In fact, gibberellins may be the hormone that plays the most important role in regulating the elongation of internodes in intact plants. Light also plays a role in regulating elongation; however, we confine our attention here to the role of auxin in cell elongation as a model of plant cell growth regulation. Although the basic mechanism of auxin action in the regulation of cell elongation has not been completely worked out, an examination of what we have learned about how this hormone regulates cell elongation illustrates one mechanism by which plant cells grow.

### The Cytology and Physiology of Plant Cell Enlargement

In most cases enlarging cells become highly vacuolate and there may be a comparatively modest increase, or even no net increase, in cytoplasmic mass during cell enlargement. Membrane surface area increases, as does the total amount of cell wall material, but the total protein content of the cell may not increase as the volume of the cell increases. Before enlargement begins, cells may have several small vacuoles, and only a small fraction of the total cell volume is occupied by these vacuoles. As enlargement occurs, the total volume of the vacuoles increases dramatically. Eventually the vacuoles fuse to form one central vacuole, which may constitute more than 90% of the total volume of the cell (Fig. 7.16).

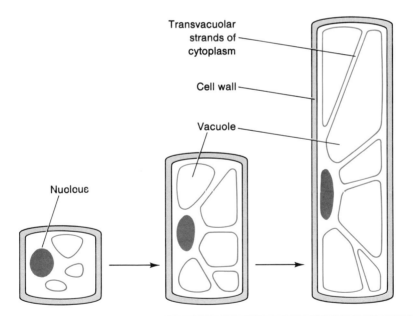

**Figure 7.16 Plant cell elongation and vacuole volume**
Plant cell elongation is accompanied by a large increase in vacuole volume. The expansion of the cell surface area and vacuolar volume during cell enlargement is not accompanied by a similar increase in cytoplasmic mass. Redrawn with permission from Alberts *et al.* (1989).

## Target Tissues and Passive versus Active Growth

The epidermal tissue of the coleoptile may control the elongation rate of this organ and function as the auxin-responsive target tissue. This can be demonstrated by cutting the coleoptile longitudinally and placing one half in water and the other in a solution containing an auxin. The coleoptile half in water curls outward, showing that the epidermis was under tension and that it limited the expansion of the organ. In the presence of auxin, the coleoptile half curls inward as a result of the rapid expansion of the epidermal cells (Fig. 7.17). If the epidermis

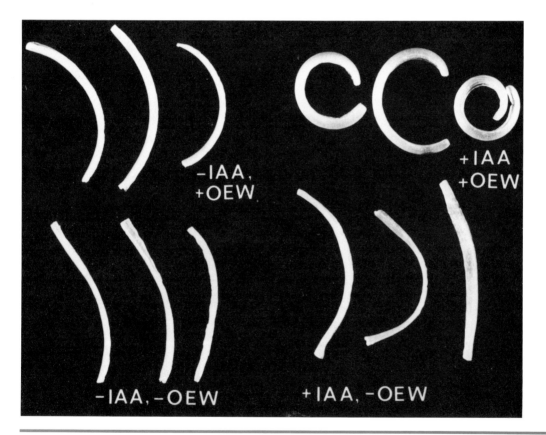

Figure 7.17 **Control of maize coleoptile expansion by epidermal tissues**
Maize coleoptiles were split in half longitudinally and placed either in water or in solutions containing indole-3-acetic acid (IAA) at 10 $\mu M$. The segments in the upper row retained their outer epidermal cells (+OEW) and were positioned so that the side of the segment halves without the epidermis faced toward the right. The outer epidermal cell layer was removed from the segments in the bottom row (−OEW). Reprinted with permission from Kutschera *et al.* (1987).

is removed from a maize coleoptile, the remaining tissues do not respond to auxin, although they may expand slowly when placed in water (Fig. 7.18). In dicot stems, the tissues that control the rate of internode elongation appear to be a layer of cortical cells just under the epidermis. Longitudinally split pea internodes behave in the same way as split coleoptiles. In water, they curl outward as the inner tissues take up water by osmosis and expand; the epidermal and subepidermal cortical cells do not expand. When auxin is added to the bathing medium, however, the epidermal and subepidermal cells expand rapidly and the split internode halves curl inward. The degree of inward curvature is proportional to the auxin concentration, and this phenomenon has been used as a bioassay for auxin in unknown solutions. As is the case with the coleoptile, the bulk of the inner tissues of this organ, the cells of the cortex and pith, grow passively. The outer tissues of both the maize coleoptile and the pea internode are target tissues for auxin, and it is the regulation of the rate of elongation of these cells by auxin that controls the elongation rate of the organ as a whole.

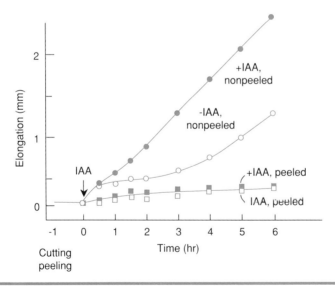

**Figure 7.18 An intact epidermis is necessary for auxin stimulation of maize coleoptile segment elongation**

Even in the presence of an osmotic stress (a mannitol solution with an osmotic concentration of 6 bar), 10 $\mu M$ indole-3-acetic acid (IAA) stimulated the elongation of segments containing an intact epidermis ("nonpeeled"), but had no effect on the growth of segments from which the epidermis had been removed ("peeled"). Redrawn with permission from Kutschera et al. (1987).

## Auxin Receptors

Auxin-binding proteins have been identified in plant cells. Initially, auxin-binding proteins were isolated from membrane fractions corresponding to the plasma membrane, the vacuolar membrane, and ER vesicles. Recently, a cDNA for an auxin-binding protein was cloned and sequenced. It encodes a 20-kDa, largely hydrophilic protein with a 38-amino-acid-long amino-terminal hydrophobic sequence that probably acts as signal peptide. It also has a glycosylation site and its carboxyl terminus ends with the sequence lysine (K)–aspartic acid (D)–glutamic acid (E)–leucine (L). This protein has been designated the auxin-binding protein (ABP) and it may be the auxin receptor. Because it has a signal peptide, it is most likely synthesized on rough ER and co-translationally inserted into the lumen of the ER, after which the signal peptide is cleaved. The sequence KDEL represents an ER localization signal. Proteins terminating with the KDEL sequence reside within the lumen of the ER, probably attached to a receptor that recognizes this sequence. Approximately 80 to 90% of the ABP is localized within the ER lumen.

Antibodies to the ABP have been used to show that most of it is localized in the epidermis of the maize coleoptile. It is not found in the internal tissues of this organ. Furthermore, these anti-ABP antibodies block auxin-induced maize coleoptile elongation. Apparently, auxin does not have to enter the target cells to increase their rate of elongation, at least over the short term. Instead, auxin binding to the ABP receptors triggers a chain of events that leads to increased elongation. There is a problem with this hypothesis, however, as most of the ABP is localized within the ER lumen rather than at the cell surface. Possibly, the KDEL signal is cleaved from the ABP and it enters the secretion pathway, resulting in the presentation of the ABP on the outer surface of the plasma membrane. There the ABP could bind with auxin to initiate the sequence of events leading to cell elongation (Fig. 7.19). Although the details of the signal transduction pathway triggered by auxin binding to its receptor are unknown, auxin-induced elongation is preceded by an increase in inositol trisphosphate and cytoplasmic calcium, perhaps in a fashion analogous to the signal transduction pathway induced by the binding of animal hormone surface receptors (see Chapter 5).

## Turgor Pressure and Cell Elongation

Because cell elongation is driven largely by water uptake, we must briefly consider water relationships in elongating and nongrowing cells. The cell contains an osmotically active solution, surrounded by a semipermeable membrane which is itself elastic, offering little or no resistance to the expansion of the cell. The membrane is in turn surrounded by the cell wall, which will resist the expansion of the cell. The solution inside the cell is rather concentrated, whereas the solution outside the cell is very dilute. As the cell membrane is semipermeable, water will

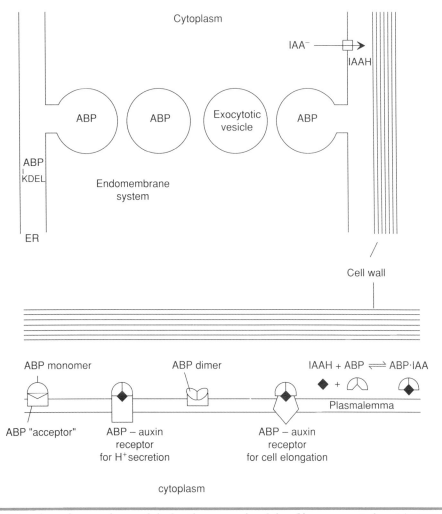

**Figure 7.19 Mechanisms by which the auxin-binding protein (ABP) could act as a plasma membrane auxin receptor**
IAA, indole-3-acetic acid. Reprinted with permission from Klambt (1990).

tend to enter the cell by osmosis. Osmotic water uptake increases the volume of the cell. The cell wall is not infinitely elastic, however. It will resist the increase in volume resulting from osmotic water uptake. If a nongrowing cell is placed in a hypotonic solution, water will enter the cell at a rate proportional to the difference between the water potential of the cell and that of the external medium. As the cell takes up water, hydrostatic pressure begins to be exerted on the cell wall. Assuming that the cell wall is not capable of expansion, an equilibrium will be reached in which the hydrostatic pressure, or **turgor pressure** as it is called, will equal the net osmotic pressure ($P = w - c$, where $P$ = turgor pressure, $w$ = water potential of the tissue, and $c$ = osmotic potential of the solution inside the cell).

The water potential of a given tissue or organ can be determined by placing comparable, weighed pieces of this tissue in solutions of increasing concentrations of an inert solute such as mannitol. After an interval of time sufficient to allow the tissue to come to equilibrium with the external solution, the tissues are again weighed. A graph of tissue weight as a function of mannitol concentration will show increases in tissue fresh weight in mannitol solutions whose osmotic potential was below that of the cells and decreases in tissue fresh weight in solutions whose osmotic potential was greater than that of the tissue. At a particular solute concentration, the tissue will neither gain nor lose weight. At this point the water potential of the cells is equal to the osmotic potential of the external solution.

Similarly, the osmotic potential of the tissue can be determined by placing comparable tissues in solutions of increasing mannitol concentration and examining the cells under the microscope to determine when incipient plasmolysis occurs (Figs. 7.20 and 7.21). Incipient plasmolysis is defined as the condition in which the plasma membrane just begins to pull away from the cell wall in 50% of the cells, as the protoplast shrinks because of the loss of water from the vacuole. The osmotic potential of the cells will equal the osmotic potential of the solution that produced incipient plasmolysis. The turgor pressure is zero at incipient plasmolysis. As $P = w - c$, the turgor pressure can readily be calculated from the values determined for $w$ and $c$.

The enlargement of a plant cell is described by

$$dV/dt = m(P - Y)$$

where

$dt$ = an interval of time
$dV$ = the change in the cell volume over this time interval
$P$ = the turgor pressure of the cell
$Y$ = the yield threshold of the cell wall (i.e., the turgor pressure that must be exceeded before the wall will expand)
$m$ = wall extensibility.

There are two components of wall extensibility: a plastic component and an elastic component. The plastic component is the ability of the wall to be deformed irreversibly; the elastic component of deformation is reversible after the deforming force is removed.

Theoretically, cell elongation could be regulated in one of the following ways:

*Regulated changes in turgor pressure.* An increase in cell turgor pressure resulting from the movement of water into the cell would cause a stretching of the wall and its expansion.

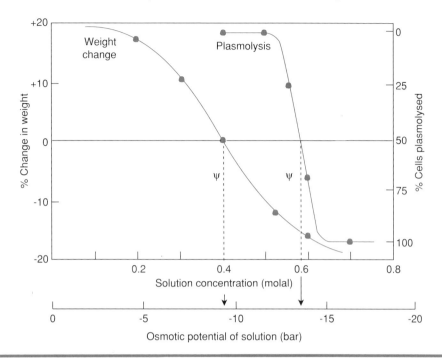

**Figure 7.20 Experimental determination of the water potential and osmotic potential of beet root tissue**
The water potential was determined by placing weighed pieces of tissue in solutions with different osmotic potentials and reweighing them an hour later. The change in weight is plotted against osmotic concentration. The osmotic potential of the tissue was obtained by placing the tissue in solutions of different osmotic potential. The osmotic potential of the tissue is equal to the osmotic potential of the solution that brings about incipient plasmolysis (see Fig. 7.19). Reprinted with permission from Galston *et al.* (1980).

> *Regulated changes in the physical properties of the cell wall.* If there was a decrease in the strength of the cell wall, the cell would expand and the wall would stretch as a result of the cell's turgor pressure.

To put this another way, the energy for cell elongation must come from turgor pressure. Turgor pressure could power cell elongation either through an increase in turgor pressure itself or a decrease in the resistance of the cell walls to turgor pressure. Although changes in either turgor pressure or the properties of the walls theoretically could result in elongation, it is a change in the mechanical properties of the wall that allows the turgor pressure-driven elongation. Auxin does not bring about an increase in turgor pressure, and turgor pressure does not increase during auxin-induced elongation. In some cases, turgor pressure

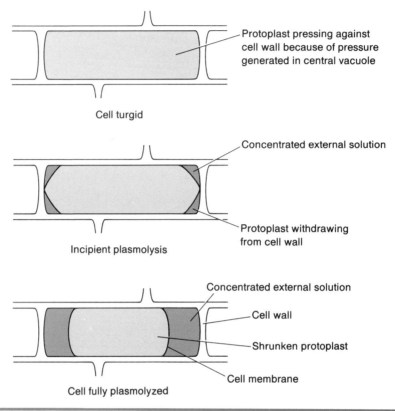

**Figure 7.21 Plasmolysis, the shrinking of the protoplast away from the cell wall**

Turgor pressure ordinarily keeps the plasma membrane tightly appressed to the cell wall. If, however, the cell is placed in a solution whose osmotic concentration is greater than that of the cell, water leaves the cell, principally from the vacuole, and protoplast volume decreases. The drawing illustrates the stages in plasmolysis. Reprinted with permission from Galston *et al.* (1980).

actually decreases during auxin-induced elongation. As a result, the physical properties of the cell wall must control cell elongation.

 The physical properties of the cell wall control elongation    The following experiment demonstrates that the cell wall is loosened by auxin treatment so that it offers less resistance to turgor pressure (Fig. 7.22A). Nongrowing segments from decapitated oat coleoptiles can be fastened in a horizontal position and weights hung from them. The weight will deform the cell walls, causing the segment to bend. When the weights are removed, the segments spring back, but not completely to their original position. The weight brought about both plastic (irreversible) and elastic (reversible) deformations of the coleoptile, but in non-growing coleoptiles the amount of plastic deformation is small. In contrast, when segments are taken from elongating coleoptiles that have been treated with auxin, the weight still causes both plastic and elastic deformations of the tissue; however, the plastic component of the deformation now is much greater. Auxin treatment increases the degree to which the coleoptiles can be deformed irreversibly (Fig. 7.22B). This experiment demonstrates that growth is accompanied by an increase in cell wall plasticity. It is usually simply stated that cell walls are "loosened" during growth. The next question we must consider is how this loosening takes place. Changes in wall plasticity could come about either as a result of the new cell wall synthesis or as a result of the enzymatic breaking of bonds within the wall framework.

Cell wall is synthesized during growth    There is a tendency for the wall thickness to remain constant as growth takes place. This is because synthesis and deposition of cell wall macromolecules ordinarily accompany cell elongation. These new wall materials could rupture chemical bonds between macromolecules and weaken the wall if the new wall materials were inserted into the existing framework of wall macromolecules, a process known as intussusception. In normally growing internodes, cell wall synthesis parallels cell elongation. Synthesis of all wall components occurs during elongation and is stimulated by auxin in proportion to the degree of growth stimulation.

Does new wall synthesis result in a change in the physical properties of the cell wall?    Most of the new cell wall material is laid down by apposition during cell elongation. That is, it is laid down on top of the wall material already present. This type of wall deposition cannot weaken the structure of the existing wall. If anything, it would have just the opposite effect. It would tend to increase the strength of the wall. Some new wall material is added deep within the existing wall matrix by intussusception, and such wall synthesis could disrupt the existing polysaccharide framework, possibly leading to broken bonds between the cellulose microfibrils and the matrix of the cell wall. This cannot, however, be the primary mechanism by which the cell wall is loosened during elongation, because wall synthesis can be blocked experimentally and auxin treatment still loosens the wall (Fig. 7.23). Although wall deposition normally accompanies cell elongation, this is not the primary mechanism for cell elongation.

The degree of order in the cell wall can be determined by the ability of ordered cellulose microfibrils to rotate a beam of polarized light. The ability of a substance to rotate a beam of polarized light is known as *birefringence,* and it is a measure of the degree of order in the arrangement of a crystalline, optically active substance. Plasmolyzed *Avena* coleoptiles show negative birefringence when examined in polarized light. As coleoptiles are stretched, however, the birefringence first becomes zero and then positive as the force applied increases. This indicates that most of the cellulose microfibrils initially are perpendicular to the

**A**

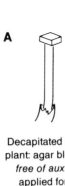

| Decapitated oat plant: agar block *free of auxin* applied for 2 hours | Block removed; coleoptile excised at base | Coleoptile section mounted on rack | Weight hung on section; curvature a result of plastic and elastic bending | Weight removed; plastic (irreversible) bending small |

| Decapitated oat plant: agar block containing *indoleacetic acid* applied for 2 hours | Block removed; coleoptile excised at base | Coleoptile section mounted on rack | Weight hung on section | Weight removed; plastic (irreversible) bending large |

**B**

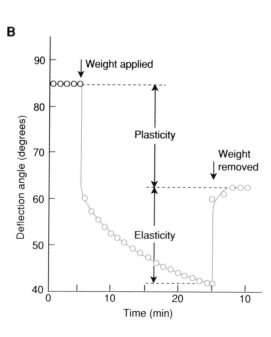

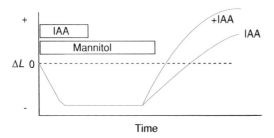

**Figure 7.23 Relationship between auxin-induced changes in wall plasticity and cell wall deposition**
Auxin-induced changes in wall plasticity are not a consequence of cell wall deposition. Oat coleoptile segments are placed in a solution of mannitol that induces incipient plasmolysis. Some of the segments are exposed to a growth-stimulating concentration of auxin during the first half of the mannitol treatment. Auxin has no effect on the elongation of the coleoptiles while they are exposed to the mannitol solution; however, when returned to water, the cells that were exposed to auxin during the mannitol treatment grew faster. Auxin treatment led to the loosening of the cell walls, even though they did not grow or deposit additional wall materials while exposed to the hormone. IAA, indole-3-acetic acid. Redrawn with permission from Tagawa, T., and Bonner, J. (1957). Mechanical properties of the *Avena* coleoptile as related to auxin and to ionic interactions. *Plant Physiol.* **32**, 207–212.

long axis of the cell and that as the tissue is stretched, microfibrils become reoriented so they are mostly parallel with the long axes of the cells. When the tissue is pretreated with auxin, the amount of reorientation is greatly reduced at any given strain. As the amount of reorientation of the microfibril is proportional to the amount of crosslinking of the cellulose microfibrils, this demonstrates that auxin loosens walls by reducing the amount of crosslinking between the microfibrils.

Auxin-induced cell wall plasticization appears to involve the hydrolytic cleavage of hemicelluloses that are linked to the cellulose microfibrils Plant cell walls contain enzymes, including enzymes capable of breaking down

**Figure 7.22 Changes in oat coleoptile cell wall plasticity as a result of auxin treatment**
(A) Simple, somewhat hypothetical, method to determine the amount of irreversible (plastic) deformation of the oat coleoptile cell walls and the effect of auxin treatment on this plastic deformation. (B) Graphic representation of the results of an experiment designed to measure the plasticity and elasticity of oat coleoptile cell walls. (A) Redrawn with permission from Bonner, J. F., and Galston, A. W. (1952). Principles of Plant Physiology. Freeman, San Francisco; and (B) Redrawn with permission of Tagawa, T., and Bonner, J. (1957). Mechanical properties of the *Avena* coleoptile as related to auxin and to ionic interactions. *Plant Physiol.* **32**, 207–212.

cell wall polysaccharides. Purified maize coleoptile cell walls undergo slow autolysis. That is, the wall polysaccharides are degraded in isolated cell wall preparations and this degradation results in a loss of about 10% of the wall substance over an 8-hour period. The soluble products produced on autolysis are glucose monomers, and it appears that it is the hemicelluloses of the wall that are degraded by these enzymes. Auxin has no direct effect on the activity of these enzymes, but when coleoptiles are incubated in a solution containing indole-3-acetic acid before the wall preparations are made, the activity of these hydrolytic enzymes is increased.

$H^+$ ion excretion weakens the cell wall    Auxin treatment of elongating stem tissue induces the excretion of $H^+$ ions from the cells, and this $H^+$ ion excretion may be the agent that weakens the cell wall (Box 7.1). The flux of protons from a plant cell can be measured by determining the pH of the external medium bathing the cell. When the proton efflux is determined for maize coleoptiles during auxin-induced growth, both the timing and magnitude of the growth response are very similar to those of the proton efflux (Fig. 7.24). The difference between an elongating cell and a static cell of the same developmental age is that the elongating cell has a weaker, or more plastic, cell wall. According to the acid growth theory, the difference in the biophysical and mechanical properties of the cell wall is due to the $H^+$ ions excreted by the auxin-treated, elongating cell.

The acid growth theory does not explain how the wall is loosened in response to proton excretion. There are, however, several possibilities. Chemical bonds between cellulose and other cell wall polysaccharides are broken during growth. The broken bonds either are acid-labile or are broken enzymatically. It is some-

**Box 7.1**

### Acid Growth Theory

Auxin
↓
$H^+$ ion secretion
↓
Broken bonds in cell wall matrix
↓
Loosening of cell wall
↓
Turgor pressure-driven cell expansion (elongation)

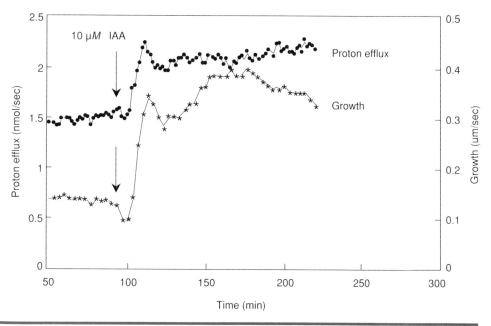

**Figure 7.24 Time courses of auxininduced growth and proton efflux coincide**

Maize coleoptile segments were treated with indole-3-acetic acid (IAA) at the time indicated by the arow, and the subsequent changes in growth (asterisks) rate and the proton efflux rate (dots) were determined and plotted as a function of time. Redrawn with permission from Luthen *et al.* (1990).

what unlikely that a pH of 4 would rupture bonds important to the integrity of the wall matrix, however. The activity of some cell wall hemicellulases could be sharply pH dependent, with an acidic pH optimum. The activity of such an enzyme could be dramatically increased by the acidification of the wall by proton excretion.

$H^+$ ion secretion occurs as a result of a membrane protein that has been termed a *proton pump*. The proton pump is an ATPase and the secretion of $H^+$ from the cell into the wall matrix occurs as the result of energy released in the hydrolysis of ATP by the pump (Fig. 7.25). Auxin has been shown to regulate the activity of the proton pump. Treatment of tobacco protoplasts causes hyperpolarization of the plasma membrane within seconds, and antibodies to the auxin receptor block this response. Furthermore, derivatives of auxin that are unable to enter the cell because of their size are able to bind to the auxin receptor and stimulate proton secretion. Auxin does not bind to the ATPase proton pump itself; rather, auxin binds first to its membrane receptor. This, in turn, must trigger events in a signal transduction pathway that ultimately leads to increased proton excretion. Perhaps auxin binding to its receptor activates a G protein

structure, but both are long rodlike molecules with a globular head at one end and a hook at the other (Fig. 8.2). The hook is inserted in the flagellar membrane, and the rod projects outward where it may interact with an agglutinin of the opposite mating type (Fig. 8.3). Perhaps these gametes adhere to each other by means of the molecular equivalent of Velcro, although with a degree of specificity much higher than that of the common garment fastener.

### Alternation of Generations

The life cycles of most other plants are much more complicated that that of *Chlamydomonas*. The life cycle of nearly all plants is composed of two **alternating generations.** Typically, the life cycle consists of a haploid phase, known as the **gametophytic generation,** that alternates with a diploid phase, known as the **sporophytic generation.** The gametophyte is the sexual generation. It reproduces by means of **gametes,** or sex cells, but does not reproduce itself directly. Instead, after the gametes fuse, the resulting zygote develops into the sporophyte. Similarly, the sporophyte does not reproduce the sporophyte directly, but makes asexual reproductive cells known as **spores,** which develop into a gametophyte. The sporophytic and gametophytic generations alternate and reproduce each other (Fig. 8.4).

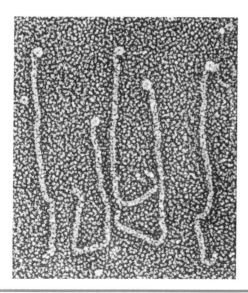

**Figure 8.2  *Chlamydomonas* agglutinins**
Electron micrograph of an *mt*$^+$ agglutinin. Reprinted with permission from Goodenough *et al.* (1985), *Journal of Cell Biology,* **101,** 924–941, ©The Rockefeller University Press.

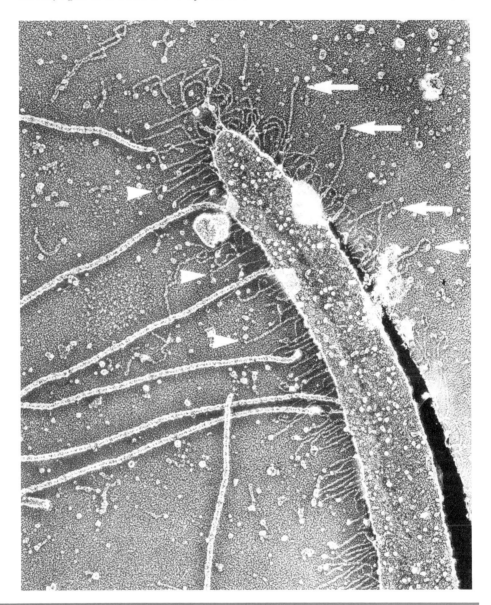

**Figure 8.3**  **Electron micrograph of a flagellum from an mt$^-$**
***Chlamydomonas***
Arrows and arrowheads point to agglutinin chains with terminal hooks. Reprinted with permission from Goodenough *et al.* (1985).

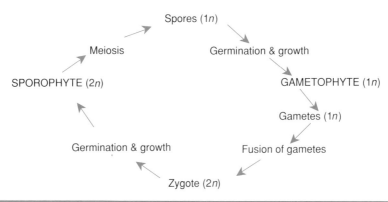

**Figure 8.4    Alternation of generations**

In many algae these two generations are independent, free-living plants, often with a startlingly different appearance. In other cases, one generation is much reduced in size and in complexity, and the other generation is dominant.

### Life Cycle of a Fern

The fern gametophyte is an inconspicuous, leafy green plant that looks nothing like the sporophyte. The plants most people readily recognize as ferns actually are the sporophytes; the gametophyte may be completely overlooked as green fuzz growing on the side of the pot. The sporophyte produces spores in special structures known as **sporangia.** The ferns are divided into two different groups, the eusporangiate and the leptosporangiate ferns, which differ radically in the developmental processes that lead to spore formation. Here we describe only the development of leptosporangiate ferns, where groups of sporangia are formed on the underside of the leaf in specialized regions known as **sori** (Fig. 8.5). Each sporangium is initiated from a single cell, which divides and grows to become a highly differentiated, multicellular structure. At maturity, it consists of a some-what flattened sphere of cells, attached to the leaf by a slender, multicellular stalk. One cell at the center of the sphere becomes differentiated as the primary archesporial cell, which will divide several times by mitosis to produce from 12 to 16 **spore mother cells.** Each spore mother cell then undergoes meiosis to form a tetrad of haploid spores (see Fig. 8.5). At maturity, the spores separate from each other and a thick wall forms around them.

The spores are forcibly discharged from the sporangium by contraction of the thick-walled cells of the annulus (see Fig. 8.5). Spores that settle onto a moist environment will germinate almost immediately. As the spore germinates, cell division occurs in one plane to produce a filament of cells. Division will continue in a single plane unless the cells are exposed to blue light. When the cells are irradiated with blue light, the plane of division of the apical cell is altered so that

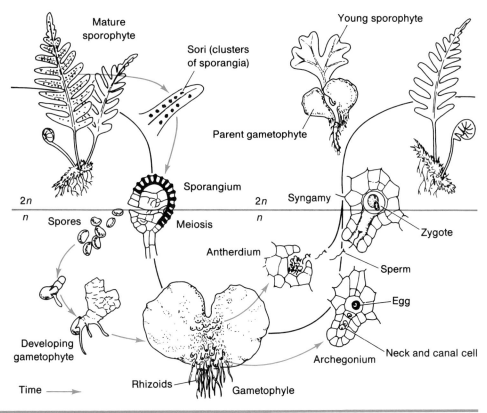

**Figure 8.5** **Life cycle of a fern**
Redrawn with permission from Jensen and Salisbury (1984).

it begins to divide in two dimensions to form a sheet of cells that will develop into the gametophyte (see Fig. 8.5). The mature gametophyte is called a **prothallus.** It is a small, flat, heart-shaped structure, approximately 1 cm in diameter, with rootlike rhizoids growing from its lower side. It is one cell thick on its outer edges, but several cells thick in the center. Except for the rhizoids, the cells of the gametophyte contain chloroplasts and are photosynthetically active. When the gametophyte is mature, female and male reproductive organs, called **archegonia** and **antheridia,** respectively, are produced (see Fig. 8.5). Antheridia form on the lower side of the gametophyte, near the growing apical notch, whereas archegonia arise on the same side of the gametophyte, but in more distal regions. Some cells within the antheridia differentiate into motile, flagellated sperm, whereas a single cell develops into an egg within the archegonium. Fertilization occurs when the motile sperm swim up the neck canal of the archegonium and reach the egg. A zygote is formed by the fusion of the sperm and egg. The zygote will undergo embryogenesis within the archegonium. Under favorable growing

conditions, the embryo continues to develop into the diploid sporophyte generation (see Fig. 8.5). Initially the sporophyte is attached to the gametophyte, and
during embryogenesis it is parasitic on the gametophyte. As it grows, it establishes
nutritional independence with the development of its own photosynthetic tissues.

Ferns have not been as successful as the angiosperms over evolutionary time.
One reason is that fern gametes are not protected. The sperm, in particular, not
only are exposed to the environment, but must swim through water to reach the
egg. This fact alone makes it difficult for ferns to become established in arid
regions. One of the reasons for the success of the angiosperms is that these
organisms have evolved protective structures for their gametes. Not only do they
provide protection for the egg, which is enclosed within an ovary, but they also
have evolved seeds and fruits, which are elaborate structures that ensure the
protection, dispersal, and nourishment of the embryo and seedling.

## Life Cycle of an Angiosperm

In angiosperms, the sporophytic generation is dominant, as it is in ferns, but the
angiosperm gametophytic generation is even more reduced than it is in ferns
and is parasitic on the sporophyte (Fig. 8.6). Angiosperms actually have two
morphologically different gametophyte generations, both of which are contained
within floral organs (see Fig. 1.13). The sporophytic generation begins with the
fertilization event that results in the formation of the zygote. The zygote undergoes embryogenesis and, along with some additional maternal tissues, forms a
seed. Both the seed and the plant that develops from it are part of the sporophyte
generation. The gametophytic generation is contained completely within the
flower, and it begins when specialized cells within the male and the female
parts of the flower undergo meiosis to produce haploid spores. The female
reproductive structures within the flower are known as **carpels.** Each carpel
consists of a stigma, which is attached to an ovary by a stalk known as a style.
The **embryo sac** is the female gametophyte and it is contained within an ovule.
The male reproductive structures within the flower are known as **stamens.** Each
stamen consists of an **anther,** which is connected to the flower by a slender stalk
known as a filament. The male gametophyte is confined to the **pollen grain,**
which is formed within the anther.

The flowers of maize (Fig. 8.7) are either male or female, never both. The
**tassel** is the male inflorescence, formed at the top of the plant. Male flowers
produced in the tassel are known as **spikelets. Ears,** the female flowers, are
formed in the axils of the leaves in the middle of the plant. Each ear contains
hundreds of ovules. The **silks** are the styles and stigmas of these ovules.

The male gametophyte forms within the anther   The anther typically
is composed of four elongate sacs which are fused together and attached to the
filament. At the center of each sac is a column of cells known as the **microspore
mother cells,** surrounded by a tissue known as the **tapetum** (Fig. 8.8). As the

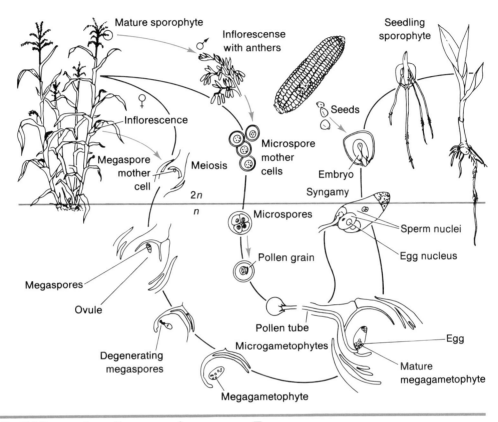

**Figure 8.6**   **Life cycle of an angiosperm, *Zea mays***
Redrawn with permission from Jensen and Salisbury (1984).

anther develops, the microspore mother cells undergo meiosis to produce four
haploid microspores, each of which develops into a pollen grain. At maturity the
pollen grain is surrounded by a thickened, elaborate, many-layered cell wall. The
innermost layer, known as the **intine,** is similar in composition to most other
primary cell walls. It is covered by an outer structure known as the **exine** (Fig.
8.9), which contains a number of unusual macromolecules, including sopropol-
lenin. The tapetum appears to contribute to the formation of the exine, whereas
the intine is a product of the pollen grain.

The pollen grain is the male gametophyte generation. It contains two cells:
the vegetative cell, from which the pollen tube develops, and the generative cell,
which will produce the sperm. The vegetative and generative cells arise as a
result of a mitotic division of the microspore. The generative cell is unusual
in several respects, including the fact that it usually lacks mitochondria and
chloroplasts. The cytoplasm was partitioned unequally during the mitotic division
of the microspore so that most of the cytoplasmic organelles, including all of the

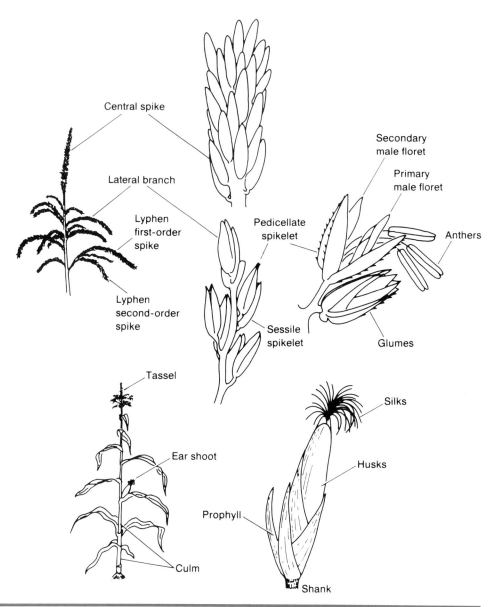

**Figure 8.7  Reproductive structures of a maize plant, a grass (monocot)**
Redrawn with permission from Poethig (1982).

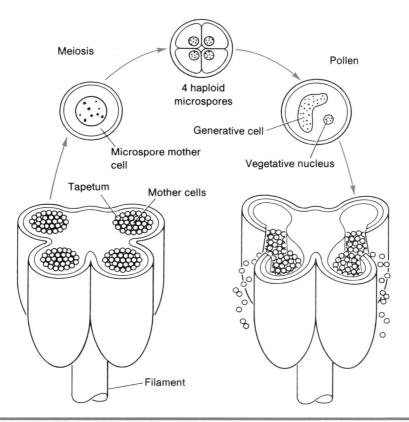

**Figure 8.8** **Diagram of an anther showing the development of pollen grains**
Redrawn with permission from Jensen and Salisbury (1984).

mitochondria and plastids, remained with the vegetative cell. Consequently, the generative cell is comparatively small. It is separated from the vegetative cell by a very thin wall, or possibly only a plasma membrane (Fig. 8.10A). As the generative cell lacks mitochondria and chloroplasts, the sperm also will lack these organelles. This is the basis for the maternal inheritance of chloroplast and mitochondrial genomes that occurs in nearly 90% of all angiosperm species. When it is first formed, the generative cell is roughly lens-shaped and pressed against the intine wall of the pollen grain (Fig. 8.10A). As it develops, the generative cell detaches from the intine wall and becomes completely surrounded by the vegetative cell cytoplasm. As it matures, the shape of the generative cell changes from spherical to a highly elongate crescent or cigar shape (Fig. 8.10B).

At some point in pollen development, the generative cell divides by mitosis and each of the two daughter cells differentiates into sperm cells. This occurs at

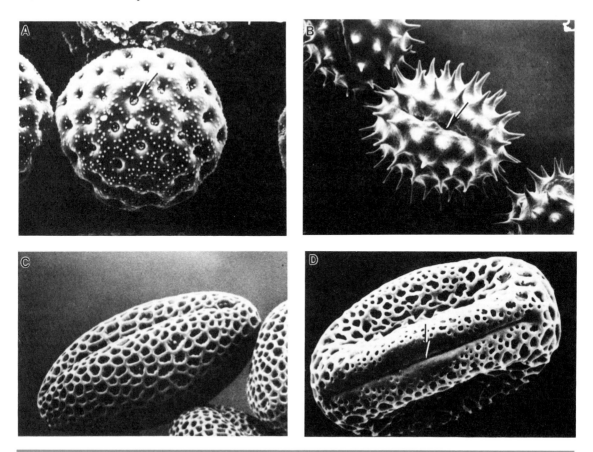

**Figure 8.9    Structure of some pollen grains as viewed by scanning electron microscopy**

(A) The wind-transported pollen of the weed *Chenopodium album* has approximately 70 pores (arrow) through the relatively smooth exine layer. ×2000. (B) Sunflower pollen grains (*Helianthus annuus* L.) have a spiny exine and are insect transported. The arrow points to one of the germination pores in a longitudinal furrow. ×1000. (C) Pollen of the turnip (*Brassica napus* L.) has a thick, netlike exine and is insect transported. ×2000. (D) The elongate, furrowed pollen grain of ivy (*Hedra helix* L.) has a reticulate exine. The pollen grain emerges from the furrow (arrow). ×2000. Reprinted with permission from Cresti *et al.* (1992).

different times in different species, but commonly it takes place during the growth of the pollen tube down the style. The elongated or crescent shape of the mature generative cell is retained by the sperm cells it will produce. This shape is imposed on the generative cell by cytoskeletal microtubules, although this shape is gradually lost when the generative cell is isolated from the pollen (Fig. 8.11). The sperm of all angiosperms are nonmotile and they are passively transported to the embryo sac by growth of the pollen tube. The crescent shape of the generative cell and sperm may be important for their transport during pollen tube growth.

The mitochondrial genome makes a contribution to the development of the male gametophyte, although the exact nature of this contribution has not been determined. As presented in Chapter 3, many plants exhibit cytoplasmic male sterility (cms) as a result of genes carried by the mitochondrial genome. The mitochondrial genes that determine cms are not essential for mitochondrial function, and these genes, when they are present, are expressed throughout plant development without affecting the growth or differentiation of other plant organs. The products of these genes, however, somehow block male reproductive development. There is no single stage or point in male reproductive development in which the cms trait acts. In some cases the anthers fail to develop, whereas in others, the anthers are formed but pollen development is abnormal. In cases where the cms trait blocks normal pollen develoment, the mitochondrial genes appear to affect the formation of the pollen wall. While they are undergoing meiosis, the microsporocytes synthesize a special cell wall consisting of the polysaccharide callose. After meiosis is completed and synthesis of the intine and exine walls of the pollen grain begun, the callose wall is broken down by an enzyme, a callase, that is synthesized and secreted by tapetal cells. The degradation of the callose wall is essential to the release of the pollen as single entities into the locule of the anther. In petunia there is a type of cms in which the timing of callase synthesis and secretion is disturbed so that it occurs either too soon or too late. This disturbs the deposition of the exine wall, and apparently pollen grains with abnormal walls cannot perform their function in pollination.

The female gametophyte forms within the ovule   The ovary contains a cavity lined with an epidermal layer. Ovules develop from these epidermal cells and are contained within the cavity of the ovary, attached to its inner surface by a short stalk called the **funiculus** (Fig. 8.12). The ovules are attached to a region of the ovary wall known as the **placenta.** With the growth of the ovule, one cell is set aside as the **megaspore mother cell** and will undergo meiosis. The meiotic division produces four haploid cells. Frequently, only one of these haploid cells develops further; the other three disintegrate. The surviving cell undergoes three mitotic divisions to form the **embryo sac,** the female gametophyte (Fig. 8.13). At maturity the embryo sac consists of seven cells containing eight haploid nuclei. The antipodal cells are not involved directly in embryogenesis.

Pollination is the transfer of pollen from a stamen to an ovule   A great many different mechanisms have evolved to increase the probability that pollen will be transferred to the stigma. In the simplest case, the pollen is wind-borne; in other cases the pollen is carried by specific animal pollinators, including birds, insects, and bats. The pollination of a flower begins when a ripe pollen grain lands on a compatible stigma. There are two types of stigmas, wet and dry. In plants with wet stigmas, the stigma consists of a loose aggregate of secretory

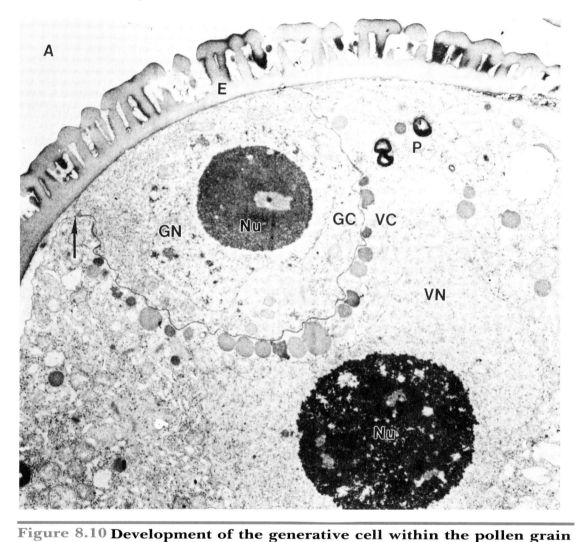

**Figure 8.10** **Development of the generative cell within the pollen grain**
(A) Transmission electron micrograph of immature *Euphorbia dulcis* pollen. The
generative cell is lens-shaped and pressed against the pollen grain wall shortly after it
is formed as a result of the mitotic division of the microspore. At the stage shown here
it is beginning to detach from the wall (arrow) and assume a spherical shape, and it
will be completely surrounded by the vegetative cell, a cell within a cell. ×33,500.
(B) Transmission electron micrograph of the mature pollen of *Tradescantia virginiana*.
This species is typical of many angiosperms in that the mature pollen grain contains
both a generative cell and a vegetative cell. The vegetative cell nucleus is highly
irregular in shape and takes up most of the mass of the pollen grain. The generative
cell is much smaller and totally contained within the vegetative cell. The overall
morphology and the nucleus of the generative cell highly elongate. ×62,000. VN,
vegetative cell nucleus; VC, vegetative cell cytoplasm; GN, generative cell nucleus; GC,
generative cell cytoplasm; Nu, nucleus; P, plastid; I, intine wall; E, exine wall.
Reprinted with permission from Cresti *et al.* (1992).

B

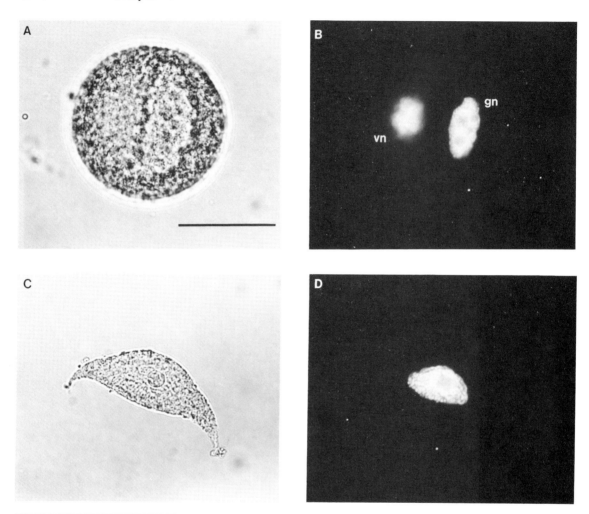

**Figure 8.11 Isolation of generative cells of tulip pollen**
(A,B) The pollen cell wall was removed enzymatically to produce the wall-less protoplast. (C,D) The pollen protoplast then was ruptured to liberate the generative cell. The photos on the left were taken with bright-field microscopy; those on the right were taken with a fluorescence microscope after the cells were stained with DAPI, a fluorescent dye specific for DNA. gn, generative nucleus; vn, vegetative nucleus. Reprinted with permission from Tanaka *et al.* (1989).

cells that produce a fluid rich in glycoproteins, mucilages, and nutrients necessary for pollen germination. Dry stigmas do not secrete large quantities of glycoproteins and mucilages, but nevertheless they are a highly specialized tissue that provides an environment that permits pollen germination. In either case the pollen must adhere to the stigma before germination can begin (Fig. 8.14). The pollen grain will absorb water and nutrients from the stigmatal surface and begin to germinate (Fig. 8.15). Pollen growth involves the formation of the pollen tube which emerges from one of the pollen pores. The pollen tube grows down through the stigma and into the style, carrying the sperm to the egg.

There are two types of styles, open and closed. Open styles have a central cavity whose inner epidermal surface is coated with mucopolysaccharides, lipoproteins, and glycoproteins. These molecules are believed to serve as a nutrient medium for the growth of the pollen tube. In addition, they appear to play a role in directing the growth of the pollen tube toward the embryo sac. In some cases, pollen tubes actually grow through the walls of the stigma papillae to enter the stylar cavity (Fig. 8.16). Here they grow along the surface of the stylar cells, in a direction determined by molecules on their surface. A plant homolog of the animal extracellular matrix molecule vitronectin has been shown to be present on the surface of these stylar cells, and perhaps it provides guidance for the growing pollen tubes. In plants with closed styles, the cells of the style are embedded in an extracellular matrix similar in composition to that found on the inner surface of the open style. Pollen tubes grow through this extracellular matrix, apparently deriving both guidance and nutrition from these matrix macromolecules.

Pollen tubes grow only at their tips   The tubes must extend for up to many centimeters to reach the embryo sac. The cell wall of the pollen tube lacks cellulose and instead contains a different polysaccharide known as callose, which is a $\beta1\rightarrow3$-linked glucan. Callose is synthesized by the Golgi apparatus and transported to the extreme tip of the pollen tube by Golgi-derived vesicles. Fusion of these vesicles with the plasma membrane helps expand the cell membrane of the elongating tube, while the content of the vesicles expands the wall (Fig. 8.17). The cytoplasmic face of the membrane of these vesicles appears to be coated with myosin, and the vesicles are transported to the tip along actin filaments.

Total cytoplasmic volume does not increase as the pollen tube grows. Instead, the bulk of the cytoplasm is in close proximity to the growing tip and continues to move with the tip as the tube elongates. It appears to remain at the tip for three reasons: First, a distal vacuole expands as the tube elongates and it helps keep the cytoplasm appressed to the tip. Second, the cell forms periodic callose cross-walls at the most distal region of the cytoplasm so that only the terminal portion of the tube retains living cytoplasm (Fig. 8.18). Finally, the cytoskeleton of the pollen tube continually transports organelles, the generative or sperm cells, and the vegetative nucleus toward the growing tip.

Many higher plants are self-incompatible   The female tissues of many higher plants will reject pollen from the same plant. Whereas in a compatible interaction between pollen and stigma, water is rapidly taken up by the pollen grain and it extends a tube within about 2 hours, in an incompatible interaction water uptake is slow. Usually the pollen grain does not produce a pollen tube,

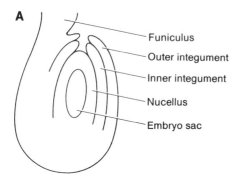

A

Funiculus

Outer integument

Inner integument

Nucellus

Embryo sac

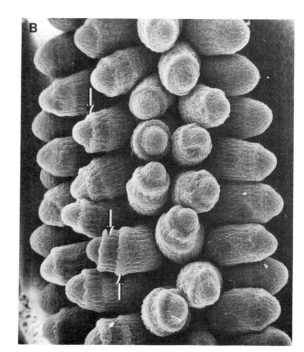

B

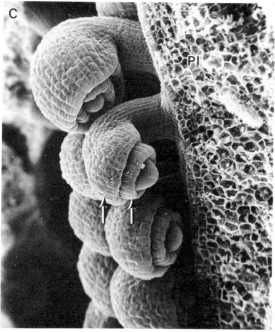

C

Pl

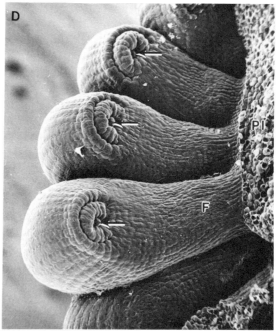

D

Pl

F

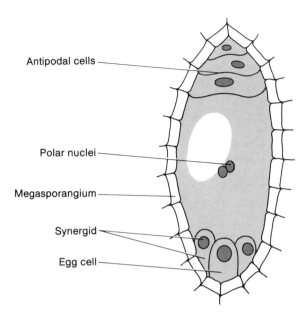

Antipodal cells

Polar nuclei

Megasporangium

Synergid

Egg cell

**Figure 8.13 Diagram of a mature female gametophyte**
Redrawn with permission from Bold and LeClaire (1987).

but if it does, the tube grows poorly and is unable to penetrate the stylar tissues, or if it does penetrate the style, the tip swells and bursts (Fig. 8.19). One of the most widespread and thoroughly studied mechanisms blocking self-fertilization is gametophytic self-incompatibility. It occurs when a gene product of the male gametophyte interacts with a gene product of the female tissues of the pistil. This interaction is controlled by the products of a single genetic locus, known as the *S* locus. There may be 40 or more *S* alleles in a natural population.

**Figure 8.12 Development of the ovule**
(A) Diagram of a mature ovule. The megaspore mother cell differentiates from the surrounding nucellar tissue early in ovule development and undergoes meiotic division to form the megaspores. Typically, one megaspore develops into the embryo sac while all others disintegrate. The tissues surrounding the megaspore mother cell and embryo sac also grow and develop. (B) The inner (arrow) and outer (arrowhead) integuments begin as ridges of tissue early in ovule development, as seen in this scanning electron micrograph of the ovule primordia of *Passiflora racemosa*. ×200. (C) Differential growth of the ovules of many species causes them to curve so that the micropyle is bent around to the funiculus (F) and the placenta (Pl), as illustrated by this scanning electron micrograph of *Passiflora vespertilio* ovules. ×150. (D) The inner and outer integuments completely overgrow the nucellus, except for the micropyle (arrow), through which the sperm will enter, as seen in this scanning electron micrograph of *Agrostemma gracilis* ovules. ×200. Reprinted with permission from Cresti *et al.* (1992).

**Figure 8.14 Pollination**

A flower of *Hibiscus rosa-sinensis,* which has a dry stigma with many pointed stigma papillae, has been pollinated and the pollen grains (PG) are attached to the tips of the stigma papillae cells. Reprinted with permission from Cresti *et al.* (1992).

**Figure 8.15 Initiation of pollen germination by water uptake**

(A) Mature pollen is dehydrated and relatively inactive metabolically. Pollen germination is initiated largely by water uptake, resulting in the rehydration of the grain. Other molecules that can influence germination or control pollen tube growth may be taken up from the stigma along with water. (B) Pollen of *Impatiens glandulifera* have four apertures, or pores (arrows). These widen as the grains are rehydrated and the pollen tube (PT) will emerge from one of them. ×2300. Reprinted with permission from Cresti *et al.* (1992).

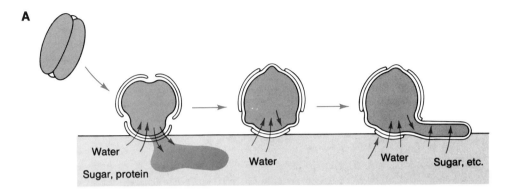

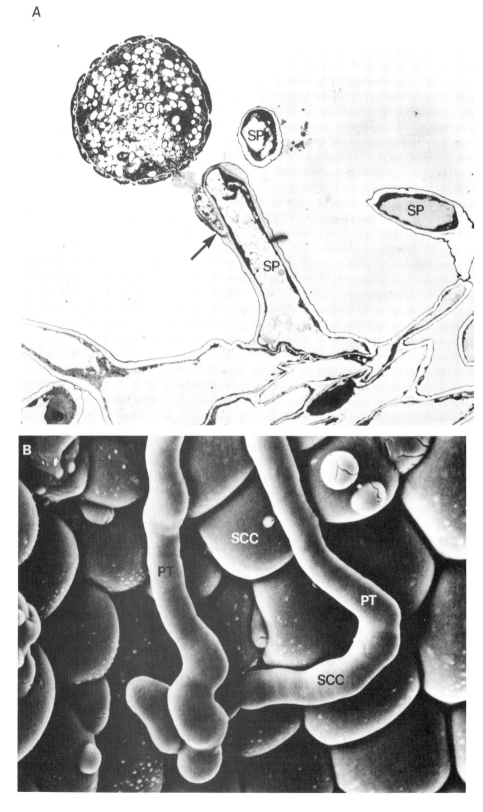

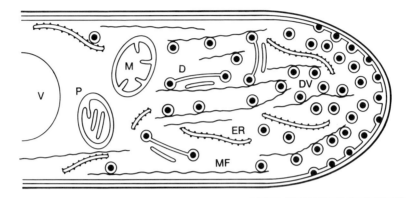

**Figure 8.17 Diagram illustrating some features of tip growth in a pollen tube**

Dictyosome (Golgi)-derived vesicles (DV) containing cell wall polysaccharides are transported to the tip by microfilaments. The vesicles fuse with the plasma membrane at the tip of the tube to expand the wall and plasma membrane as the tube grows. D, dictyosomes (Golgi); ER, endoplasmic reticulum; M, mitochondria; P, plastid, MF, microfilament. Reprinted with permission from Sievers and Schnepf (1981).

Although we do not know exactly how the *S* alleles interact, it is known from genetic studies that if the *S* alleles in the pollen and female tissues are the same, the pollen tube will not develop normally. The protein products from the *S* alleles expressed in the stigma and styles of tobacco are relatively basic glycoproteins secreted into the extracellular matrix of the transmitting tract. Comparison of the gene products of several of these *S* alleles from two species of tobacco demonstrates that they contain only limited regions of identical sequence (Fig. 8.20), but one region of all of these proteins has sequence similarity to fungal RNases. Apparently the RNase activity of the *S* proteins destroys the pollen RNA after an incompatible reaction, but not after a compatible pollen–stigma reaction. It has also been shown that expression of a foreign gene encoding an RNase will bring about male sterility when the gene is expressed in pollen.

**Figure 8.16 Growth of the pollen tube through the stigma and style**

(A) Transmission electron micrograph of a germinating pollen grain (PG) of spinach in which the pollen tube has penetrated the cuticle (arrow) of the stigma papillae (SP) and begun to grow through the papillae cell wall on its way to the ovary. ×2300. (B) Growing pollen tubes (PT) in the stylar canal of lily. Lily has an open style in which the pollen tubes grow over the surface of the stylar canal cells (SCC). ×660. Reprinted with permission from Cresti *et al.* (1992).

in other species, particularly monocots, the endosperm becomes the reserve food storage organ.

Embryogenesis is summarized in Box 8.1.

*The origin of the different tissues and organs of the seedling plant can be traced to specific cells early in embryogenesis* Because of the precise division patterns during the formation of the globular-stage embryo, it is possible to recognize specific cells or groups of cells at this stage that will give rise to the three main tissue types: epidermal, ground, and vascular (Fig. 8.25C). That is, in addition to the protoderm, the origin of both the vascular tissue and cortical tissues can be traced to a small group of cells within the globular embryo. Similarly, in the heart-stage embryo, the regions giving rise to the different regions of the seedling, cotyledons, hypocotyl, and radicle, also are evident; however, it probably would be a mistake to say that the fate of any of the cells is precisely determined during early embryogenesis. Plant cells appear to be plastic in their development, and specialized characteristics of these different tissues and organs probably is determined by the interactions of physical factors and cell–cell interactions that come into force only as the embryo develops.

*Endosperm is formed in gametic embryogenesis* The formation of endosperm tissue in gametic embryogenesis is initiated from the product of the other fertilization event that occurs after the pollen tube ruptures in the embryo sac and the second sperm fuses with the two polar nuclei to form the triploid endosperm nucleus. An endosperm tissue usually develops from the endosperm nucleus during embryogenesis. In some dicots, such as soybean, the

**Box 8.1**

## Chief Events of Embryogenesis

1. Establishment of the precursors, or initials, for the dermal, ground, and vascular tissues within the plant body: These are differentiated in a radial pattern within the embryo. This occurs at least by the globular stage, and the formation of this basic pattern may actually occur even earlier during the octant stage.

2. Establishment of the apical–basal polarity of the embryo which will persist throughout the life of the plant. This occurs during the transition from the globular to the heart stage and probably is responsible for this transition.

3. Establishment of the root and shoot apical meristems. Although the primary meristems do not become active until after germination, they are established in the heart stage.

endosperm tissue surrounds the embryo during the early stages of embryogenesis, but it disappears during seed maturation and the cotyledons become the storage organs of the mature seed. In other dicots such as tobacco and castor bean, however, the endosperm persists throughout seed formation and there is a significant amount of endosperm in the mature seed, where it represents an important site for the deposition of the stored food reserves (Fig. 8.26). In monocots, such as maize, the mature seed is composed of a large mass of endosperm tissue which may make up 70% of the seed. The monocot endosperm tissue actually is composed of two regions: an outer region known as the **aleurone,** in which the storage proteins are deposited, and an inner **starchy endosperm,** which contains relatively little storage protein.

The endosperm and suspensor contribute to embryogenesis   It is often assumed that both the suspensor and endosperm play important roles in the development of the embryo, in addition to any role that the endosperm may play as a storage organ for the seed. This is suggested by experiments in which attempts have been made to culture very young embryos. The mature embryo usually is fully autotrophic. It is capable of photosynthesis and of synthesizing all organic molecules it requires for its growth. Immature embryos, however, are dependent on the parent plant, not only for the basic photosynthate, but also

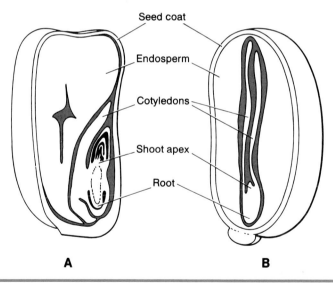

**Figure 8.26 Diagram of a seed and a fruit, both of which have well-developed endosperm tissues surrounding the embryo at maturity**
(A) The maize kernel technically is a type of fruit, known as a caryopsis, which contains a seed, (B) the castor bean (*Ricinus communus*), a dicot, is a seed. Both are shown as they would appear in longitudinal section. Reprinted with permission from Jensen (1972).

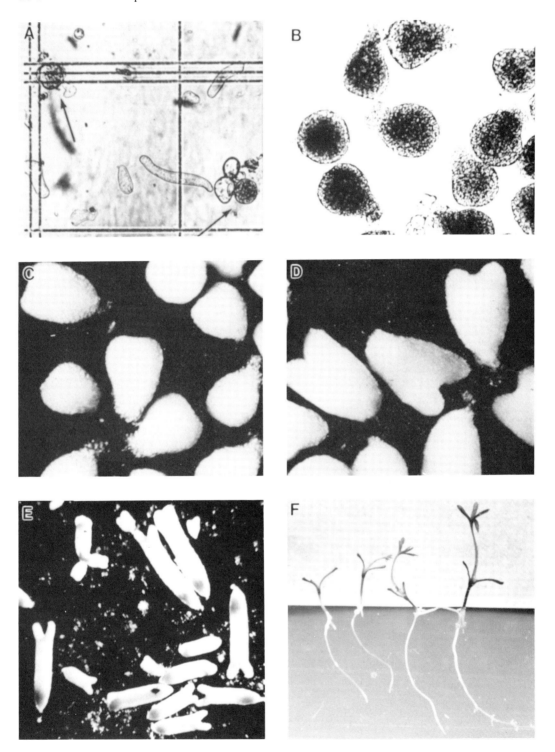

for an array of other nutrients. In some cases, immature embryos need to be in a solution with a high osmotic potential if they are to proceed with embryonic development. Although mature embryos will germinate and grow when cultured on a simple mineral salt solution, immature embryos frequently require sucrose, amino acids, and also phytohormones to continue to grow and carry out the events of embryogenesis. This suggests that the early embryo is not capable of synthesizing many of the molecules it needs for its growth, and other tissues, possibly the endosperm and the suspensor, are essential, not only to provide these nutrients, but also to provide the environment essential for embryogenesis. Contradictory evidence has come from the fact that somatic cells can undergo embryogenesis in culture, without either a suspensor or an endosperm, which suggests that the development of the embryo may be autonomous, requiring only nutrients from the parent plant.

Somatic cells of many plants undergo embryogenesis in special circumstances such as tissue culture    In a few plants, somatic embryogenesis is part of the normal vegetative reproductive strategy of the plant. Plants that will undergo somatic embryogenesis in tissue culture include soybean, maize, and carrot, but it is in the carrot that this phenomenon has been most thoroughly studied. When placed in a culture medium containing basic minerals, sucrose, and the synthetic auxin 2,4-dichlorophenoxyacetic acid (2,4-D), the cells proliferate actively, forming a tissue known as callus, which is largely undifferentiated and lacks recognizable organs. When the callus tissue is transferred to medium lacking 2,4-D, the cells continue to proliferate, but now they do so in a highly organized fashion, with the developing aggregates proceeding through the stages of embryogenesis (Fig. 8.27). It is not clear what actually initiates somatic embryogenesis. Although 2,4-D inhibits embryogenesis in these cultured carrot cells, some as yet unidentified internal signals may initiate this developmental pathway. Whatever it is, it can be lost after the cells have been in culture for a prolonged period. Newly cultured cells produce a very high percentage of embryos, whereas after a year or more in culture, most carrot isolates lose the capacity to form embryos altogether or have a greatly reduced capacity. The stages of carrot somatic embryogenesis, shown in Fig. 8.27, are very similar morphologically to those of zygotic embryogenesis. There are, however, some differences. Although single cultured cells can form embryos, most somatic embryogenesis is initiated from small clumps of cells, rather than from a single cell, as is the case with the fertilized egg. Also, these somatic embryos are formed without any accompanying

**Figure 8.27** **Stages of somatic embryogenesis in carrot suspension cultures**
(A) Cultured cells with proembryogenic cell masses (arrows). (B) Globular embryos. (C) Beginning of the heart stage. (D) Heart-stage embryos. (E) Torpedo-stage embryos. (F) Carrot plantlets resulting from the germination of the somatic embryos. Reprinted with permission from Halperin (1966).

differentiation of, or association with, endosperm tissue, and they lack a suspensor.

Somatic embryogenesis should be a very useful system for determination of the biochemical and molecular events that regulate embryonic development. Studies to understand the molecular basis of embryogenesis, using carrot somatic embryogenesis, are being conducted employing several approaches. For example, the proteins produced by carrot somatic embryos can be compared with those produced by other carrot tissue, using the technique of two-dimensional polyacrylamide gel electrophoresis. Of the approximately 800 proteins that can be resolved by this technique, only about 1 to 2% are embryo specific. Similarly, the screening of cDNA libraries has permitted the identification of a small number of genes whose expression is unique to embryos. At present, the nature of these genes and the functions of the proteins they encode has not been determined. Although the number of stage-specific gene products found in carrot somatic embryogenesis would be consistent with the possibility that these acted as master regulatory proteins, in fact we do not know what role they play in plant embryogenesis, if any. The cDNAs and polypeptides visualized by two-dimensional gel electrophoresis from carrot somatic embryos most likely represent the products of the more strongly expressed genes.

## Embryogenesis and Seed Development

The development of the soybean embryo and seed has been characterized at both the morphological and molecular levels. In the soybean cultivar Forest, completion of embryo and seed development takes 120 days and can be divided into five stages (Fig. 8.28). Stages I, II, and III are the globular, heart, and torpedo stages of embryogenesis. These are completed within the first 30 days of fertilization. By the end of stage III all of the cells of the embryo are in place, along with all of the structural organs of the embryo; however, the embryo is only a fraction of its final size and weight. Storage protein synthesis and deposition have not yet begun. The bulk of the storage proteins is synthesized and deposited in the cotyledon cells during stage IV, the **midmaturation phase,** which results in a rapid increase in the size and weight of the cotyledons. This also can be called the seed development phase as the other structures that characterize the seed, such as the seed coat, are formed at this stage. The seed can increase in fresh weight by a factor of 100 or more during this period. At the end of the maturation phase, the seed enters the dormancy phase, during which transcription of most genes and protein synthesis stop and the embryo loses most of its water. A sequence of events similar to this occurs in other species, although the time frame may be very different. In *Arabidopsis,* the entire process is completed in 7 days. Also, in some species, the storage organ is not the cotyledons, but rather the endosperm.

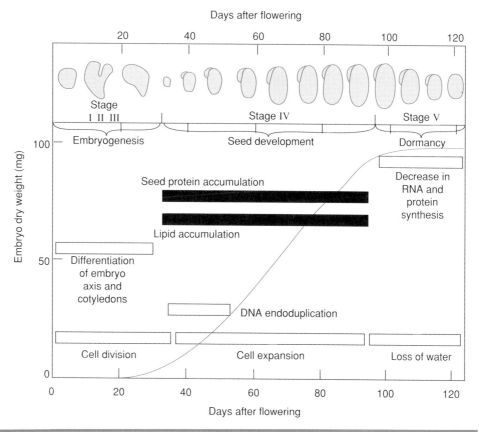

**Figure 8.28 Stages of soybean embryogenesis and seed formation and their duration**
The top row illustrates the appearance of the embryo or seed at the times indicated, but these are not drawn to scale. The early stages of embryogenesis would not be visible at the scale with which the seeds are drawn. Reprinted with permission from Goldberg *et al.* (1989).

## Gene Expression during Embryogenesis

Goldberg and his colleagues examined the mRNA complexity at various stages of soybean embryogenesis. They hybridized cDNAs prepared from mRNA extracted from cotyledon-stage (III) and midmaturation (stage IV) embryos to single-copy soybean genomic DNA. The hybridization kinetics showed that both stages contained 14,000 to 18,000 different mRNAs. In other words, approximately the same number of genes was being expressed in both stages. This is about the same number of genes expressed in a mature leaf or stem. Thus, plant embryogenesis is not accompanied by a reduction in the number of genes expressed as embryogenesis progresses. The abundance of some of the mRNAs was very different in the cotyledon and midmaturation embryos, however. At the cotyledonary stage, there were two abundance classes. One was a class of moderately abundant messages, consisting of about 180 different mRNA sequences, each present about 1000 times per cell. Most of the mRNAs belong to the second abundance class in which each mRNA was present in only a few copies per cell. The midmaturation stage contained, in addition to these two classes, a new class of six or seven superabundant mRNAs, each of which were present in about 100,000 copies per cell. These superabundant mRNAs were transcripts from the genes encoding seed storage proteins.

Quantitative changes occur in gene expression during embryogenesis    There are approximately 15,000 different mRNAs present in midmaturation embryos of soybean. Most of these (about 90%) are also found in the cotyledon-stage and mature embryos. In fact, they persist in the dry seed and are also found in the postgermination cotyledons and in the mature leaf. Only a small number of genes are expressed exclusively in embryos. Some of the embryo-specific genes also are stage specific. There are a group of genes that are expressed only in early embryogenesis, another set expressed only in the midmaturation phase, and another set expressed in late embryogenesis. The last include genes that may be involved in seed desiccation, whereas the genes expressed in the midmaturation phase include those encoding the seed storage proteins. A few genes, perhaps as many as 100 are expressed during the entire seed development period and a few are also exclusively expressed in embryogenesis and seed development (Fig. 8.29). The function of most of these genes is unknown. Some of them may be regulatory genes whose function is to initiate the transition from one stage to the next. The most interesting genes expressed during embryogenesis are those that regulate the major events of embryogenesis. These are difficult to identify in a plant such as soybean, both because the soybean genome is relatively large and because it has a long life cycle, making a genetic approach to the identification of these genes much more time consuming.

How are the genes responsible for the embryogenic program identified?    *Arabidopsis,* with its small genome and short life cycle, is much more amenable to these kinds of studies, and considerable progress toward a molecular–genetic characterization of embryogenesis has been made recently using this organism. The most promising method for the identification of genes responsible for the key events of embryogenesis is the generation of mutations that affect embryogenesis. Although some embryonic lethal mutations could be expected to affect common processes essential for life, in other cases lethal mutations may inactivate genes specifically required for the completion of an important stage of embryogenesis. Additionally, mutations that result in abnormal embryos or seedlings may identify genes that must act at a particular stage of embryogenesis if normal embryonic development is to occur.

Two approaches have been used to generate mutations affecting embryogen-

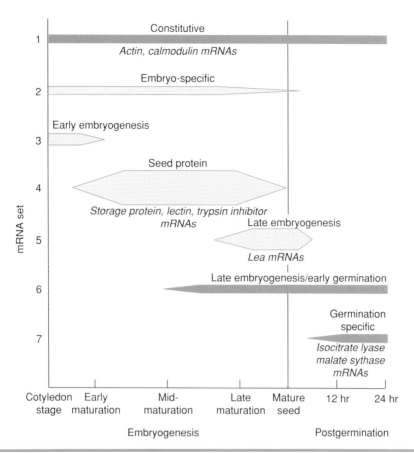

**Figure 8.29 Changes in abundance of different groups of mRNAs during seed development and germination**
Seven different groups, or sets of mRNAs, are depicted as they appear during seed development and subsequent germination. The relative thickness of each line indicates the abundance of that group of mRNAs. Reprinted with permission from Goldberg *et al.* (1989).

esis in *Arabidopsis:* treatment of seeds with chemical mutagens such as ethyl methanesulfonate (EMS), and insertion of T-DNA from the *Agrobacterium tumefaciens* Ti plasmid randomly into genes (see Chapters 3 and 9). With the latter method, seeds are treated with *Agrobacterium* bearing a modified Ti plasmid in which the oncogenes have been deleted and a gene encoding an antibiotic resistance gene has been inserted between the T-DNA borders. Some of the cells of the embryo of the treated seeds will be transformed by the *Agrobacterium*, and T-DNA containing the antibiotic resistance gene will be inserted into some of their genes. EMS mutagenesis usually results in point mutations, whereas T-DNA insertional mutagenesis disrupts the controlling elements or coding sequences of genes. The advantage of T-DNA insertional mutagenesis is that the mutated

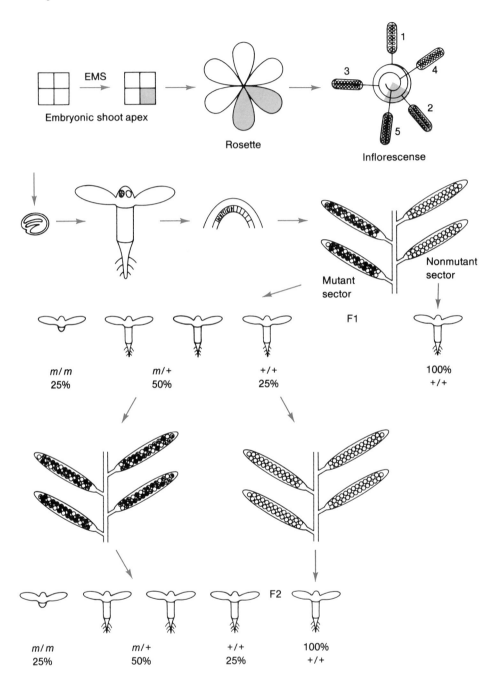

genes are tagged as a result of the insertion of the T-DNA into them, and subsequently they can be found more readily for sequencing by screening for the T-DNA or for the antibiotic resistance gene.

With either technique, only some of the cells of the embryo will contain mutated genes. If the mutation is to be transmitted to future generations, it must occur in a cell that will contribute to the formation of gametes. This means that the mutation must occur in one of the cells of the shoot apical meristem whose progeny participate in flower formation. If it occurs in one of the apical initials, the sector of the plant derived from the shoot apical initial containing the mutated gene will also carry the mutation (Fig. 8.30). In most cases, the mutations affecting embryogenesis have no effect on the characteristics of the plant derived from the mutagenized seed (known as the M1 generation). The progeny of the M1 generation will, however, show the effects of the mutation if it affects a gene essential for embryogenesis. The seeds produced by the M1 generation contain embryos which are the M2 generation, so the effects of the mutation on embryogenesis can be determined directly by examining the developing seed. Usually only one of the two alleles for any given gene will be mutated and the mutated gene will be recessive. As *Arabidopsis* is self-fertile, embryos derived from flowers formed in the mutant sector will either be wild type, heterozygous for the mutant, or homozygous for the mutant gene. The effects of the mutation can be seen directly in the 25% of the seed that are homozygous for the mutation, and, if the mutation is lethal, it can be carried as a heterozygote.

Nine genes have been identified that function in determining the major events of embryogenesis    Mutations in four of these regulatory genes affect aspects of the basic apical–basal pattern of the plant, three genes affect organ size, and two specify the pattern of tissue differentiation. Mutations in the four axis-specifying genes result in a deletion of a major part of the axis. Mutations in the gene *gurke* delete the cotyledons and the shoot apical meristem. Mutations in the *fackel* gene result in the lack of the central part of the axis, so that the seedling consists of cotyledons attached directly to the root (Fig. 8.31). Mutations in the *monopteros* gene result in seedlings lacking both a hypocotyl and a root, and mutations in the *gnom* gene give seedlings lacking roots and cotyledons (see

**Figure 8.30** **Strategy for isolating mutations in genes controlling** ***Arabidopsis* embryogenesis**

Seeds were treated with the chemical mutagen ethyl methanesulfonate at 0.3% for 8 hours and then germinated. In some cases the mutagen will have induced mutations in initial cells in the shoot apical meristem that will ultimately give rise to the seeds. In the F1 generation, one-quarter of seed will be homozygous for the mutant gene and the phenotypic effect of the mutation can be observed as embryo lethality or embryonic deformations. Half of the progeny from the same cross will be heterozygous for the mutant gene. As most of these mutations are recessive, the heterozygotes show no phenotypic effect of the mutation. Reprinted with permission from Jurgens *et al.* (1991).

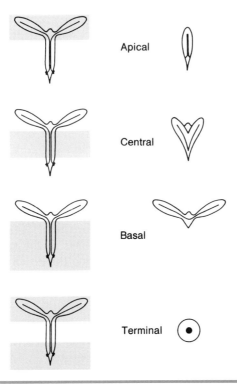

Apical

Central

Basal

Terminal

**Figure 8.31 Mutations in genes controlling *Arabidopsis* embryogenesis that lead to deletions of major portions of the embryo and seedling**

Mutations in the *gurke* gene result in deletions of the apical meristem and cotyledons, whereas mutations in the *fackel* gene bring about deletion of the central (hypocotyl) region, and mutation of the *monopteros* gene deletes the root and root apical meristem. Deletions of both the apical and basal portions of the embryo and seedling result from mutations in the *gnom* gene. Reprinted with permission from Mayer *et al.* (1991).

Fig. 8.31). In the most extreme form, *gnom* mutant seedlings are spherical and lack axial polarity entirely. These genes are responsible for the determination of the apical–basal pattern of the plant, but they have relatively little to do with the determination of the basic tissue types or cell differentiation. For example, *gnom* mutants contain epidermal, ground, and vascular tissue, although the organization of these tissues may be disrupted. The *gnom* mutants differentiate tracheary elements, but these elements are not connected to form a vascular system. Mutations in the genes *knolle* and *keule* result in an altered pattern of tissue differentiation. Specifically, seedlings homozygous for mutations in these genes lack an epidermis. Thus, they are involved in determining the radial pattern of tissue differentiation.

An *Arabidopsis* gene responsible for cotyledon differentiation has been identified    Screening seed produced by *Arabidopsis* plants transformed by *Agro-bacterium* T-DNA led to the identification of a *lec* (for leafy cotyledon) mutant in which the cotyledons had the characteristics of true leaves. The cotyledon primordia are initiated during the transition from the globular to the heart stage of embryogenesis, and not from the apical meristem. Although cotyledons superficially resemble leaves, there are a number of differences between them, in addition to their origin. Cotyledons are storage organs (see below) and are specialized for the storage of proteins and lipids (Fig. 8.32A), whereas leaves are specialized for photosynthesis. Cotyledons of the *lec* mutant lack lipid and protein storage bodies, but contain plastids (Fig. 8.32B). In addition, the pattern of vascular tissue in the *lec* mutants more closely resembles that of leaves than it does the simple vascular pattern of cotyledons (Fig. 8.33). Cotyledons of wild-type *Arabidopsis* also lack trichomes, whereas trichomes are formed from the epidermal layers of true leaves and on the leafy cotyledons of the *lec* mutant as well. Thus,

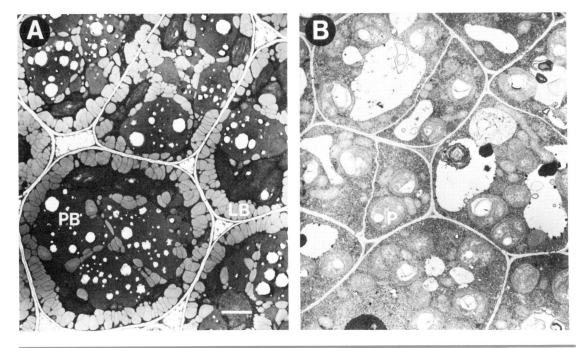

Figure 8.32 **Comparison of cellular structure in *lec* mutant and wild-type *Arabidopsis* cotyledons**
(A) Wild-type cotyledon cells with protein bodies (PB) and lipid bodies (LB). Transmission electron micrograph of a leafy cotyledon of a seed homozygous for the *lec* mutation. Cells show numerous plastids (P) but no protein or lipid bodies. Bar = 2 $\mu$m. Reprinted with permission from Meinke (1992).

  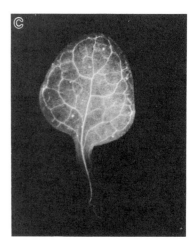

**Figure 8.33 Comparison of the pattern of vascularization in cleared leaves and cotyledons from 14-day-old *Arabidopsis* seedlings**

(A) Wild-type cotyledons. (B) Leafy cotyledons from a plant homozygous for the *lec* mutation. (C) Vascular pattern in the first true leaf of the wild type. Reprinted with permission from Meinke (1992).

a functional *lec* gene appears to be necessary for the development of cotyledons from the cotyledon primordia. When the *lec* gene is nonfunctional, organs develop from the primordia, but they are the wrong organs.

At present, we do not know exactly how these regulatory genes function to specify the major events of embryogenesis. By most definitions of the term *homeotic*, the *lec* gene would be considered to be a homeotic gene because it controls organ identity. When it is mutated, the wrong organ is formed in the wrong place; leaves are formed instead of cotyledons. Clearly the genes identified as essential for the establishment of the radial tissue pattern and axial patterns are important regulatory genes and, by stretching the definition, also could be considered homeotic genes. By analogy with the homeotic genes found in other organisms, it is most likely that these genes encode transcription factors that regulate the expression of other genes. For example, the radial pattern of tissue differentiation within the globular-stage embryo appears to be determined by the *knolle* and *keule* genes. These could be expressed in an opposite pattern within the octant-stage embryo, perhaps with the *keule* gene expressed most strongly in the innermost cells, and the *knolle* gene expressed most strongly in the outermost layer. This could establish a gradient of transcription factors within the embryo that would lead to the expression of genes specific for the program of epidermal cell differentiation in the outer cell layer and a different program within the inner cells. This possibility can be evaluated after these genes have been cloned and sequenced.

## Seed Storage Proteins

Seed storage proteins play an exceedingly important role in the reproduction and survival of angiosperms. In addition to their importance for seedling survival, seed storage proteins are interesting because they are encoded by genes that are very strongly regulated by development. That is, these genes are expressed only during a limited part of the life cycle of the plant and only in cotyledons or endosperm, depending on the species. As a result, they can be considered a model for those genes that determine the functional characteristics of a given cell or tissue type. As considerable progress has been made toward understanding how the expression of these genes is regulated, we will examine these genes and their regulation in some detail.

The seed storage proteins are a unique class of proteins synthesized in response to developmental cues during seed formation. They are proteins that act as a food reserve for the germinating seedling while it is growing heterotrophically, before it can photosynthesize at rates sufficient to sustain itself. They are a source of stored nitrogen and carbon, as well as amino acids. Most of the seed storage proteins of dicots are **globulins;** that is, they are proteins that are insoluble in water, but soluble in concentrated salt solutions. In grasses, the seed storage proteins typically are **prolamines.** Prolamines are proteins that are insoluble in water or concentrated salt solutions, but soluble in alcohol or in detergent solutions. Typically the seed storage proteins are rich in arginine, glutamine, glutamate, and asparagine. Seed storage proteins are often given trivial names, usually derived from the name of the species or genus of the plant in which they were first characterized. For example, the main seed storage globulins of soybean are known as glycinins (after *Glycine max*), whereas the prolamines of *Zea mays* are called zeins.

The expression of the soybean storage protein genes is initiated as the maturation phase begins and their mRNAs come to represent at least 20% of the total mRNA found in the midmaturation embryo. As the embryo enters dormancy, their expression decreases. Although some storage protein gene mRNA may persist in the dormant seed, it disappears during seed germination and is virtually undetectable in the organs of the mature plant. *In situ* localization studies have shown that the storage protein mRNAs are localized in the cotyledons and the embryo axis (Fig. 8.34). They were not detectable in the surrounding nonembryonic tissues or in any of the tissues of the mature plant. One apparent exception is the product of the lectin gene. Lectins are glycoproteins that bind certain sugar residues and have the capacity to stimulate cell division in animal cells. They have been very useful to cell biologists studying the animal cell surface, but their function in the plant is largely unknown. Lectins are produced during seed formation, and they are stored along with the more abundant seed storage proteins in the cotyledons of dicots, such as soybean. In soybean, lectins proteins are found in root epidermal cells, and the lectin protein is synthesized in these cells as a result of the activity of the lectin gene, although at levels much lower than observed in cotyledons during the maturation phase of seed development.

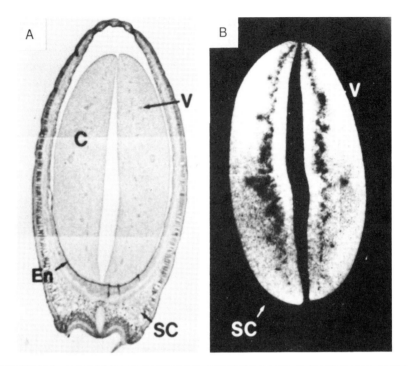

**Figure 8.34 Localization of storage protein mRNA within the cotyledon cells of a developing soybean seed**

(A) Light micrograph of a cross section of a midmaturation embryo. En, endosperm; SC, seed coat; C, cotyledon; V, vascular tissue. (B) Dark-field micrograph of the seed shown in (A) after *in situ* hybridization with a labeled probe for a soybean storage protein mRNA. Reprinted with permission from Goldberg *et al.* (1989). ©Cell Press.

Storage proteins are synthesized on the rough endoplasmic reticulum and deposited in protein bodies   In all cases examined so far, they are first synthesized as a larger precursor which is then processed to the smaller storage protein in the ER. The gene encodes a signal sequence as well as the polypeptide itself. The signal sequence is necessary for the transport of the protein across the ER membrane and is cleaved from the protein during transport. The dicot storage proteins are glycosylated in the ER by the same mechanism that operates in mammalian ER, and this mannose-rich chain of sugar residues is subsequently modified in the Golgi. That is, these storage proteins have the characteristics of secreted proteins. They are transported to the tonoplast and secreted from the cytoplasm into the vacuole. In other cases, such as in grasses, the storage proteins accumulate in membrane-bound vesicles within the cytoplasm. The storage proteins accumulate in dense, crystalloid structures called **protein bodies.**

*Maize storage proteins are prolamines known as zeins*  Zeins are soluble in 70% ethanol alone or in 70% ethanol plus a reducing agent. Proteins extracted by ethanol alone belong to a group known as zein 1. This group is a mixture of proteins with masses of 22,000 and 19,000 Da. These make up the bulk of the storage proteins in the maize kernel. The zein 2 group is extracted by alcohol and reducing agents and includes proteins with masses of 22,000, 10,000, 15,000, and 10,000 Da. The 22- and 19-kDa zeins have an amino acid composition that is 20% glutamine, 20% leucine, 14% alanine, 9% serine, and 7% serine. The 15- and 10-kDa zeins have a similar composition, but also contain a high content of sulfur amino acids, methionine, and cysteine. Box 8.2 outlines zein structure.

Region III is a tandem repeat of a block of 20 amino acids, starting with a string of glutamines. Each of these repeats has an $\alpha$-helical structure, with the polar amino acids distributed in three symmetrical sites. If the repeated $\alpha$ helices are folded back on one another in an antiparallel fashion, the polar groups can form $H^+$ bonds with the adjacent helixes. The nine helices would thus form a cylinder. This probably is very important for the packing of the zein proteins into the protein body.

*Seed storage proteins are encoded by several different multigene families*  Their expression is very highly developmentally regulated. There are more than 20 different zein proteins. Southern genomic analysis has shown that these 20 proteins are encoded by about 65 different genes. The reason there are more genes than proteins is that not all of the genes are expressed. The 22-kDa zein family is highly homologous. There is about 90% sequence homology among the approximately 25 members of this gene family. There is less homology among the 19,000-Da zeins and also more genes. At least 55 different genes encode this group of zeins. There appear to be only about one or two genes each for the 15,000- and 10,000-Da proteins. None of these genes have introns. Zeins are made between 10 and 50 days postpollination, with a peak occurring at about day 35. Synthesis occurs exclusively within the endosperm tissue and primarily within the aleurone layer of the endosperm. These genes are expressed nowhere else in the plant.

---

**Box 8.2**

**Zein Protein Structure**

| Region I | Region II | Region III | Region IV |
|----------|-----------|------------|-----------|
| ATG–signal peptide – | amino terminus – | glutamine repeat – | carboxy terminus |
| 21 amino acids | 36–37 amino acids | 9–10 × 20 amino acids | 10 amino acids |

Several genetic loci have been identified that have a dramatic effect on the synthesis of the zein storage proteins in maize. Several of these loci confer what is known as the opaque phenotype on the maize kernel. Normally, the aleurone layer is translucent where the storage proteins are deposited. The opaque phenotype exhibits an opaque aleurone, primarily as a result of the decreased deposition of the zein storage proteins in this tissue.

Zein is poor in two amino acids essential for human nutrition, lysine and tryptophan. The opaque phenotype was first identified because it increased the lysine and tryptophan contents of the maize seed protein, relative to other amino acids. Further analysis demonstrated that opaque mutants increased the relative percentages of these amino acids because they decreased zein synthesis. The seeds of plants with the opaque phenotype have a very low zein content. The lysine and tryptophan contents of the total seed proteins are increased because the synthesis of other seed proteins is unaffected by the opaque mutations and other proteins have a more normal distribution of amino acids.

Although maize with the opaque mutations are of limited value for human nutrition, they have given us an opportunity to examine some of the factors that regulate zein gene expression, for the opaque mutants are in genes that regulate the synthesis of the zein storage proteins. Mutations in the *opaque*-2 gene result in partial to complete repression of the 22-kDa class of zein proteins, whereas mutations in the *opaque*-7 and *flowery*-2 genes reduce the overall zein content of the seed. The magnitude of the effect of these mutations depends on the genetic background in which the mutants occur. These regulatory genes map to parts of the genome different from that of the zein genes, although in the case of *opaque*-7, the locus is near some of the zein genes. Recently the *opaque*-2 gene was cloned by transposon tagging. Analysis of the amino acid sequence of the protein encoded by the *opaque*-2 gene indicated that it was a transcription factor containing a basic domain and a leucine zipper (bZIP). The *opaque*-2 protein, O2, has a DNA binding domain that binds to the promoter region of the 22-kDa zein genes with a high degree of specificity. The O2 protein also has a leucine zipper dimerization domain, but at present it is not known if the O2 protein binds as a homodimer or a heterodimer. As other genes are known to regulate zein protein synthesis, specifically *opaque*-7 and *flowery*-2, it is possible that the protein product of one of these genes forms a heterodimer with O2 to regulate zein gene transcription, but the nature of these proteins has not been determined at this time.

Gene transfer experiments have been useful in understanding how seed storage protein gene expression is regulated    A number of seed storage protein genes not only have been cloned and sequenced, but have also been transferred to other plants by means of *Agrobacterium tumefaciens* Ti-mediated transformation. Usually tobacco or petunia is used as the host plant for these experiments because both are highly susceptible to transformation by this organism and because they are easily regenerated in tissue culture. For example, the soybean lectin gene has been introduced into tobacco cells and plants have been regenerated from these transformed cells. When the pattern of expression of the soybean lectin gene was examined in the transgenic tobacco plants, soybean lectin mRNA was confined to the developing tobacco seeds in stages of their development analogous to those of the soybean seed where the gene ordinarily is expressed. The foreign gene also was expressed in the root of the tobacco plant at much reduced levels, much as it was in the soybean plant itself. The role of *cis*-acting elements in the regulation of these genes has been investigated by deleting portions of the 5′ promoter region of the genes and transferring them into tobacco, where the effects of these deletions on the expression of the genes is evaluated. This has led to the identification of sequences necessary for the regulation of their expression. For the lectin gene, two regions were found to be involved in the regulation of the expression of these genes. The first 77 bp upstream of the ATG start codon has been shown to be essential for the correct developmental order of the expression of the lectin gene, whereas another region about 1 kb upstream of the start codon appears to be involved in regulating the level of expression of the gene.

Although the specific nature of the mechanisms responsible for the developmental regulation of storage protein gene expression is still being worked out, it is clear that these genes contain specific DNA sequences to which transcription factors bind. The transcription of these genes by RNA polymerase II requires the binding of specific transcription factors to these regulatory sequences. Some of these transcriptional factors may be proteins that are found only within the cotyledons or endosperm. Alternatively, it may be that a unique combination of several more general transcription factors is required for the transcription of these genes. In this event, this particular combination of transcription factors would be present only in these tissues.

## Seed Dormancy

The onset of dormancy in seeds is part of the normal developmental pathway for seed formation. If the seed is to serve its purpose of reproducing and propagating the plant, it must attain a high degree of resistance to cold, drought, and other environmental challenges. The development of dormancy progressively shuts down the cellular metabolic processes or reduces them greatly. This includes most gene transcription and the translation of mRNAs into proteins. In addition, the onset of dormancy is accompanied by the differentiation of protective structures such as the seed coat. Thus, the preparation for dormancy is an active process that involves the transcription of new genes; some of these genes will in turn shut off gene transcription more generally, whereas others may be involved in the formation of the specialized structures of the seed coat.

Seed dormancy is triggered by abscisic acid, which accumulates in the seed. For many species there exist viviparous mutants in which the seeds do not become dormant, but instead proceed directly into germination while still attached to the mother plant and continue their development to become seedlings. In many cases these mutants have been shown to fail to produce abscisic acid. There also is an herbicide known as Fluoridon which is an inhibitor of abscisic acid biosynthesis. When developing seeds are treated with Fluoridon, the seeds become viviparous. These studies strongly indicate that dormancy is triggered by the hormone abscisic acid. We do not know exactly where abscisic acid is produced in the plant, although there is evidence that it is synthesized in the mature leaves and transported into the developing seed, where it accumulates toward the end of seed development.

A group of genes have been identified that are expressed only during the onset of seed dormancy. These have been termed the **late embryogenesis-associated (*lea*) genes.** Abscisic acid has been shown to regulate the expression of many of these *lea* genes. At present we do not know what any of these genes are, but it seems likely that they regulate the events leading to dormancy (Box 8.3). For example, some of the proteins encoded by *lea* genes may be general DNA-binding proteins that turn off the transcription of other genes in a fairly nonspecific fashion. It is somewhat paradoxical that some genes that turn off other genes are being actively expressed, but many examples of this type of regulation have been observed in prokaryotic systems.

### Seed Germination

Seed germination involves the breaking of dormancy and the resumption of the growth of the embryo; that is, the processes that lead to dormancy are reversed. Gene transcription is resumed, protein synthesis begins again, and the rates of respiration and intermediary metabolism increase dramatically. In short, life resumes after a state of suspended animation.

*Germination is triggered simply by imbibition of water in many plants* The immediate cause of germination often is simply the imbibition of water by the seed. There may, however, be other barriers to germination that must be overcome as well. In some species seeds have a photoperiodic requirement for germination. In other species, a hard seed coat must be breached by scarification before water can enter the seed and germination begin. In yet other cases seeds

---

**Box 8.3**

### Steps in the Development of Dormancy in Seeds

1. Loss of water (desiccation)
2. Differentiation of the seed coat
3. Cessation of gene transcription and protein synthesis
4. Reduction of respiration and other activities of intermediary metabolism

contain built-in germination-inhibiting compounds which must be leached away
before germination can begin.

Early seedling growth is dependent on metabolizable substrates stored
in the seed   The early growth of the seedling is heterotrophic. The seedling
initially either cannot photosynthesize, because the chloroplasts have not yet
differentiated, or does not photosynthesize at rates sufficient to sustain its growth.
The stored food reserves represent the source of carbon, nitrogen, and metabolizable substrates that can be used for the energy and biosynthetic needs of the
germinating seedling.

Seeds contain stored food reserves   These food reserves commonly are
oils, proteins, and starch, which must be hydrolyzed to be made available to the
seedling for its growth. In some plants these food reserves are not in the embryo
per se, but rather in adjacent tissues such as the endosperm. In most dicots,
however, the cotyledons are the storage organs. The cotyledons store not only
proteins, but oils and starch as well. They have become specialized as storage
organs. In cereal grains the food reserves are stored in the endosperm during
embryogenesis, but the endosperm cells are dead at maturity. These food reserves
are mobilized during germination. Stored food reserves are complex, water-
insoluble molecules that must be digested where they were deposited. Thus,
the mobilization of the stored food reserves requires hydrolytic enzymes. These
hydrolytic enzymes include proteases, lipases, and amylases. In monocots, these
hydrolytic enzymes are produced and secreted by the aleurone layer of the
endosperm, whereas in dicots they are made by the cotyledon cells during germination.

Hormonal signals regulate the synthesis and secretion of hydrolytic
enzymes by the aleurone   Gibberellins play a key role in the germination of
barley grains and many other seeds. Gibberellins are synthesized in the germinating seedling, or they are released from storage as an inactive conjugate, in response to the imbibition of water. The appearance of active gibberellin in the
aleurone tissues of the seed triggers the *de novo* synthesis of a number of hydrolytic
enzymes (Table 8.1). These enzymes must be actively secreted. The food reserves
are stored in the endosperm, and the hydrolytic enzymes that will break down
these reserves are synthesized in the aleurone layer which surrounds the endosperm.

Synthesis and secretion of these hydrolases require the development
of ER in aleurone and this also is gibberellin-dependent   Barley grains
can be cut in half so that one half contains the embryo and the other contains
only the aleurone layer, surrounding the endosperm. If the half lacking the
embryo is incubated on moist sand, the hydrolases are not synthesized and the
food reserves are not digested. If, however, these same embryoless halves are

## Table 8.1    Hormonal regulation of gene expression in barley aleurone layers

| Gene | Hormone treatment | | | |
|------|------|------|------|------|
| | None | GA[a] | ABA | GA + ABA |
| α-Amylase | | | | |
|   High p*I* | − | + + + + + | − | − |
|   Low p*I* | + | + + + + + | + | + |
| Thioprotease | − | + + + | − | − |
| Nuclease (RNase + DNase + 3′-nucleotidase) | + | + + + | − | ND[b] |
| β-1,3–1,4-Glucanase | − | + + + | − | − |
| Actin | + + | + + | + + | + + |
| Alcohol dehydrogenase | + | − | + | + |
| GA suppressed[c] | + + + | − | + + + | + |
| ABA-induced p29 | + | − | + + + | + |
| ABA-induced p36 (glcNH$_2$-lectin-?) | − | − | + + | + |

[a] GA, gibberellic acid; ABA, abscisic acid.
[b] ND, Not determined.
[c] Unidentified cDNA clones.
Reprinted with permission from Ho *et al.* (1987).

treated with gibberellic acid (GA), the hydrolases are synthesized and the food reserves are digested. The hydrolases are actively secreted from the aleurone cells, and they accumulate in the medium in which the embryoless halves are incubated (Fig. 8.35). Secretion of proteins from cells involves a highly conserved process in which specific messenger RNA molecules are translated on ribosomes that become attached to the endoplasmic reticulum. The newly synthesized proteins accumulate inside the lumen of the ER, where they will be processed, largely through the attachment of carbohydrate chains and the cleavage of an amino-terminal amino acid sequence known as the leader sequence. Before they have been stimulated with gibberellic acid, cells of the aleurone layer have a very poorly developed ER network. One of the effects of gibberellic acid is that the architecture of the aleurone cells is dramatically altered, with the extensive development of an ER network.

Gibberellin and abscisic acid act in the regulation of α-amylase synthesis by barley aleurone cells    The regulation of metabolite mobilization by gibberellic acid is a complex process that involves not only the synthesis and organization of the protein export machinery, but also the initiation of the expression of a number of genes encoding hydrolytic enzymes. For example, at least four different α-amylases are produced by the aleurone cells in response to the hormone. These α-amylases have been the most extensively investigated of the

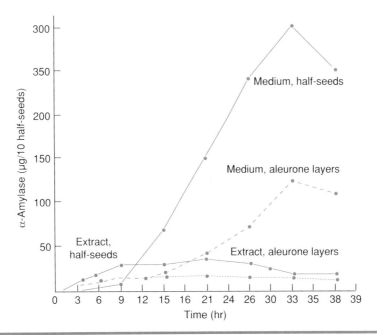

**Figure 8.35 Time course for the appearance of amylase activity in barley half-seeds and in isolated aleurone layers, after adding gibberellic acid at 1 $\mu M$**

Data from Varner, J. E., and RamChandra, G. (1964). Hormonal control of enzyme synthesis in barley endosperm. *Proc. Natl. Acad. Sci.U.S.A.* **52,** 100–106.

hydrolases produced by germinating barley seeds. The four $\alpha$-amylases differ in their isoelectric point (p$I$) and in the time of their maximum appearance after gibberellic acid stimulation. The two high-p$I$ $\alpha$-amylases reach their maximum abundance approximately 12 hours after gibberellic acid treatment, whereas the two $\alpha$-amylases with comparatively low p$I$ values show a slower development and do not reach their maximum expression until 24 to 32 hours after gibberellic acid stimulation. $\alpha$-Amylase activity continues to accumulate in the medium for a long period when aleurone cells or protoplasts are treated with gibberellic acid. The appearance of the $\alpha$-amylase enzymes can largely be blocked by another hormone, abscisic acid (Fig. 8.36).

Expression of the $\alpha$-amylase genes is regulated at the transcriptional level    The genes encoding both the high- and low-p$I$ $\alpha$-amylases are expressed in aleurone tissue during germination and in response to gibberellin stimulation.

# 9

# Apical Meristems and the Formation of the Plant Body

Most plant development occurs after embryogenesis. With the exceptions of the cotyledons and, in some cases, the first leaves, the major tissues and organs of the plant are not formed during embryogenesis. Embryogenesis establishes only a rudimentary plant axis, with the shoot and root apical meristems at either end. The body of the plant is generated after embryogenesis as a result of the activity of the meristems. Although the root and shoot apical meristems are established during embryogenesis, they do not become active until after embryogenesis is complete and seed germination has begun. Then the plant body is constructed by the repetitive divisions and subsequent differentiation of cells produced by these unique structures. Meristems are cell populations that retain some of the properties of embryonic cells, principally the capacity to divide, long after embryogenesis is over. The primary shoot and root apical meristems formed during embryonic development may endure throughout the life of the plant, exhibiting cyclic periods of activity and quiescence. In addition, secondary meristems, such as the vascular cambium, may develop later and contribute additional tissues to the plant body. The developmental events that lead to the formation of almost all of the tissues and organ systems of the plant are a consequence of the activities of the meristems, rather than the unfolding of structures that were formed during embryogenesis. Clearly, the key to under-standing plant development lies in comprehending the nature and activity of meristems.

# The Shoot Apical Meristem

The vegetative shoot apical meristem is a population of small, isodiametric cells with "embryonic" characteristics, located at the extreme end of the shoot axis. The meristematic cell population typically contains a few hundred to a thousand cells, although the *Arabidopsis* shoot apical meristem has only about 60 cells. The shoot apical meristem repetitively produces lateral organs (leaves and lateral buds) as well as stem tissues, but in the course of doing so, it retains its embryonic characteristics. The apical meristem also continuously regenerates itself. Although the shoot apical meristem is at the extreme tip of the shoot, it is surrounded by immature leaves that fold over and cover it. These leaves were produced by the activity of the meristem. It is useful to distinguish the shoot apex, the apical meristem plus the most recently formed leaf primordia, from the apical meristem proper, which refers to the embryonic cell population only and not any of its derivative organs.

If we dissect away the immature leaves that ordinarily obscure the apex and examine it with a dissecting microscope or a scanning electron microscope, the upper surface of meristem itself can be viewed directly (Fig. 9.1A). Often it is a flat or slightly mounded region, 100 to 300 $\mu$m in diameter, surrounded by the most recently formed leaf primordia and young leaves. The size of the shoot apical meristem varies not only among the different species of plants, but also within a given species, depending on its activity at the moment it is examined. The shoot apical meristem is a dynamic structure that changes during its activity cycle and during the seasons. The size and structure of the apex change during its cycle of leaf and stem tissue formation. Additionally, many plants exhibit seasonal growth. They may grow rapidly in the spring, enter a period of slower growth during the summer, and become dormant in the fall. The size and structure of the apex change with this seasonal activity.

The structure of the shoot apical meristem has been studied extensively in sectioned meristems prepared for light or electron microscopy. These approaches show that the meristem is composed mostly of small, thin-walled cells, with a

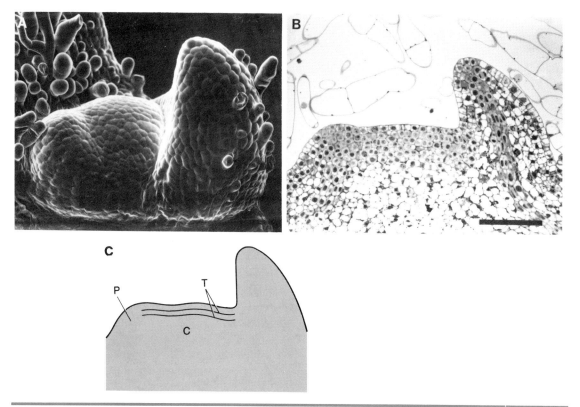

**Figure 9.1   Shoot apical meristem**
(A) Scanning electron micrograph of the tobacco (*Nicotiana tabacum*) shoot apical meristem, after removal of the larger leaves. The apex is flanked by the most recently formed primordia, $P_1$ and $P_2$. ×240. (B) Longitudinal histological section through a tobacco shoot apical meristem with two leaf primordia at approximately the same developmental age as those seen in (A). Bar = 100 $\mu$m. (C) Diagram of the apex illustrated in (B) showing the location of the different tissue regions: C, corpus; P, leaf primordium; T, tunica layer. (A and B) Reprinted with permission from Poethig and Sussex (1985) and (C) from Steeves and Sussex (1989).

dense cytoplasm, lacking large central vacuoles (Fig. 9.1B). Angiosperm meristems usually have a highly stratified appearance in which the cells nearest the surface are organized into two or more distinct layers in which all the cells have the same orientation. These constitute the **tunica** layers of the meristem and are designated L1, L2, and sometimes L3, with the L1 layer outermost (Fig. 9.2C). All cell divisions are **anticlinal** in the tunica layers; that is, they occur so that the new cell wall separating the daughter cells is oriented at right angles to the surface of the meristem. **Periclinal** cell divisions are those in which the new cell wall is formed parallel to the surface of the meristem, or parallel to the axis of

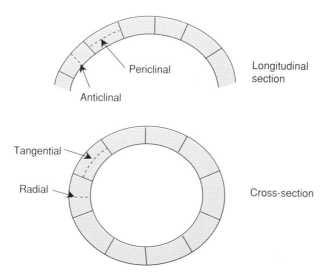

**Figure 9.2    Diagram illustrating the terminology used to describe the orientations of the planes of cell division found in plants**

the organ in which they occur (Fig. 9.2). The tunica layers overlay more internal cells, known as the **corpus,** in which cell divisions occur in other planes. Cell divisions in the corpus may be either anticlinal or periclinal, or they may be oriented indiscriminately with respect to the surface.

Active apical meristems have an additional organizational pattern superimposed on the tunica–corpus-type organization. This has been called **cytohistological zonation** because several more or less distinct zones can be recognized in the apical meristem. Each zone is composed of cells which may be distinguished not only on the basis of their division planes, but also on the basis of differences in size and degree of cytoplasmic density. The center of the meristem is occupied by a cluster of relatively large cells known as the **central zone** (Fig. 9.3). This is flanked by the **peripheral zone,** a doughnut-shaped region of smaller cells that surrounds the central zones. Leaf initiation occurs within the flanks of the meristem, in the peripheral zone (leaf initiation will be discussed in greater detail later). Underneath the central cell zone is the **rib meristem,** which gives rise to the pith. The cytohistological zonation pattern is a reflection of the differing rates of mitotic activity of cells in different regions of the meristem. The duration of the cell division cycle in the different regions of the shoot apical meristem can be determined by allowing the apex to take up tritiated thymidine. Cells undergoing DNA replication will incorporate the radioactive thymidine into their nuclear DNA, and the presence of the radioactivity in the nucleus can be detected by autoradiography of histological sections. By determining the time required for labeled cells to enter mitosis, it is possible to estimate the duration of the

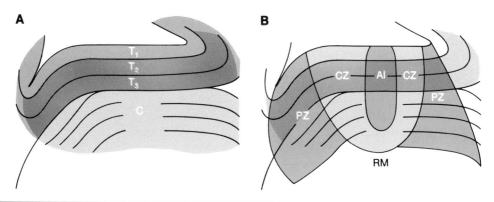

**Figure 9.3    Cytohistological zonation of the shoot apical meristem**
These diagrams of the structure of the ivy (*Hedera helix*) shoot apical meristem show its (A) tunica–corpus organization and (B) the position of the cytohistological zones in relation to the tunica and corpus in the same meristem. C, corpus; T, tunica; CZ, central zone; PZ, peripheral zone; RM, rib meristem; AI, apical initials. Redrawn with permission from Rogers and Bonnet (1989).

cell division cycle in different regions of the apical meristem. This approach demonstrates that cells in the peripheral zone complete their cell cycle more rapidly than those in the central zone of the apex (Table 9.1). As a consequence central zone cells tend to be larger than those of the peripheral zone. When the apical meristem becomes quiescent, as with the onset of dormancy, mitotic activity slows or ceases, and cytohistological zones become difficult to distinguish, or the pattern disappears altogether, leaving only the tunica–corpus organization.

**Table 9.1    Duration of the cell cycle in the various zones of the shoot apical meristem of several angiosperms**

|  | Apical region | |
| Species | Central zone | Peripheral zone |
| --- | --- | --- |
| *Rudbeckia* | >40 | 30 |
| *Pisum* | 70 | 28 |
| *Datura* | 76 | 36 |
| *Trifolium* | 108 | 69 |
| *Solanum* | 117 | 74 |
| *Chrysanthemum* | 140 | 70 |
| *Coleus* | 237 | 125 |
| *Sinapis* | 288 | 157 |

Reprinted with permission from Lyndon (1990).

The vegetative apical meristem is self-perpetuating. Not only does it continue to produce stem tissue and lateral organs, but it also generates itself. An apical meristem can continue its seasonal activity for many years, possibly even hundreds of years in the case of trees. This is possible because some meristematic cells always retain the capacity for cell division and are not committed to any particular developmental pathway, at least as long as the meristem remains vegetative. These uncommitted cells are designated the **apical initials.** When an initial cell divides, one of the daughter cells inherits the characteristics of the initial while the other may enter a developmental pathway. The apical initials reside among the slowly dividing cells of the central zone and represent the ultimate source of all the cells in the meristem and the entire stem.

## The Root Apical Meristem

Unlike the shoot apical meristem, the root apical meristem is not terminal; rather, it is covered by a multicellular structure derived from the root apex known as the **root cap.** The root cap protects the apical meristem as the root forces its way through the soil, but its outer layers are continuously sloughed off as the root grows. The cells of the root cap are not a permanent part of the root. When viewed in longitudinal section, the root is seen to be made up of files of cells, which can be traced to a small group of cells below the root cap that represent the meristem proper. Although it is difficult to define the extent of the meristem exactly in most cases, angiosperm roots often have one to four layers of cells from which all the cells in the body of the root are derived. These layers contain the root apical initials. In some cases, such as radish (*Raphanus sativus*), the layers are readily seen and cell lineage is easily established. Here, the outermost layer of initials produces cells that become the epidermis and the root cap, whereas the middle layer of initials produces the cortex, and the derivatives of the innermost layer become the vascular cylinder. In grasses such as maize there is a separate initial layer for the root cap and the second layer of initials produces derivatives that become both the epidermis and the cortex (Fig. 9.4).

The pattern and frequency of cell division within the root apical meristem have been determined by studying the incorporation of a radioactively labeled precursor of DNA, such as tritiated thymidine. When a root that has taken up tritiated thymidine is sectioned and autoradiographed, the pattern of DNA synthesis is revealed by the presence of reduced silver grains in the photographic film over the nuclei. In a growing root, the cells of the central region of the root apex, including the apical initial cells, are labeled infrequently, indicating that they divide relatively infrequently. This region of relatively infrequent cell division has been termed the **quiescent center** of the root meristem (Fig. 9.5). The term is somewhat of a misnomer as the cells of the so-called quiescent center actually do divide, but with much longer cell cycle times than the surrounding

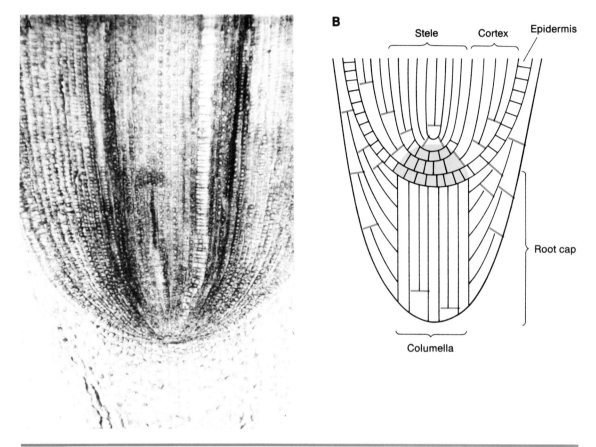

**Figure 9.4   Root tip of maize**
(A) Electron micrograph of the root tip of maize (*Zea mays*), showing the root cap and root apical meristem in longitudinal section. (B) Interpretation of the structure of the root tip, in which files of cells are shown along with the periclinal division that established each file (heavy line). Walls formed by anticlinal divisions are shown only for the epidermis and apical initials. The cells of the vascular cylinder, cortex, and epidermis can be traced back to four layers of initial cells in the meristem (shaded). Reprinted with permission from Lyndon (1990).

cells. For example, the cells of the bind weed (*Convolvulus arvense*) quiescent center have a cell cycle time of 430 hours, whereas those of the root cap have a cell cycle time of 13 hours. This means that the root cap cells will have divided 33 times, with the potential of producing approximately $8.5 \times 10^9$ cells, before cells of the quiescent center will have completed a single division. As a result of this difference in cell proliferative rates, the initial cells must be seen as the ultimate source of cells for the body of the root and not the factory that produces them. The size of the quiescent center varies with the activity of the root, in

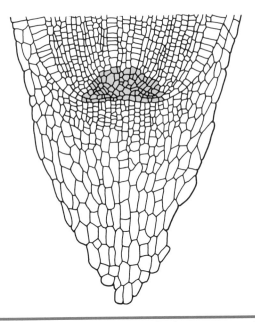

**Figure 9.5**  **Quiescent center of the root apical meristem**
The long cell cycle times of the apical initials create a zone within the meristem with a low tritiated thymidine labeling index, in which relatively few mitotic figures are seen. This region, shown in green, has been termed the quiescent zone. × 150. Reprinted with permission from Clowes (1964).

many cases. For example, when the roots are growing at their maximal rates, they have a large quiescent center that extends considerably beyond the region of the initials; however, when they are examined just as growth is initiated after a dormant period, there is virtually no quiescent center.

## Meristems with a Single Apical Cell

Our understanding of the mechanics of meristems has been greatly aided by investigations of the apical meristems of ferns and other lower vascular plants. The root and shoot apical meristems of these organisms have a single apical cell, which is the ultimate source of all the cells in these meristems and thus the root or shoot body. The apical cell is large and resides at the center of the apex. All the cells in the fern root can be traced to the apical cell through a precise pattern of division of the apical initial cell and its derivatives. Initially the apical cell may divide rapidly, but as growth continues its rate of cell division declines, and in some instances, division may stop altogether. In those instances in which the apical cell stops dividing at a particular point in development, the growth of the organ also stops; it is determinant in its development. The growth and development of the root in the water fern *Azolla* are determinant. The *Azolla* root apical cell is shaped something like a pyramid, with four more or less triangular sides. It divides a total of 55 times, cutting off derivative cells in each of its four planes, before it stops dividing entirely (Fig. 9.6). The derivatives produced from the

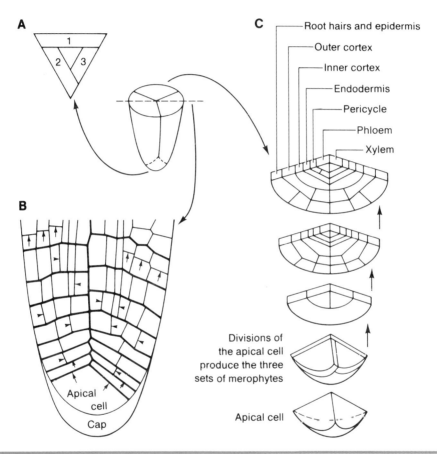

**Figure 9.6** **Tissue formation in the root of the water fern *Azolla*, which has a meristem with a single apical cell**
(A) The apical cell has three faces from which derivatives are formed to construct the three phytomeres of the root. (B) Two of the three phytomeres are shown in longitudinal section (delineated by thick lines). Each phytomere is formed by one of the three derivatives of the apical cell. Horizontal arrowheads point to the cell plates of the formative divisions, which establish cell lines that differentiate with a particular function. Proliferative divisions that increase the numbers of cells of a particular type are shown as vertical arrows. (C) Series of sections through one phytomere at progressively greater distances from the apex, showing the development of cell types. Redrawn with permission from Gunning (1982).

distal face of the apical cell become the root cap, so they do not contribute to the main body of the root. Divisions in the other three planes produces derivatives, each of which divides again a number of times to form one-third of the root axis. The planes and numbers of divisions of each of these apical cell derivatives are very precise. Periclinal divisions of the apical derivatives establish the files of cells that form the various tissues of the root, whereas anticlinal divisions increase the numbers of cells in each file.

## Repetitive Activity of Vegetative Meristems

Plant development often is highly repetitive. This is especially true of the vegetative shoot apex in which meristematic activity results in the formation of a given structure, such as the leaf, and then it repeats this activity again and again. The modular unit constructed by the vegetative shoot apex is not just the leaf, but also the node to which the leaf is attached, the internode below the leaf, and the lateral bud that forms in the axil of the leaf where the leaf joins the stem. These structures are produced as an integrated unit in the flanks of the shoot apical meristem. This modular unit has been termed a **metamere** or **phytomere** (Fig. 9.7). The metamere is an integrated unit that undergoes coordinated development and tissue differentiation. This is particularly evident from the pattern of vascular tissue differentiation within the leaf and stem. The metamere formed by the root apical meristem is less evident, but the primary root meristem cells repetitively divide, often in a very precise pattern, to produce the tissue pattern found in the root axis. The activity of the root apical meristem is highly recursive in the sense that it repetitively constructs the same tissues in the same pattern as long as it remains active and undisturbed.

## Formation of Leaves

Although the phytomere produced by the shoot apical meristem is an integrated unit, most studies on the meristematic activity leading to the formation of the phytomere have focused on the leaf. This is because the leaf primordia can be readily observed as they protrude from the surface of the apex, whereas the nodal and internodal tissues are more difficult to distinguish from the adjacent meristematic tissue, particularly in the early stages of phytomere development. Phytomeres are initiated discontinuously by the shoot apex. At regular time intervals and at specific places in the flanks of the meristem, increased cell proliferation marks the initiation of the leaf primordium and subjacent stem tissue. Cells both in the surface layers and in underlying tissues divide to form the phytomere; that is, phytomere initiation is a multicellular event (see Fig. 9.1). The outermost tunica layer continues to divide anticlinally and its derivatives become the epider-

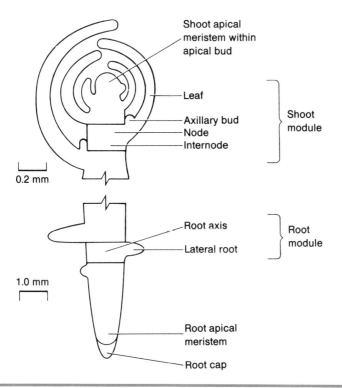

**Figure 9.7 Repetitive construction of the phytomere or metamere by apical meristems**

The phytomere formed by the shoot apical meristem comprises a leaf, axillary bud, node, and internode. The phytomere formed by the root apical meristem comprises the root axis and lateral root. Redrawn with permission from Lyndon (1990).

mis of the leaf and internode. The internal tissues of the leaf are derived mainly from the remaining tunica layers, whereas the nodal and internodal tissues are derived from both corpus and tunica cells. The leaf primordium first becomes visible as a slight protrusion from the apical surface. This outgrowth, which will grow to become the leaf, is known as the leaf buttress. Usually it is formed away from the center of the apex and slighly above and between two older primordia.

Leaves are formed in a highly characteristic, genetically determined pattern, or **phyllotaxy.** Most plants have one of three different patterns of leaf placement. Leaves may be initiated singly from the apex so that there is one leaf at each node and the phytomere consists of a single leaf, node, and internode. Alternatively, more than one leaf may be initiated at a time so that the phytomere consists of a node with two or more attached leaves. Plants with two leaves per node are said to be **decussate;** plants with **whorled** phyllotaxy have three or more leaves per node. All members of the mint family (Labiatae) have decussate phyllotaxy and each successive leaf pair is formed at right angles to the previous pair. As a result the adult leaves, as well as the primordia, are stacked in four vertical rows, called **orthostichies,** along the stem.

Plants that produce a single leaf at each node initiate leaf primordia in a pattern that results in the arrangement of the leaves in a helix around the stem. This is known as the **generative spiral.** In addition, if a sufficiently large segment of the stem is examined, additional regular rows of leaves can be distinguished because the positions in which leaves are formed within the helix recur with a highly regular pattern. Two different sets of intersecting rows of leaves can be distinguished. These are the **contact parastichies.** One set spirals clockwise down the axis, and the other set spirals in the opposite direction. As the phyllotactic pattern of the mature stem is a reflection of the pattern of leaf initiation by the shoot apical meristem, this pattern also can be determined by examining the arrangement of leaf primordia around the apex (Fig. 9.8). In a given species there are a set number of contact parastichies in one direction and a different number in the opposite direction. The phyllotaxy of a given species can be expressed as the number of contact parastichies in each direction. For example, the phyllotaxy of potato is 2 + 3 because it has two parastichies in the clockwise direction and three in the counterclockwise direction, whereas the fern *Dryopteris*

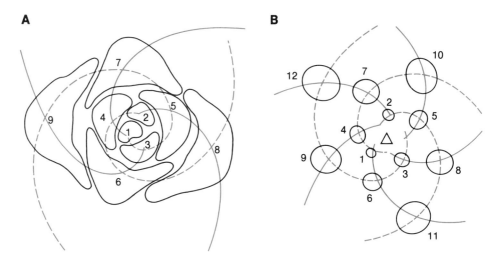

**Figure 9.8** **Phyllotaxy in the shoot apex of (A) the potato (*Solanum tuberosum*) and (B) the fern *Dryopteris dilatata***
There are two parastichies in the clockwise direction (dashed color lines) and three in the counterclockwise direction (continuous color lines) in the potato shoot apex, so the phyllotaxy is 2 + 3. The fern has three parastichies in the counterclockwise direction and five in the clockwise direction, so the phyllotaxy is 3 + 5. Redrawn with permission from Steeves and Sussex (1989).

*dilatata* has 3 + 5 phyllotaxy (see Fig. 9.8). Phyllotactic patterns occur in numbers that are part of the regular mathematical progression known as the Fibonacci series: 1 + 1 + 2 + 3 + 5 + 8 + 13 + 21 etc. In many cases, the number and placement of leaves by the apical meristems are very stable characteristics of a species. In fact, there are whole plant families in which all species have the same phyllotactic pattern in their vegetative organs. In other cases, the phyllotaxy varies with the size and activity of the apex and during normal vegetative development (Fig. 9.9). Also, it is usually the case that the phyllotaxy of floral organs differs from that of vegetative shoots.

## Apical Initial Cells, Stem Cells, and Cell Lineages

### Concept of the Stem Cell

Some of the cells within a vegetative meristem appear to have many of the properties of the "stem cells" that are the progenitors of various animal cell lineages. All the cell types found in blood, for example, are ultimately derived from a small population of stem cells found in the bone marrow in adult mammals. The stem cells are capable of dividing indefinitely and are pluripotent, meaning that their derivatives can differentiate as red blood cells, white blood cells, granulocytes, and platelets, but they cannot form other cell types in the body. These stem cells are relatively few in number and they divide very slowly. Their derivatives, in contrast, divide relatively rapidly, but their capacity for division is limited, and ultimately the derivatives stop dividing altogether. The derivatives of the stem cells also are more limited in the range of cell types they can form.

### Apical Initial Cells

Apical initials, superficially at least, are similar to stem cells. They divide infrequently, they are capable of regenerating the meristem if it is damaged, and they are the ultimate precursors of all the other cells of the root or shoot. Presumably this is precisely what many investigators meant when they described certain cells within apical meristems as apical initial cells; however, this designation implies that we have more knowledge about the potentials of these cells than we actually do. Historically there has been considerable debate about both the number and even the existence of the apical initials. To what degree are the cells that have been called apical initials equivalent to the "stem" cells that are the precursors for different animal cell lineages? Before we accept apical initial cells as the plant equivalent of stem cells, we must address three questions:

1. Do the apical initial cells retain the capacity to divide indefinitely, while their derivatives may divide a limited number of times, but then differentiate, never to divide again during normal plant development?

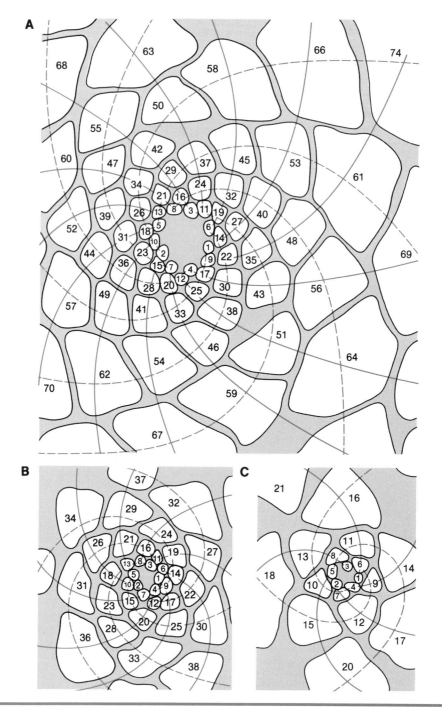

**Figure 9.9** **Variation in phyllotactic patterns among different branches on the same individual of the star pine (*Araucaria excelsa*)**

Each of these phyllotactic patterns is characteristic of a particular level of branching: (A) The main shoot has an 8 + 13 pattern of parastichies. (B) The primary branches have a 5 + 8 pattern. (C) The secondary branches have a 3 + 5 pattern. Redrawn with permission from Steeves and Sussex (1989) as redrawn from Church (1904).

2. If the apical meristem contains true initials in the preceding sense, are they pluripotent in the sense that their fate is partially determined and that they produce derivatives that are already committed to differentiate as particular cell types?

3. If each initial can only form a limited number of different cell types, how many different initials are there and what is the range of cell types that each type of initial can produce?

## Determinant Meristems

Basically these questions are about cell lineage. If there are permanent apical initial cells that give rise to specific tissue and cell types, it should be possible to demonstrate this by determining the cell lineage patterns within the plant. Can we demonstrate that specific tissues and/or cell types are always derived from specific apical initial cells? The conclusions obtained from these studies differ radically depending on whether the species examined exhibits determinant or indeterminate growth. A **determinant** plant is one that produces a limited amount of vegetative growth before it flowers. Flowering often ends vegetative growth, so the activity of the vegetative shoot apical meristems of a determinant plant do not continue indefinitely. Annual plants usually are determinant. Investigators working with determinant plants, such as sunflower and maize, have concluded that these plants lack permanent initials. Although the fate of the apical cells is not strictly determined at the time of seed germination in determinant species, generalized fate maps can be constructed for the cells within the meristem of the embryonic meristem, with concentric rings of cells giving rise to progressively more apical structures as the plant matures (Fig. 9.10).

## Indeterminate Meristems

In contrast, plants exhibiting prolonged **indeterminate** growth, such as ivy and juniper, have a small number of initials in each layer of the meristem, perhaps three. These initials are not necessarily permanent in the strict sense of the word. Shoot apical meristems have one of two different kinds of initials, depending on the species. In lower vascular plants, whose apical meristems have a single apical initial cell, division of the initial cell produces two very different daughter cells. One is large and retains the function of an initial; the other daughter is much smaller and lacks the capacity to divide indefinitely. If the meristem is indeterminant, after every division of the initial, the larger daughter always acts as an initial and thus the initial cell line may be permanent. Of course, there are some meristems in which the apical cell stops dividing after a specific number of divisions, as it does in *Azolla* roots. In these cases the meristem is determinant and the apical cell is an impermanent initial, but this is a somewhat unusual case. More often, meristems with a single apical cell are indeterminate and the apical cell is a permanent initial.

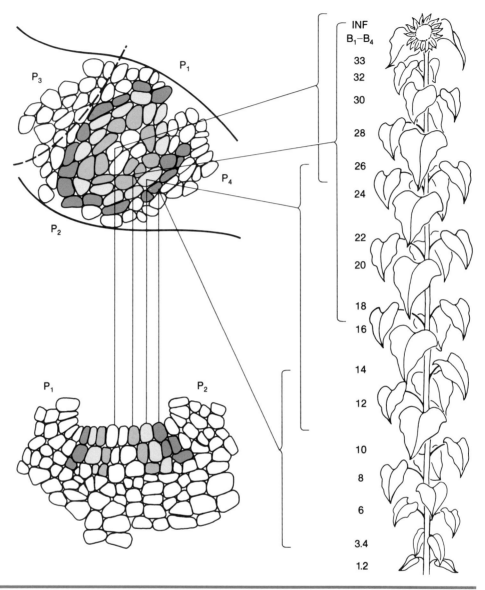

**Figure 9.10 Fate map for the cells of the shoot apical meristem that are present in the seed of sunflower before germination**

The mature sunflower plant is shown on the right, with the nodes numbered. INF, inflorescence; $B_1-B_4$, involucral bracts around the inflorescence. The cells of the meristem giving rise to the different tissues and organs of the plant were determined by the analysis of radiation-induced sectorial chimeras. The meristem was found to contain concentric rings of cells, each of which produced the tissues and organs of progressively more apical portions of the plant. The shoot apical meristem of the dry seed is shown in the two drawings on the left in a surface view (upper drawing) and in transverse section (lower drawing). The four concentric rings of meristematic cells are indicated by different degrees of shading, with lines drawn to indicate the portions of the plant derived from each of these layers. The meristem is composed of three tunica layers, all of which contribute to the formation of the tissues and organs found in each section of the plant. $P_1-P_4$, leaf primordia present in the meristem of the dry seed. Reprinted with permission from Jegla and Sussex (1989).

In contrast, in most angiosperms the apical meristems contain several initial cells and these cells exhibit what is known as stochastic division kinetics; that is, the initial cells divide to produce a population of daughter cells, some of which are selected as the next generation of initials and others that differentiate. The selection of daughter cells to act as initials is, however, random. It is possible that one, both, or neither of the progeny of a particular initial cell will continue as an initial cell. As there is a finite probability that both progeny of a given apical initial will lose the capacity to act as a stem cell, no cell can be permanent initial; however, a given cell may remain as an apical initial for a substantial period of time. Plants with indeterminate growth may be said to have persistent apical initials.

## Chimeric Meristems

*Chimeras* were monsters of Greek mythology with the body of a goat, tail of a serpent, and head of a lion. A chimeric meristem is composed of layers or sectors of cells of different genetic composition, making it possible to identify these cells and their derivatives. Plants with chimeric meristems have been very valuable in cell lineage studies. Chimeras can be produced by a variety of means. The alkaloid colchicine can be used to create chimeric apical meristems in which some layers or regions are polyploid as a result of a colchicine-induced increase in the number of chromosomes in the cells that happened to be dividing at the time of colchicine treatment (Fig. 9.11). Chimeras also can be created by treating plants with ionizing irradiation, such as x-rays, at levels that induce mutations in some but not all cells.

Many variegated plants, including common cultivars of ivy and geranium, are green–white chimeras in which the white or yellow sectors are produced by mutant cells containing defective genes for chloroplast development. The white tissues arise from a different part of the meristem than the green tissues, so the cell lineage of specific structures can be determined. Mutations leading to a deficiency of chloroplast development are particularly useful, as it is possible to determine whether or not the cells are green (normal) or white (mutant) by visual inspection of the plant (Fig. 9.12).

Two types of chimeras are useful for these studies; periclinal and sectorial (Fig. 9.13). Periclinal chimeras occur or can be induced in angiosperms because their shoot apical meristems are composed of concentric cell layers, in which divisions are oriented in one plane. As discussed earlier, although the number of layers varies somewhat, it is common to find meristems with two or three distinct tunica layers, designated $L_1$, $L_2$, and $L_3$. The $L_1$ layer gives rise to the epidermis, whereas the other layers produce the more internal tissues of the stem. Periclinal chimeras are mosaics in which the genetic composition of one or more of these meristem layers differs from that of the other layers. Mericlinal chimeras affect only a sector of one tunica layer and the derivatives of these cells. Sectorial chimeras are similar, except that a portion of the cells in several layers

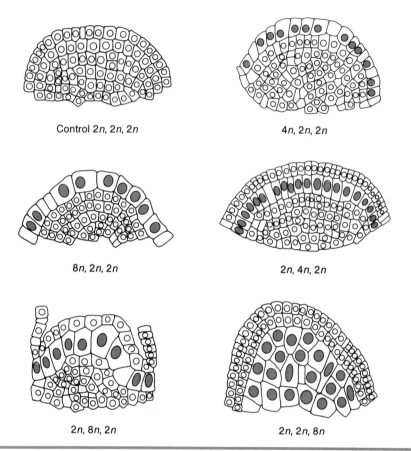

Control 2n, 2n, 2n

4n, 2n, 2n

8n, 2n, 2n

2n, 4n, 2n

2n, 8n, 2n

2n, 2n, 8n

**Figure 9.11 Polyploid cells forming periclinal chimeras in the shoot apices of *Datura* after colchicine treatment**
The ploidy of each layer of cells within the apex is shown. Diploid nuclei are depicted as open circles; polyploid nuclei are shown as filled circles. Redrawn with permission from Satina *et al.* (1940).

differs genetically from the remainder. In either case, the lineages of cells derived from mutant cells can be determined.

## Cell Differentiation and Cell Lineage

The organization of apical meristems indicates that one or more initial cells are the ultimate source of cells in any particular tissue. Specific initials usually can be identified for each of the three root tissues, epidermal, vascular, and ground, and the epidermal tissues in the shoot obviously are derived from a small number of initial cells in the $L_1$ layer. In other cases it is more difficult to identify the

**Figure 9.12 A sunflower plant, with albino tissues forming a mericlinal chimera**
Irradiated dry seeds induced a mutation in a cell occupying the $L_2$ tunica layer. The mutation prevented normal chloroplast development so the tissues derived from the mutated cells appear white. The mutated cell gave rise to a sector of the plant that extends from the 18th node through the inflorescence. Reprinted with permission from Jegla and Sussex (1989).

initials, but the pattern of tissues produced by the meristem suggests they are derived from a small group of cells, some of which probably are its initials. What is the significance of the fact that each plant tissue has its origin in a specific initial or small group of initials? There are at least two different possible explanations for this relationship:

1. The initials may be pluripotent "stem" cells with a rigidly fixed range of developmental possibilities. That is, initials of the outermost tunica layer may be

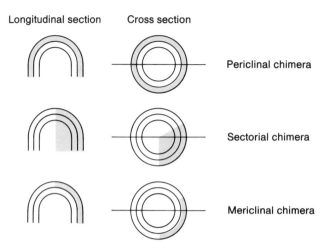

**Figure 9.13** **Diagram of the shoot apex in chimeric plants**
The cells with a different genetic composition are shown in green and the position
these cells would occupy within the apex is shown in longitudinal and cross section.
The concentric rings represent the tunica layers of the meristem or their derivatives in
the plant body. Redrawn with permission from Steeves and Sussex (1989).

determined so that their derivatives can only differentiate as epidermal, stomatal,
or trichome cells, whereas the initials of other layers are determined to form a
different range of cell types.

2. The initials may be multipotent stem cells, producing derivatives whose
fate has not yet been determined. The initials of each layer are only generally
determined to be shoot or root cells, and the specific fate of each derivative is
determined later in development by its chemical and physical milieu, including
its interaction with its neighbors. In other words, the fate of the cell is determined
by its position within the plant body and it is determined gradually as a result
of a prolonged exposure to the morphogenetic stimuli.

The bulk of the evidence supports the latter view. Even in those plants with
persistent apical initials, it is clear that cell differentiation is not dependent on
cell lineage. As the pattern of cell division within the apex is very predictable,
the different tissues within the mature plant can be traced to specific layers or
sometimes even to specific cells within the meristem. Even though a particular
cell layer such as the $L_1$ of the tunica usually does give rise to epidermal cells,
there is no evidence that the derivatives of the $L_1$ layer are committed to become
epidermal cells before they actually begin to synthesize epidermal waxes and
take on the other characteristics of this tissue. The plane in which a cell divides
determines the position of its daughter cells within the plant, and this in turn
plays the most significant role in deciding the fate of the daughters. The strongest
evidence for the importance of position in determining a cell's ultimate fate
comes from an examination of the fate of cells that are displaced from their

usual position so that they come to occupy a different layer. The vast majority of the divisions in the tunica layers are anticlinal, which of course is responsible for creating the layer in the first place. Nevertheless, occasional periclinal divisions occur so that one derivative comes to occupy the next tunica layer. This does not alter the composition of the tissue derived from this layer. Instead, the derivatives assume a function appropriate for a cell occupying that location. Figure 9.14, for example, shows a leaf from a periclinal chimera of tobacco that had a mutation in the apical initial cells of the $L_2$ tunica layer so that none of its derivatives could form chloroplasts. Chloroplasts are able to develop in cells derived from the $L_1$ and $L_3$ layers, however, as these cells lacked the mutation. The majority of cells of photosynthetic mesophyll tissue are derived from the $L_2$ layer, so the leaves of this chimeric tobacco contain a great deal of white tissue; however, they are not totally white, but rather they are variegated, with sectors of green tissue which must be derived from $L_1$ or $L_3$ layers. Occasional periclinal divisions in the $L_1$ and $L_3$ layers early in leaf development established clones of normal cells that differentiate as a green mesophyll cells. This tells us that plant cell differentiation is not dependent on cell lineage. The fate of a cell during development is determined by the position it occupies in the plant body. It also tells us that plant development is highly regulative.

## Determinant versus Indeterminate Development

Leaves, stems, petals, stamens, and most other plant organs exhibit determinant growth; that is, they grow to some genetically determined maximum size and then cease growing. Even though there may be considerable variation in the size a particular organ of a given species may attain, there is a maximum limit that is not exceeded. These determinant organs are initiated by a group of dividing cells in the flanks of a vegetative or floral meristem. At some point these cells begin to behave as an integrated unit, with an altered axis of growth and a more restricted developmental potential than the apical meristem proper. At least some of the cells of the vegetative meristem proper retain the capacity to divide indefinitely, perhaps for many years; however, the derivative cells that form the leaf or internode will divide only a limited number of times before growth ceases. The cells of a primordium divide in a characteristic and complex pattern to form an organ such as a leaf, but at some point all mitotic activity contributing to the body of the leaf ceases, and any subsequent growth occurs by cell enlargement only.

### When Does a Meristem or Primordium
### Become Determined?

The difference between the determined state and the indeterminate state can be demonstrated by examining the developmental potential of apical meristems and the primordia derived from them after they have been removed from the

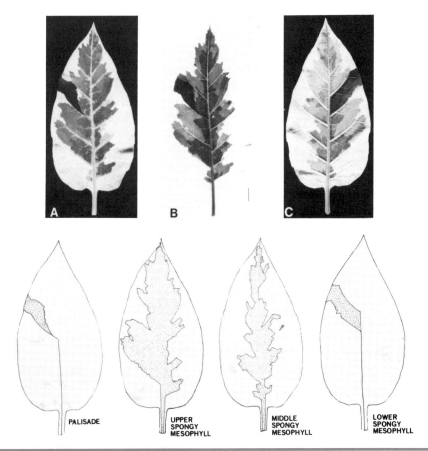

**Figure 9.14 Contribution of different layers of the meristem to the formation of tissues within a tobacco leaf**

X irradiation induced a mutation in the $L_2$ layer of the meristem so that the cells derived from this layer were unable to form chloroplasts. As a result, the origin of the tissues of the leaf could be determined from this green/white/green periclinal chimera. (A) Upper surface of a leaf photographed with reflected light. (B) Same leaf photographed with transmitted light. (C) Underside of the leaf photographed with reflected light. The white areas are tissues in which chloroplast development did not occur and thus were derived from the $L_2$ tunica layer. The dark regions (green) were derived from either the $L_1$ or $L_3$ layer. The differences in shading occur because the leaf tissues that are nearer to the camera appear darker when photographed in reflected light (A,C). In transmitted light the darkest regions result from overlapping layers of green cells in the upper and middle spongy mesophyll. The lower series of drawings depicts the distribution of green tissue (stippled) in the four different mesophyll layers. The upper and lower epidermis is derived from the $L_1$ layer and would contain green guard cells, but the chloroplasts do not develop in the bulk of the epidermal cells in either the wild-type or mutant plants. Reprinted with permission from Poethig (1987).

plant and placed in tissue culture. In a number of cases, the isolated apical meristem, without any lateral organs, will grow and develop in tissue culture to form a complete shoot (Fig. 9.15). Also, the quiescent center of maize roots will develop in culture to form a complete root. The isolated tissue of the quiescent center directly regenerates the root apical meristem, which then grows into a root (Fig. 9.16). Similarly, cultured leaf primordia develop into mature leaves. We do not know precisely when a primordium becomes committed to be a leaf, but very young leaf primordia in some cases are uncommitted; that is, they develop into vegetative shoot meristems instead of leaves when cultured *in vitro*. Older primordia clearly are determined and develop into normal leaves when cultured in isolation from the rest of the plant (Table 9.2).

## Changing the State of Meristems

Entire apical meristems also can be determinant in their development. The best common example of this is the floral meristem, which produces successive whorls of primordia, each of which develops into a different kind of organ (sepals, petals, stamens, and carpels), before its meristematic potential is exhausted and growth ceases. Additionally, some vegetative meristems are determinant, particularly in the lower vascular plants. We have already discussed one of these, the root apical meristem of the water fern *Azolla*, in which the apical cell divides only 55 times before the root reaches its maximum length and growth ceases. Derivatives of the apical cell also divide during root development, but at maturity all cell division stops, including division of the apical cell.

As vegetative meristems may be transformed into floral meristems (see later), it is clear that a meristem can lose the indeterminant state. There are not many examples of determinate meristems reverting to the indeterminate state, however. One well-documented example of such a reversion is shown in Fig. 9.17. It is surprising that there are not more examples of such transformations given the facts that virtually all living plant cells are totipotent and highly differentiated plant cells can, in some circumstances, dedifferentiate to produce tissues, from which whole plants will regenerate in tissue culture. Apparently, although retaining the potential for a high degree of developmental plasticity, once the cells of a primordium or a whole meristem have had their developmental fate determined, this state is highly stable and, although it can be reversed on cell division in tissue culture, it is difficult to reverse in the normal course of development.

# Gene Expression Regulating Meristem Function

We know very little about the molecular mechanisms responsible for meristematic activity in plants. A number of different classes of genes could be expected to be expressed preferentially in the functioning meristem. First, there are genes

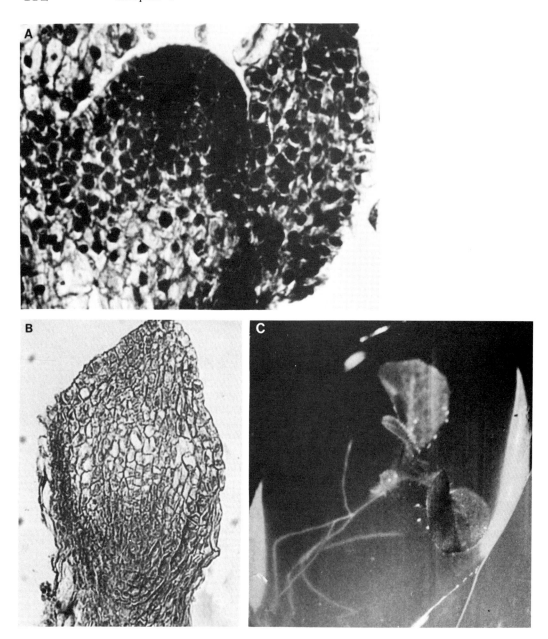

**Figure 9.15 Development of a plant from the cultured apical meristem of tobacco (*Nicotiana tabacum*)**

(A) Meristem immediately after its excision (×395). (B) The meristem after 12 days in culture, showing a leaf primordium in the upper right (×130). (C) A plant with roots that has developed from a cultured meristem. (D) Rotated plant that developed from a cultured meristem. Reprinted with permission from Smith and Murashige (1970).

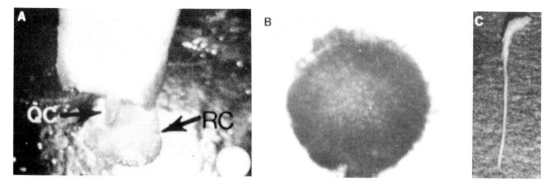

**Figure 9.16** **Regeneration of a root from an isolated maize root quiescent center in tissue culture**
(A) Maize root with a partially removed root cap (RC), showing the location of the quiescent center (QC), which was isolated and placed in tissue culture. (B) The isolated quiescent center. (C) Root regenerated from the cultured quiescent center. Reprinted with permission from Feldman and Torrey (1976).

involved in cell proliferation. These would include genes that regulate the cell cycle, such as the plant equivalents of the yeast *cdc* and cyclin genes, as well as genes encoding proteins required for DNA replication and the formation of the mitotic apparatus. A second group of genes would be those involved in determining cell fate. These would be the plant equivalents of homeobox genes and

**Table 9.2** **Development of leaf primordia of the fern *Osmunda* during culture on a simple medium containing sugar and mineral salts[a]**

| Primordium designation | Number forming leaves | Number forming shoots | Number not growing | Total number tested |
|---|---|---|---|---|
| $P_1$ | 2 | 7 | 11 | 20 |
| $P_2$ | 2 | 12 | 6 | 20 |
| $P_3$ | 4 | 10 | 6 | 20 |
| $P_4$ | 4 | 11 | 5 | 20 |
| $P_5$ | 8 | 11 | 1 | 20 |
| $P_6$ | 12 | 8 | 0 | 20 |
| $P_7$ | 16 | 4 | 0 | 20 |
| $P_8$ | 17 | 1 | 1 | 19 |
| $P_9$ | 19 | 0 | 1 | 20 |
| $P_{10}$ | 20 | 0 | 0 | 20 |

[a] The youngest visible primordium is designated $P_1$, and the oldest is $P_{10}$. The numbers of cases in which these developed into leaves or shoots are indicated. In some cases the primordium did not grow (Steeves, 1966).
Reprinted with permission from Steeves (1966).

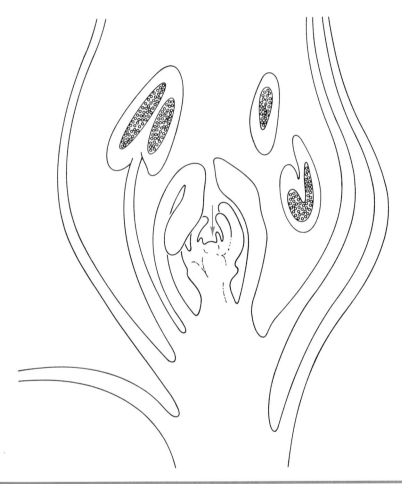

**Figure 9.17 Reversion of a floral bud of *Impatiens balsamina* to vegetative development after it had initiated stamens, but before the development of carpels**
The arrow points to the vegetative apex, which has begun to initiate leaves. Redrawn with permission from Krishnamoorthy and Nanda (1968).

other transcriptional regulators. Finally, there would be a class of genes whose expression is limited to specific tissues and organs and whose products are involved in the function of these structures. Some of these meristem-specific genes have been cloned and characterized from a subtraction cDNA library prepared from developing cauliflower heads. One of the problems in working with meristems at the molecular level is getting enough material to work with, as meristems normally represent an extremely small part of the total plant mass. Cauliflower, however, provides an answer to this problem. Cauliflower is a mutant cultivar of *Brassica oleracea* in which there is an abnormal proliferation of meristems. Instead of forming leaves, each meristem produces additional meristems in its flanks. The cauliflower head is composed of perhaps as many as $10^6$ meristems, so it represents an unusually abundant source of these structures. Although their development is abnormal, they have many of the important properties of meristems.

## Genes Expressed in Meristems

A number of meristem-abundant cDNAs have been isolated from cauliflower meristem cDNA libraries. The sequences of most of the meristem-specific or meristem-abundant cDNAs have not revealed any clues as to their function, but one of them, designated *meri-1*, encodes H3 histone. As histone gene expression and protein synthesis are closely tied to the cell cycle, this would be an example of a gene whose activity we expected to be coordinated with the proliferative functions of the meristem. The cauliflower meristem-abundant cDNAs have been shown to have homologs in *Arabidopsis* where they are encoded by small multigene families. The promoter regions of some of the genes encoding these meristem-specific cDNAs have been fused to a promoterless *GUS* reporter gene and this construct has been introduced into tobacco plants by *agrobacterium tumefaciens*-mediated transformation. The promoter of one known as *mer-5* directed *GUS* gene expression in the apical dome of the transformed tobacco, and the promoter of *meri-1*, the H3 histone gene, directed *GUS* expression primarily in the flanks of the meristem, which of course represents the region with the highest rates of cell proliferation.

## Determination of Cell Fate in the Functioning Meristem

The determination of cell fate apparently involves the acquisition of new functions that enable cells to specialize, as well as an eventual loss of the ability of the cells to divide. The genes for cell proliferation are turned off as other genes that specify the differentiated characteristics of the cell are turned on. The cessation of cell proliferation may not occur immediately on the initiation of a particular pathway of organ or tissue development, but the proliferative potential is limited, at least while the organ occupies its normal position on the plant. Recently a gene that is involved in cell fate determination was cloned and partially character-ized from maize. The maize mutant known as *knotted* (*Kn*1) is a dominant mutation that has a dramatic effect on leaf morphology. Plants carrying the *Kn*1 mutation have grotesquely distorted leaves as a result of abnormal cell divisions along the veins of the leaf. The *Kn*1 mutations affect a region of the maize leaf known as the ligule. The ligule is a tissue outgrowth at the juncture between the blade and the sheath, the specialized basal region of the leaf that wraps around the stem. Abnormal cell divisions that result in knot formation, however, occur primarily within the vascular bundles of the veins as they run through the ligule region.

Not only do these abnormal cell divisions occur after normal cell proliferation has ceased in the leaf blade, but the division planes also are abnormal. This late cell division distorts the vascular tissues and forms knots that protrude from the leaf surface (Fig. 9.18). Cell differentiation is relatively normal in the mutant

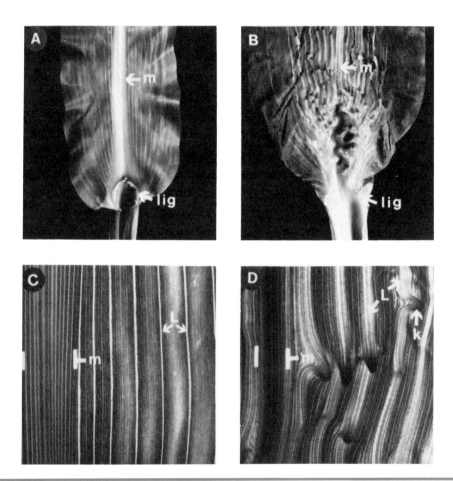

**Figure 9.18  Leaves of plants exhibiting the knotted (*Kn*1−0) phenotype, as compared with the leaves of wild-type maize**
(A) The eighth leaf of a wild-type plant near the point where the ligule (lig) separates the sheath and leaf blade. The veins of the wild-type normal leaf are all parallel to the midrib (m). (B) A comparable leaf from a plant carrying the dominant *Kn*1−0 allele. Cell divisions late in leaf development have distorted the veins, sheath, and ligule. (C,D) Closeups of the leaves shown in (A,B), respectively. There are 8 to 10 intermediate veins between the lateral veins (L) in the wild type. There are knots (k) sporadically throughout the leaf in the mutant. Reprinted with permission from Vollbrecht *et al.* (1991).

phenotype, except in the vicinity of the knots. As cell division normally continues in the sheath after it has stopped in the blade, the *Kn*1 mutants are in some ways exhibiting sheathlike characteristics in the blade. The *Kn*1 mutations are gain-of-function mutations. That is, the mutant phenotype results from the incorrect expression of the gene rather than the loss of its expression in normal development. One of the mutant *Kn*1 alleles, known as *Kn*1-0, which has been cloned and sequenced, is the result of a tandem duplication of a 17-kb region containing the gene. Other *Kn*1 mutations result from alterations in noncoding regions of the gene that would affect the tissues in which the gene is expressed, without interfering with the ability of the protein product of the gene to function. This is consistent with the fact that the *knotted* phenotype results from gain-of-function mutations in which the gene is expressed at the wrong time in development.

## Homeodomain Proteins Determining Cell Fate

The *Kn*1 gene contains a conserved region known as the **homeobox** that is present in genes that control important events in the development of other eukaryotes, including *Drosophila,* mouse, humans, and yeast. Homeobox-containing genes encode a class of transcription factors known as **homeodomain proteins.** The homeodomain is a region of the protein that consists of 64 amino acids with a helix–turn–helix structure and is capable of binding to DNA in a sequence-specific manner (Fig. 9.19). The DNA binding homeodomain is encoded by the homeobox. The regions of these homeodomain proteins outside the homeobox are not conserved. Homeodomain proteins regulate the expression of genes that play an important role in specifying cell fate and organ identity, apparently in all eukaryotic organisms. When mutated, these genes often have homeotic effects. Two other maize homeobox genes have been identified using the *Kn*1 homeobox sequence as a probe. These are designated *ZMH1* and *ZMH2*. They encode proteins that are similar at 57 of the 64 residues of the *Kn*1 homeobox, although they share little sequence similarity outside the homeodomain. This suggests that there is a family of homeobox genes present in maize, and most likely in all other plants, whose expression determines cell fate.

   The wild-type *Kn*1 gene is expressed in developing vascular tissue in the vegetative shoot apex and in floral meristems. Neither the *Kn*1 mRNA nor the *Kn*1 protein has been detected in wild-type leaves at any stage of their development, although both are present in leaf vascular tissue of *knotted* mutants in the region giving rise to knots. It is likely that the wild-type *Kn*1 gene regulates the expression of genes necessary for vascular tissue formation. As ectopic expression of the *Kn*1 gene results in abnormal proliferation within the vascular tissue and a distortion of the veins, the normal function of the *Kn*1 gene may control the proliferation of vascular tissue as well as the determination of the fate of some of the cell types in the xylem or phloem. Other homeodomain genes may be involved in determining cell and tissue differentiation in the functioning meristem.

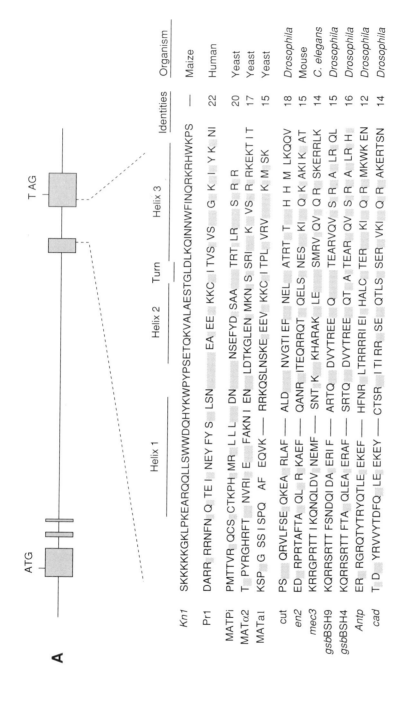

# Flowering

Flowering is the most dramatic example of a programmed developmental change in higher plants. A typical angiosperm flower consists of four whorls of organs: sepals, petals, stamens, and carpels. These are produced in successive order by the activity of the floral meristem. This involves the differentiation and growth of the floral organs from the reproductive meristem as a result of the expression and interaction of sets of gene encoding transcription factors, which control other genes whose products are used for the unique functions of these organs. Before the floral meristems are able to differentiate floral organs, the fundamental nature of the plant must change from vegetative development to reproductive. This process has been termed **floral evocation.** Not only does it involve the formation of the floral meristems, but often many other changes take place in the structure and physiology of the plant as well.

## Developmental Regulation of Floral Evocation

The shift from vegetative to reproductive development is controlled by several interacting mechanisms. Floral evocation is triggered by developmental and environmental signals. Many plants are induced to flower when they experience a photoperiod of a certain duration (described later), but they can do so only after they have attained a particular size. The importance of developmental age in floral evocation can be illustrated by considering flowering in maize. Many maize varieties flower only after they have produced 16 to 17 vegetative nodes. There are at least two possible explanations for this developmental pattern. First, it is possible that a flowering signal is synthesized by mature leaves and transported to the vegetative apical meristems, transforming them into reproductive meristems. Floral induction would occur only when sufficient quantities of the floral-inducing

**Figure 9.19 Sequences of homeodomain proteins from maize and other organisms**

(A) The upper diagram represents the *Kn*1 gene, encoding the homeodomain protein that, when mutated, causes the knotted phenotype. The colored regions indicate the region that encodes the homeodomain, in which the protein would have the helix–turn–helix structure characteristic of this group of transcriptional regulators. The amino acid sequence of the homeodomain region (amino acid residues 263–326) is shown for the Kn1 protein and it is aligned with comparable regions of several other known homeodomain proteins. The amino acid residues that are identical to those of the Kn1 protein are shaded, and asterisks indicate amino acid residues that are invariant in these eukaryotic homeodomains. (B) The amino acid sequence of a larger segment of the homeodomain of the *Kn*1 gene has been aligned with the comparable regions of two maize cDNAs encoding other homeodomain-like proteins. ZMH1 and ZMH2. Reprinted with permission from Vollbrecht *et al.* (1991).

substance arrived at the shoot apical meristem and a critical number of vegetative leaves would be necessary to synthesize enough of the flowering stimulus. Alternatively, it is possible that the meristems of these plants actually are determinant and that they are programmed to initiate floral development after completing a finite amount of vegetative growth. Fate maps of the apical meristem in the dry seed of annual plants such as maize show that the upper, reproductive parts of the plant tend to be derived from different cells than the vegetative, basal regions (see Fig. 9.10). Perhaps certain cells of the apical meristem are preprogrammed for floral development. This question can be answered by examining the development of isolated apical meristems in tissue culture. Apical meristems are cultured from plants at different developmental ages to see how many more nodes they will produce before they flower. If the meristems are determined very early in their development, plants regenerated from older apices should flower much sooner and produce fewer vegetative nodes than apices taken from young plants. When these experiments were done with maize, however, the regenerated plants flowered after producing 16 to 17 nodes, regardless of the age of the meristem at the time they were placed in culture for regeneration (Table 9.3). The maize shoot apical meristem is truly indeterminate until it is transformed by a developmental signal emanating from the mature plant.

### Juvenility and Phase Change in Floral Evocation

Most plants exhibit distinct juvenile and adult phases and these phases occur in both a sequential spatial sequence along the shoot axis and in a specific temporal

**Table 9.3**  **Number of nodes formed by plants regenerated from cultured maize apical meristems**[a]

| Genetic background | Number of nodes at time of culture | Number of nodes at flowering | Standard deviation | Number of plants examined |
|---|---|---|---|---|
| W23 | Control | 17.1 | 1.4 | 16 |
|  | 7 | 19.5 | 3.2 | 10 |
|  | 8 | 15.3 | 2.2 | 4 |
|  | 9 | 18.8 | 2.6 | 8 |
|  | 10 | 16.7 | 2.1 | 17 |
| B73 | Control | 16.0 | 1.0 | 3 |
|  | 8 | 15.0 | 1.6 | 4 |
|  | 9 | 16.7 | 2.3 | 3 |
|  | 10 | 18.3 | 3.0 | 4 |

[a] Shoot apices were removed from plants at different stages of their vegetative development and placed in culture to regenerate plants. The regenerated plants were then grown in the greenhouse to determine how many vegetative nodes they produced before they flowered.
Reprinted with permission from Irish and Nelson (1988).

order. That is, juvenile organs are formed first, so they occupy the lower part of the stem (Fig. 9.20). At some point in time, the plant makes the transition from juvenile to adult development. Thereafter the apical meristem forms adult structures, which then occupy the upper regions of the stem. Only the adult plant can reproduce, but both vegetative and reproductive structures are produced during the adult phase, usually at different times. That is, a certain amount of adult vegetative development will precede the formation of reproductive structures. Nevertheless, the transition from the juvenile to the adult phase of development must occur before the plant can become reproductive.

In many cases the juvenile–adult phase change will affect nearly every aspect of the morphology and anatomy of the plant. English ivy (*Hedera helix*) is an example of a plant with a dramatic phase change (Table 9.4). Not only does

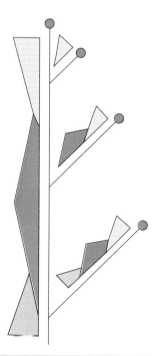

**Figure 9.20** **Poethig's model for the pattern of juvenile, adult, and reproductive structures within the shoot of an annual plant**

Juvenile, adult, and reproductive developmental programs occur sequentially during shoot development, and the structures of different parts of the shoot are determined by the interaction of these programs. The vertical line represents the primary shoot axis; diagonal lines represent the lateral branches. ▢, juvenile; ▨, adult vegetative; ▢, adult reproductive; ●, inflorescence. Reprinted with permission from Bassiri *et al.* (1992).

**Table 9.4    Comparison of some of the characteristics of juvenile and adult forms of *Hedera helix***

| Characteristic | Juvenile plant | Adult plant |
|---|---|---|
| Growth habit | Prostrate | Upright |
| Geotropism | Plagiotropic | Orthotropic |
| Phyllotaxy | Alternate | 2 + 5 spiral |
| Leaf production | One leaf/week | Two leaves/week |
| Shoot growth | Vigorous | Slight |
| Anthocyanins | Prominent | Slight to none |
| Rooting ability | Vigorous | Weak |

Reprinted with permission from Stein and Fosket (1969).

the phase change permit flowering, but it also modifies such nonreproductive characteristics as the response of the plant to gravity, its ability to form adventitious roots, leaf shape, and phyllotaxy. Phase changes alter the potential of the meristem, as well as the characteristics of the tissues and organs produced from them. A woody perennial such as a tree retains juvenile tissues and meristems after it has undergone the juvenile–adult transition, although the juvenile buds may be dormant. These dormant juvenile meristems can generate structures with juvenile characteristics if they are stimulated to grow.

Very little is known about the molecular–genetic mechanisms responsible for phase changes, although genes have been identified that affect the transformation. Four semidominant mutations in maize are known that affect the juvenile–adult transition. These are *corngrass* (*Cg*) and *Teopod* 1, 2, and 3 (*Tp1*, *Tp2*, *Tp3*). There are four components of the maize phytomere: the leaf, a bud in the axil of the leaf, the internode, and an additional leaflike structure known as the prophyll that is formed between the bud and the internode. Juvenile phytomeres have short internodes that produce adventitious roots (prop roots). They also have small prophylls and narrow, hairless leaves that are covered with epidermal wax. In some maize varieties the axillary buds of the juvenile phytomeres develop into lateral branches known as tillers. After the juvenile phase of development, a series of adult vegetative phytomeres are produced in which the internodes are longer and do not form roots, the leaves and prophylls become progressively larger, and the leaves have epidermal hairs but no surface wax. Adult vegetative development changes abruptly when the ear node is formed. A female flower, the ear, is formed in the axil of the ear node instead of the lateral bud. The meristem continues to produce phytomeres after the ear node, but these lack prophylls and have progressively shorter internodes and smaller leaves. The activity of the shoot apical meristem terminates with the formation of the male reproductive structure, the tassel (Fig. 9.21). Plants carrying the mutations *Tp1* or *Tp2* will have four or five additional juvenile phytomeres, depending on the

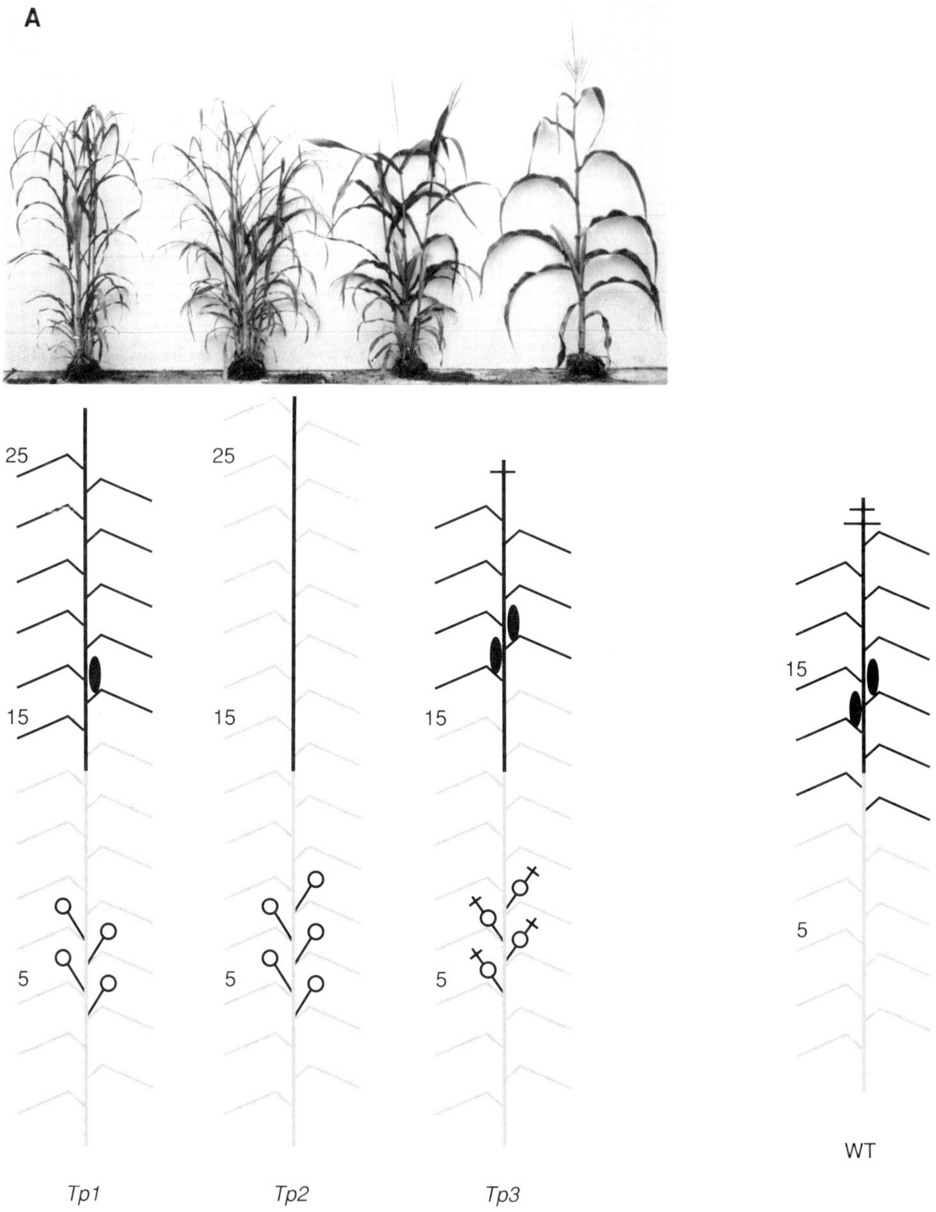

**Figure 9.21 *Teopod* mutations alter maize shoot morphology**
(A) Photograph of wild-type (WT) maize plants with the genotype *Oh51a* and the phenotypes of three different *Teopod* mutations, *Tp1*, *Tp2*, and *Tp3*, in the same genetic background. (B) Diagram of the distribution of shoot organs and their characteristics for the four plants shown in (A). The diagram does not accurately depict the relative lengths of the internodes; juvenile phytomeres are shorter than adult phytomeres. Stippled leaves represent leaves with epidermal wax; stippled internodes, internodes with prop roots. ♀, Tiller terminating in a pistillate inflorescence; ☿, tiller terminating in a mixed staminate and pistillate inflorescence; ◗, ear; ╪, branched tassel; |, unbranched tassel. Reprinted with permission from Poethig (1990).

genetic background. These additional internodes are short and have adventitious roots, while the leaves are small and narrow with epidermal wax, and more tillers are produced. The effect of the *Tp3* mutation is similar to that of *Tp1* and *Tp2*, but it is less severe. These mutations delay the transition from juvenile to adult development. As a result, they have been called **heterochronic.** A heterochronic mutation is one that affects the timing of developmental events; however, these mutations also alter nearly every other characteristic of the shoot, including the morphology of the tassel and ear. In some cases hormones, particularly gibberellins, have been shown to cause adult plants to revert back to juvenile. Gibberellins, however, do not appear to cause developmental changes similar to the *Teopod* mutations.

### Environmental Control of Floral Evocation

Not all plants have to grow up before they can flower. A few plants can flower at any age, even as seedlings, if they receive the right environmental clues. Even in plants with strong developmental controls of flowering, environmental signals may interact with these developmental controls to regulate floral evocation. The most important of these environmental signals is daylength. Many plants are sensitive to the length of the day, a phenomenon known as **photoperiodism.** Their built-in clocks tell them how long the day is in each diurnal cycle. Flowering is only one of a number of plant developmental characteristics that can be regulated by photoperiod.

Plants fall into three broad groups with respect to their response to daylength (Table 9.5). Many tropical and cultivated plants do not respond to daylength. These are known as **day-neutral** plants. Another group of plants requires short days to flower. A **short-day plant** such as cocklebur (*Xanthium pensylvanicum*), for example, will not flower unless the photoperiod is shorter than 15.7 hours in a normal 24-hour diurnal cycle. This means that when it is grown under noninductive photoperiods, it will not flower, and the shoot apical meristem continues its indeterminate growth, repetitively forming vegetative phytomeres (leaves, nodes, and internodes). On receiving a single inductive photoperiod, however, the vegetative meristem is transformed to a floral meristem and its growth and developmental potential become restricted. The cocklebur is an obligate short-day plant. Regardless of how large it grows, the plant will remain vegetative until it has received an inductive photoperiod. Many other temperate zone plants also are obligate short-day plants. Some plants, however, are facultative short-day plants. When young, their flowering is controlled by photoperiod, but after they attain a certain size they flower regardless of the photoperiod. The third group of plants with respect to photoperiod is **long-day plants,** in which the day must be longer than some critical value for the plant to flower. Once again, some plants are obligate long-day plants, whereas others are facultative.

In summary, then, long-day and short-day plants are fundamentally different in behavior. Long-day plants do not flower unless the photoperiod is longer than some critical value, whereas short-day plants require a photoperiod shorter than

**Table 9.5**     **Plants exhibiting photoperiodic control of flowering**

| Daylength requirement | Degree of control | Species |
|---|---|---|
| Short-day plants | Obligate (absolute requirement) | Chrysanthemum |
| | | Coffee |
| | | Poinsettia |
| | | Strawberry |
| | | Tobacco (Maryland mammoth) |
| | | Duckweed (*Lemma gibba*) |
| | | Cocklebur (*Xanthium*) |
| | Facultative (quantitative requirement) | Hemp (*Cannabis*) |
| | | Cotton |
| | | Rice |
| | | Sugar cane |
| Long-day plants | Obligate (absolute requirement) | Carnation (*Dianthus*) |
| | | Henbane (*Hyoscyamus*) |
| | | Oat (*Avena*) |
| | | Ryegrass (*Lolium*) |
| | | Clover |
| | Facultative (quantitative requirement) | Pea |
| | | Barley |
| | | Lettuce |
| | | Wheat (spring wheat) |
| | | Turnip (*Brassica rapa*) |
| Day-neutral plants | | Cucumber (*Cucumis*) |
| | | Tomato (*Lycopersicum*) |
| | | Potato (*Solanum tuberosum*) |
| | | Bean (*Vicia faba*) |
| | | Rose |

a critical value if they are to flower (Fig. 9.22). They actually overlap considerably with respect to the duration of the critical photoperiod. These two groups differ not so much in terms of the absolute length of the day as in their response to the changing length of day relative to night. This difference can be illustrated by the response of long-day and short-day plants to a flash of light during the dark period. Interrupting the dark period of a short-day plant grown under a noninductive photoperiod will induce flowering, whereas a flash of light in the middle of an inductive photoperiod for a long-day plant will inhibit its flowering (Fig. 9.23).

### Phytochrome Control of Signal That Induces Floral Evocation

Photoperiodic control of flowering is known to be mediated by the phytochrome system (see Chapter 6); however, almost nothing is known about the mechanism by which the perception of the photoperiod by the phytochrome system leads to floral development. The photoperiodic signal is perceived by the leaves of the

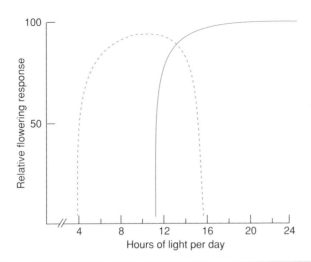

**Figure 9.22 Diagrammatic representation of the differing response of short-day and long-day plants to photoperiods of different lengths**

The graph illustrates the way an idealized short-day (broken line) or long-day (solid line) plant would respond to photoperiod. Photoperiods shorter than about 4 hours do not induce flowering because they cannot meet the photosynthetic needs of the plant. Reprinted with permission from Hart (1988).

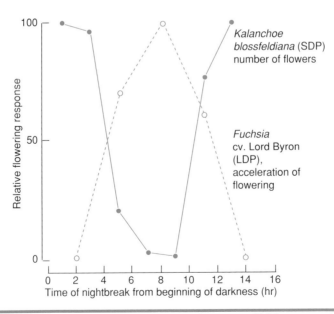

**Figure 9.23 Effect of light interruption at different times during the dark period on flowering in a short-day plant (SDP) and a long-day plant (LDP)**

Reprinted with permission from Hart (1988).

plant and is somehow transferred to the shoot apical meristem where it transforms development. It has been proposed that a flowering hormone, named florigen, is produced in leaves in response to inductive photoperiods. Once florigen is made by the leaves it is transported to the apical meristems, where it changes their developmental potential. The chemical nature and even the existence of florigen remain speculative. It is, however, clear that an inductive photoperiod alters the pattern of gene expression within the meristem. Two groups of genes have been isolated whose transcripts accumulate within the meristems after photoperiodic induction of flowering in the long-day plant *Sinapis alba*. Although we do not know how the products of these genes may function in floral evocation at present, they could be homeodomain genes whose products halt the expression of genes responsible for the vegetative program and initiate the transcription of genes responsible for the differentiation of floral organs.

# Floral Differentiation

### Inflorescence and Floral Meristems

There are two different patterns of reproductive development in the angiosperms. In the first, entire indeterminate vegetative meristems are transformed directly into a determinant floral meristem. Alternatively, the vegetative meristem is first transformed into an **inflorescence meristem,** which then generates the floral meristems. The inflorescence meristem does not produce floral organs directly, but it also does not generate leaves. Instead of leaves, it may produce bracts with floral meristems, or sometimes a mixture of floral and more inflorescence meristems, in the axils of the bracts (Fig. 9.24). Bracts may have some of the characteristics of leaves, but typically they are much reduced in size when compared with the vegetative leaves of the same plant. The inflorescence meristem may be indeterminate, as it is in *Arabidopsis* and many similar plants that produce a type of inflorescence known as a **raceme,** but in other plants it is determinant. The sunflower family (Compositae) produces a type of determinant inflorescence known as a **head** that people often mistake for a single flower. In reality, a head is an inflorescence composed of hundreds of individual flowers, called florets, each of which is borne in the axil of a bract (Fig. 9.25). Inflorescence meristems thus represent a distinct type of meristem whose characteristics and activities often differ markedly from those of either floral or vegetative meristems. These differences not only include the type of organs differentiated from them and whether or not they exhibit determinant or indeterminate growth, but also their phyllotactic pattern and the growth characteristics of the internode (Table 9.6).

The vegetative apical meristems of wild-type *Arabidopsis* form approximately 12 phytomeres with very short internodes, resulting in a basal rosette of leaves, during vegetative growth. The vegetative meristem is transformed into an indeterminate inflorescence meristem approximately 25 days after germination and

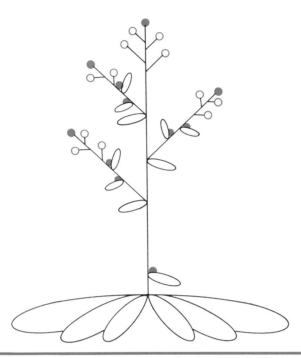

**Figure 9.24 Diagram of the morphology of the *Arabidopsis* wild-type** Floral meristems are indicated by open circles. Closed circles indicate florescence meristems. Redrawn with permission from Irish and Sussex (1990).

ceases the production of vegetative phytomeres. The inflorescence meristem produces lateral buds in the axils of cauline leaves, but later produces only buds at nodes, without cauline leaves. Cauline leaves are smaller than vegetative leaves and similar in size and structure to the bracts that subtend many angiosperm flowers. *Arabidopsis* flowers, however, lack bracts. The first three lateral buds formed by the inflorescence meristem are themselves inflorescence meristems and their activity repeats the pattern of develoment of the primary inflorescence meristem. Subsequent lateral buds, lacking cauline leaves, are determinant floral buds that form only floral organs.

### Initiation and Differentiation of Floral Organs

Floral meristems usually can be distinguished from vegetative meristems, even early in floral development, by their larger size. The transition from vegetative to floral development is marked by an increase in the frequency of cell divisions within the central zone of the meristem. The cells of this region completed their division cycle slowly and some of them act as apical initials in the vegetative meristem. The increase in the size of the meristem as floral development com-

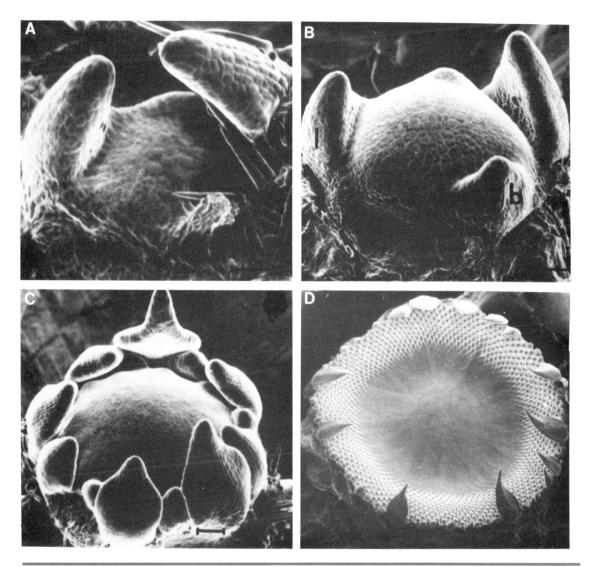

**Figure 9.25 Changes in the structure of the apical meristem during the transition from vegetative to inflorescence meristem in *Helianthus annuus***

(A) The vegetative meristem of the seedling, which initiates only leaf primordia. (B) Transition meristem of a 23-day-old plant after the central zone cells have begun to divide. The meristem has initiated the last three leaf primorida (1) and has begun to initiate the primordia for the large bracts that surround the inflorescence, known as involucral bracts, one of which is visible (b). (C) Formation of the inflorescence receptacle disk. The meristematic dome has broadened and flattened while numerous involucral bracts have been initiated from the periphery of the meristem. (D) A later stage in floral differentiation in which floret initiation has occurred from approximately one-third of the disk, starting from the periphery and progressing toward the center. Each floret primordium is subtended by a bract primordium. In all four figures, bar = 100 $\mu$m. Courtesy of D. Jegla.

**Table 9.6**   **Some characteristics of vegetative, inflorescence, and floral meristems in the snapdragon (*Antirrhinum majus*)**

|  | Meristem type | | |
|---|---|---|---|
|  | Vegetative  → | Inflorescence → | Floral |
| Determinant or indeterminate | Indeterminate | Indeterminate | Determinant |
| Phyllotaxy | Decussate | Spiral | Whorled |
| Internode length | Long | Short | Very short |

Reprinted with permission from Coen *et al.* (1990).

mences is largely a result of the increased division rate of these central cells. Four different floral organs then are initiated sequentially in the flanks of the meristem; sepals are initiated first, followed by petals, then stamens, and finally carpels (Fig. 9.26). Each set of organs is initiated as a whorl. By the time the carpels are initiated, the meristematic cells of the apex have been used up and only the primordia of the floral organs are present on the apical surface. In the wild-type *Arabidopsis* flower, the first (outermost) whorl consists of four sepals, which are green at maturity. The second whorl is composed of four sepals, which are white at maturity. The third whorl contains six stamens, two of which are substantially shorter than the other four. Finally, what is considered to be the fourth whorl is a single complex organ, the pistil (Fig. 9.27). The pistil includes an ovary that is composed of two fused carpels, each with numerous ovules, as well as a stigma capping a short style.

## Mutations Affecting Floral Differentiation

A number of different mutations are known to affect the differentiation of floral organs. People have been cultivating interesting and attractive floral variants of roses and other ornamental plants at least as far back as the ancient Greeks. A wild-type rose has five sepals and petals and many stamens. In many cultivated rose varieties, several of the whorls of stamens are replaced by extra whorls of petals. Many of these modifications of the wild-type flower structure are the result of homeotic mutations which have occurred spontaneously. A **homeotic mutation** is one that results in the formation of the wrong organ at the wrong place. In recent years the systematic investigation of homeotic floral mutants has greatly aided our understanding of the molecular mechanisms of floral determination and the differentiation of floral organs. The two plants in which the most progress has been made in this direction are *Arabidopsis thaliana* and the snapdragon (*Antirrhinum majus*). In both of these plants, flowering is a two-step process in which (1) the vegetative meristems are transformed into inflorescence meristems, and (2) the inflorescence meristems then produce the floral meristems in the axils of bracts.

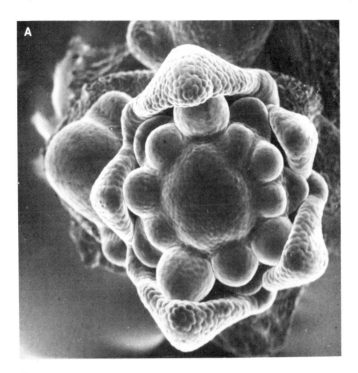

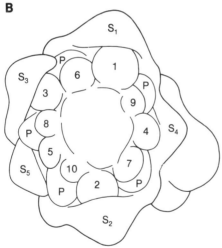

**Figure 9.26 Formation of floral organ primordia**

(A) Scanning electron micrograph of *Silene* floral but taken after the formation of the stamen primordia. The carpel primordia have not yet formed from the central meristematic dome, which remains undifferentiated. (B) Map of the floral apex shown in (A). The developing floral bud formed the five sepal primordia ($S_1$, $S_2$, $S_3$, etc.), then the five petal primordia (P), and then ten stamen primordia (numbered 1–10). Reprinted with permission from Lyndon (1990).

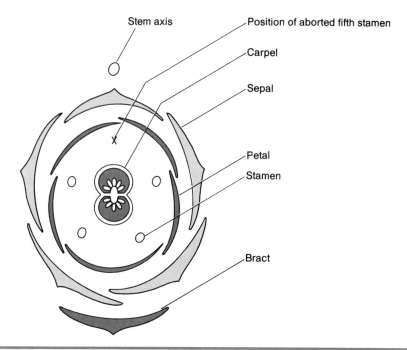

Figure 9.29 **Floral diagram of a wild-type snapdragon (*Antirrhinum*) flower**
Redrawn from Carpenter and Coen (1990).

(Fig. 9.30). Type B mutations affect the second and third whorls. Second-whorl organs develop as sepals instead of petals, and third-whorl organs become carpels instead of stamens. The *Arabidopsis* mutations *apetala3* (*ap3*) and *pistillata* (*pi*) bring about this type of transformation (Fig. 9.30C). Type C mutations affect the third and fourth whorls; instead of stamens, petals are formed at the third whorl and sepals are formed in the forth whorl. The *agamous* (*ag*) mutations of *Arabidopsis* fall into this class (Fig. 9.30B). Similar types of homeotic mutations have identified genes controlling floral organ identity in snapdragon. Mutations in the snapdragon gene *deficiens,* for example, have an effect comparable to those in the *Arabidopsis apetala3* gene.

Homeotic genes controlling *Drosophila* embryonic development are closely linked to each and occur in an invariant order on a specific chromosome. Not only are these homeotic genes conserved in animal evolution, but the gene order also is conserved. The order in which these homeobox genes occur on the chromosome apparently is important in determining the sequence in which they are expressed. This does not seem to be the case in plants. The chromosome map locations of each of the four *Arabidopsis* homeotic genes affecting floral organ identity, *amagous, apetala2, apetala3,* and *pistillata,* have been determined. Significantly, none of them are linked, and in fact, each is on a different chromosome (Fig. 9.30F).

**Figure 9.30 Homeotic mutants affecting floral development in**
*Arabidopsis thaliana*

(A) Wild-type flower of the *Landsberg erecta* strain with four sepals, four petals, six stamens, and an ovary consisting of two fused carpels. Sepals are not visible in this photo but were present. (B) The *agamous* 3 homozygous mutant lacking stamens and ovary. The reproductive organs have been converted to petals. (C) An *apetala3—1* mutant in which the organs of the third whorl have developed as carpels instead of stamens. The organs of the outer two whorls were removed to show the inner whorls. (D) An *apetala2—/apetala2—1* heterozygous mutant in which the first-whorl organs have many of the characteristics of carpels, the organs of the second whorl are absent, and the organs of the third whorl are anthers, as in the wild type, but they are reduced in number. The fourth whorl is normal. (E) An *apetala2—1* mutant in which the petals have many of the properties of stamens. The organs of the third and fourth whorls are normal. (F) (See next page) Diagrammatic representation of flower structure in the wild type and in plants homozygous for mutant alleles of several homeotic genes that control floral organ identity. The genetic map positions of each of these homeotic genes on the five linkage groups (chromosomes) of *Arabidopsis* also are shown. Reprinted with permission from Meyerowitz *et al.* (1989). ©Company of Biologists Ltd.

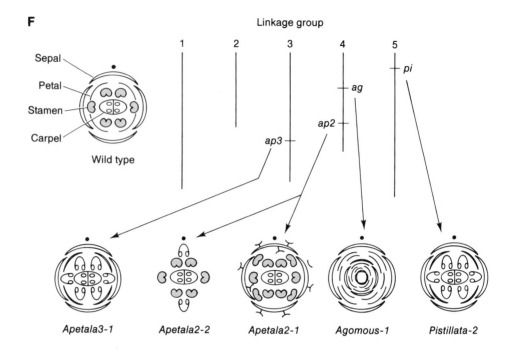

## Homeotic Genes

A number of the genes affected by these homeotic mutations have been cloned and characterized, including the snapdragon *deficiensA* gene and the *Arabidopsis agamous, apetala1,* and *apetala3* genes. They encode related proteins belonging to a class of transcription factors containing a conserved region known as the MADS domain (Fig. 9.31). In addition to the proteins controlling floral organ identity, this class of transcription factors includes the serum response factor of mammals that is responsible for serum-induced transcription of *c-fos* protooncogenes and a yeast transcription factor known as MCM-1 that is involved in mating type determination. The MADS box is the region of these genes encoding the DNA binding MADS domain of these protein transcription factors. Proximal to the MADS domain is a dimerization domain. This is a region of the protein that can bind to another protein that has a similar functional domain. As a result of the dimerization domain these transcription factors can form either homodimers, by binding to another molecule of the same protein, or heterodimers, by binding to another transcription factor containing a similar domain. As explained in Chapter 4, controlling elements in the promoters of genes often are duplicated, so transcription factor dimerization results in stronger

- - -  DNA binding ─────────────────────────────────────
────────  Dimerization  ─────────────────────

SRF      RVKIKMEFIDNKLRRYTTFSKRKTGIMKKAYELSTLTGTQCLLLVASETGHVYTFATRK
MCM1     RRKIEIKFIENKTRRHVTFSKRKHGIMKKAFELSVLTGTQVLLLVVSETGLVYTFSTPK

                                    *
DEF A    RGKIQIKRIENQTNRQVTYSKRRNGLFKKAHELSVLCDAKVSIIMISSTQKLHEYISPT
DEF H22  RGKIEIKRIENTTNRQVTYSKRRNGIMKKAKEISVLCDAHVSVIIFASSGKMHEFCSPS
DEF H23  RGKVQLKRIENKINRQVTFSKRRGGLLKKAHELSVLCDAEVALIVFSNKGKLFEYSTDS
AC       RGKIEIKRIENTTNRQVTFCKRRNGLLKKAYELSVLCDAEVALIVFSSRGRLYEYSNNS

Cons     RgKIqIkrIDN   nRqvTF KRK GI KKA ELSvLcdT vsLLV S      kV eF      s

**Figure 9.31 Amino acid sequences of the DNA binding and dimerization domains of proteins involved in the control of differentiation in mammals, yeast, and plants**

Conserved amino acids in the consensus sequence (Cons) are indicated with capital letters if the residue is present in all six species and in lowercase boldface letters if conserved in all plant sequences. Conserved amino acid substitutions are indicated with lowercase roman letters. The conserved phosphorylation site also is underlined. The conserved amino acid that is altered in the *nicotianoides* allele of the *deficiens* gene is indicated with an asterisk. Reprinted with permission from Schwarz-Sommer *et al.* (1990).

and more specific binding of the factor to these regulatory elements. MADS domain transcription factors also contain a region within the DNA binding domain that can be phosphorylated. As a result, DNA binding could be controlled by phosphorylation of the protein. The snapdragon genome contains at least eight genes whose protein products are 65 to 90% homologous to the deficiens (DEF) protein in the DNA binding and dimerization domains, but outside these domains they have little homology. The agamous (AG) protein is homologous to the DEF protein within its DNA binding and dimerization domains. The *Arabidopsis* genome also contains a family of at least six genes encoding proteins with homology to the AG protein. These have been called the *AGL* genes.

*In situ* hybridization has been used to determine the tissues and organs of the plant in which the MADS box genes are expressed. For example, the *agamous* gene is not expressed in any part of the plant except in floral meristems, where its expression first appears in the primordia that give rise to stamens and carpels. The gene continues to be expressed during the growth and differentiation of these two organs. As these are the organs most strongly affected by mutations in this gene, the results support the idea that the protein product of the *agamous* gene plays an important role in the determination of these organ types. The AG protein probably controls the expression of genes whose products are involved in the formation and/or function of these organs. The *Arabidopsis* genes encoding the AGL proteins also are expressed in various floral organs, although one of them also was expressed in vegetative meristems. This suggests that these genes are part of a family of genes encoding transcription factors that determine tissue and organ identity in both vegetative and reproductive organs.

Genetic and *in situ* hybridization studies suggest that the three classes of MADS box genes are expressed in overlapping fields within the meristem. Two whorls of organs develop in each of the fields. Type A genes are expressed and function in whorls 1 and 2, type B genes in whorls 2 and 3, and type C genes in whorls 3 and 4. A model illustrating how the products of these genes interact to control organ identity is shown in Fig. 9.32. The expression of the *apetala2* gene alone results in sepal formation, and the expression of the *agamous* gene alone results in carpel formation. Petals are formed as a result of the combined action of *AP2* and *AP3/PI*, whereas stamen differentiation requires the action of *AG* and *AP3/PI*. *AP2* and *AG* also are antagonistic. *AG* prevents the expression of *apetala2* in whorls 3 and 4; conversely, *apetala2* prevents the expression of *agamous* in whorls 1 and 2. As a result, a mutation that blocks the expression of the *AG* gene permits the expression of *apetala2* in whorls 3 and 4, resulting in the formation of petals in whorl 3 and sepals in whorl 4. Mutations blocking the expression of *apetala2* now permit the expression of *agamous* in whorls 1 and 2, resulting in the formation of stamens in whorl 2 where *AP3/PI* are also active, and carpels where only *AG* is active. The wild-type and most of the mutant phenotypes can be predicted and explained by this model. It must, however, be remembered that the model is apt to be an oversimplification of a mechanism regulating floral organ determination and differentiation.

Assuming that the different MADS domain transcription factors are active as either homodimers or heterodimers, we might expect that *AP2* acts as a homodimer to induce sepal differentiation, but as a *AP2–AP3/PI* heterodimer to induce the differentiation of petals. Similarly, *AG* presumably acts either as a homodimer to induce carpel formation or as an *AG–AP2/PI* heterodimer to induce stamen formation. The heterodimers would be able to bind to regulatory elements different from those of either of the homodimers, and the two homodimers also would differ from each other in the genetic elements they recognized. As a result, the *AG–AP3/PI* heterodimer would regulate a different group of genes than the *AG* homodimers, for example. Consequently, different classes of genes will be transcribed in each of the four whorls. The next challenge in uncovering the mechanism responsible for floral organ development is to identify the genes whose expression requires the different MADS transcription factors and then to understand how the products of these genes bring about organ differentiation.

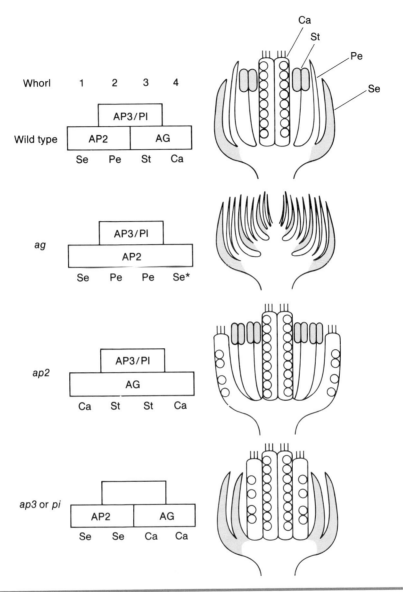

**Figure 9.32 Model for the determination of floral organ identity by the interaction of three classes of floral homeotic genes**
The three rectangular boxes on the left represent meristematic fields, each containing a pair of adjacent floral whorls, in which the three types of homeotic genes are active in the wild type: AP3 = *apetala3*, PI = *pistillata*; AP2 = *apetala2*; AG = *agamous*. Absence of letters within the field indicates that the protein product is not present in that field as a result of a mutation in the respective gene. The meristematic positions in which the four whorls of floral organs will arise are indicated at top left, relative to the fields. The organ type of each whorl is indicated below the whorl. Se, sepal; Pe, petal; St, stamen; Ca, carpel. The floral phenotypes are shown in the drawings on the right for the wild type and for the mutations *agamous* (*ag*), *apetala2* (*apa2*), and *apetala3* (*ap3*), or *pistillata* (*pl*). Redrawn with permission from Bowman *et al.* (1991).

## Summary

Plant development is mostly postembryonic. The plant body is formed through the activity of meristems, which are small populations of embryonic cells located at or near the extreme tip of the root and shoot axis. The angiosperm shoot apical meristem is highly stratified, with two or three tunica layers in which cell divisions are nearly all oriented at right angles to the meristem surface, or anticlinal. These tunica layers overlie a group of corpus cells in which the plane of cell division is not as highly ordered. Active shoot apical meristems exhibit cytohistological zonation in which the cells in the center of the meristem, the central zone cells, are relatively large and divide infrequently as compared with the surrounding peripheral zone cells, which divide more rapidly. A rib meristem underlies the central zone. The root apical meristem differs from the shoot apical meristem in at least two ways: the root apical meristem is not terminal, but is covered by the root cap, and it does not produce any lateral organs. It generates only the tissues of the primary root and, in many roots, the root cap.

Vegetative meristems are self-perpetuating. Some meristematic cells retain the capacity to divide indefinitely, or until the meristem is transformed into a floral meristem. These uncommitted cells are known as the apical initials. The apical initials are found among a group of slowly dividing cells in the central zone of the shoot apical meristem and in a region of the root apex known as the quiescent center. The apical initials have many of the properties of stem cells; they divide slowly, they retain the potential for cell division through many division cycles, and they are pluripotent. Ferns have meristems with a single apical initial cell that is the ultimate source of all the cells in the root or shoot. Angiosperm apices, however, contain a small number of apical initials, perhaps three, in each layer of the meristem and it is unlikely that any of them are permanent initials. Chimeric plants, in which the different layers of the meristem have a different genetic composition, can be created because each layer is derived from a different set of initials. Cell lineage studies with chimeric plants have demonstrated that cell differentiation is not dependent on cell lineage. Instead, plant cell differentiation is position dependent and cell fate is determined gradually.

The vegetative shoot apical meristem is highly repetitive in its activity. It produces a modular structure, the phytomere, consisting of a leaf attached to a node, a lateral bud in the axil of the leaf, and an internode subtending the leaf. Leaves are initiated in a characteristic pattern by the activity of the apical meristem. Leaf initiation begins with increased cell division in all tunica layers in the flanks of the apical meristem, resulting in the formation of the leaf primordium. Although vegetative meristems exhibit indeterminate growth, the leaf is determinant. It exhibits rapid, limited growth, but all growth ceases at maturity. The primordium becomes committed to this determinant "leaf" developmental pathway gradually after it is initiated. Homeotic genes have been identified that determine cell and tissue fate in the functioning meristem. The *Kn*1 mutations in maize have identified a homeotic gene that appears to control cell fate in vascular tissue. The *Kn*1 gene is a member of a family of genes that contain a conserved element known as the homeobox. The homeobox encodes a protein domain with a helix–turn–helix structure that binds to DNA in a sequence-specific manner. These genes contain proteins that belong to a class of transcription factors that control cell, tissue, and organ identity in eukaryotic organisms from *Drosophila* and humans to plants.

Flowering involves a fundamental change in the development of the plant. Floral evocation is an abrupt transition from vegetative to reproductive development that alters the structure and developmental character of the apical meristem. Floral evocation is brought about by both environmental and developmental factors. Before most angiosperms can flower, they must attain a certain size. In some cases plants exhibit three different developmental phases: juvenile, adult vegetative, and adult reproductive. Before they can flower, juvenile plants must make the transition to the adult phase. Genes have been identified in maize that control the juvenile–adult phase transition. Photoperiod also plays a major role in controlling floral evocation in many adult plants. Many plants have a mechanism for measuring the length of the day and will not flower unless they receive either short or long days. Actually, this mechanism measures the duration of the dark period and it is mediated by phytochrome. Long-day plants will not flower unless the dark period is longer than some critical value; short-day plants require a dark period that is shorter than a critical value for floral evocation.

In some cases, the flowering signal transforms vegetative meristems directly into floral meristems. In other cases, however, the vegetative meristem is first transformed into an inflorescence meristem, which then generates the floral meristems. The inflorescence meristem produces phytomeres in which the lateral meristems may differentiate floral organs. The floral meristem exhibits determinant growth. It initiates four successive whorls of organs before its meristematic capital is exhausted and its activity ceases. The primordia in each whorl differentiate into one of the four floral organs. The primordia of the first whorl differentiate as sepals, those of the second whorl develop into petals, those of the third become stamens, and the fourth

*continues*

whorl differentiates as the pistil. Homeotic mutations have identified genes responsible for the determination of floral organ identity. Each of these mutations affects two whorls of floral organs. The *agamous* mutations of *Arabidopsis*, for example, result in the formation of petals in the third whorl, instead of stamens, and sepals in the fourth whorl, instead of the pistil. Three types of homeotic genes have been identified: type A affects whorls 1 and 2, type B affects whorls 2 and 3, and type C affects whorls 3 and 4. These mutations identify genes encoding transcription factors containing a conserved DNA binding MADS domain. Each of these genes is expressed within a specific field within the meristem that encompasses a pair of whorls. The transcription factors interact to control the expression of genes specific for floral organ identity.

## Questions for Study and Review

1. What are the nature and function of meristems? Compare the structure and activity of the root and shoot apical meristems. How do apical meristems differ from secondary meristems, such as the vascular cambium?

2. Describe the pattern of cell division within the shoot apical meristem and explain the significance of this pattern.

3. Compare the structures of the shoot apical meristem of a typical dicot and the shoot apical meristem of a fern. What is the function of the apical cell of the fern shoot apical meristem?

4. What are the characteristics of the cells called apical initials? How are the apical initials of angiosperm meristems similar to the apical cells of the meristems of ferns and other lower vascular plants?

5. What is the function of the root cap and how is it formed? What is the developmental fate of the root cap cells?

6. What is the quiescent center of the root apical meristem? How are the cells of the quiescent center and the apical initials of the shoot apical meristem similar?

7. Plant development is highly recursive (the same event occurs repeatedly at regular intervals). Explain the recursiveness of the development of the stem during vegetative growth. In what respect is angiosperm reproductive development recursive?

8. What is meant by the term *phytomere* and how is this significant for our understanding of the functioning of the shoot apical meristem? What is the phytomere of the root?

9. If a plant has decussate phyllotaxy how are its leaves arranged around its stem? What are orthostichies and how do they differ from contact parastichies?

10. How do determinant and indeterminate meristems differ? What are some examples of determinant meristems? How can you tell when a meristem has been determined?

11. What are the characteristics of the

plant meristems said to be "chimeric"? Experimentally, how can plants with chimeric meristems be generated? What important questions might be answered by studying chimeric meristems?

12. Are the apical initial cells of angiosperm shoot apical meristems pluripotent or multipotent? Define the terms *multipotent* and *pluripotent* and provide experimental evidence that would help us decide the developmental capacity of shoot apical initial cells.

13. How might homeodomain proteins such as *Kn*1 and its alleles function in the formation of leaves?

14. Certain cultivars of maize always produce approximately 16 vegetative phytomeres before they flower. A fate map of the cells in the seedling apical meristem shows that the upper part of the plant that will ultimately produce the flowers tends to be derived from different meristematic cells than the lower, vegetative structures. How do we know that the apical meristem cells giving rise to the reproductive structures are not already determined in the seedling or even the dry seed?

15. Compare short-day and long-day plants with respect to their responses to the changing length of the day during the growing season.

16. How do inflorescence, floral, and vegetative meristems differ?

17. What is a homeotic mutation and how do homeotic mutations affect the pattern of organs initiated from a floral meristem? How does the existence of the *floricaula* mutation of *Antirrhinum* help us understand how inflorescence meristems produce floral meristems?

18. What is the difference between type A, B, and C homeotic mutations that affect organ identity?

19. What do we know about the protein product of the normal allele of the *agamous* gene and how does this help us understand how it functions to determine organ identity?

20. Explain Meyerowitz's model for the specification of floral organ identity (see Fig. 9.33).

## Further Reading

### General References

Barlow, P. W. (1978). The concept of the stem cell in the context of plant growth and development. In *Stem Cells and Tissue Homeostasis* (B. I. Lord, C. S. Potten, and R. J. Cole, Eds.), pp. 87–113. Cambridge University Press, Cambridge.

Barlow, P. W. (1989). Meristems, metamers and modules and the development of shoot and root systems. *Bot. J. Linn. Soc.* **100,** 255–279.

Clowes, F. A. L. (1961). *Apical Meristems.* Blackwell, Oxford.

Coen, E. S. (1991). The role of homeotic genes in flower development and evolution. *Annu. Rev. Plant Physiol. Plant Mol. Biol.* **42,** 241–279.

Coen, E. S., and Meyerowitz, E. M. (1991). The war of the whorls: Genetic interactions controlling flower development. *Nature* **353,** 31–37.

Drews, G. N., and Goldberg, R. B. (1989). Genetic control of flower development. *Trends Genet.* **5,** 256–261.

Gifford, E. M., Jr. (1983). Concept of apical cells in bryophytes and pteridophytes. *Annu. Rev. Plant Physiol.* **34,** 419–440.

Gifford, E. M., Jr., and Corson, G. E., Jr. (1971). The shoot apex in seed plants. *Bot.Rev.* **37,** 143–229.

Hake, S. (1992). Unraveling the knots in plant development. *Trends Genet.* **8,** 109–114.

Hart, J. W. (1988). *Light and Plant Growth.* Unwin Hynan, London.

Klekowski, E. J., Jr. (1988). *Mutation, Developmental Selection, and Plant Evolution.* Columbia University Press, New York.

Lyndon, R. F. (1990). *Plant Development. The Cellular Basis.* Unwin Hyman, London.

Medford, J. I. (1992). Vegetative apical meristems. *Plant Cell* **4,** 1029–1039.

Poethig, R. S. (1987). Clonal analysis of cell lineage patterns in plant development. *Am. J. Bot.* **74,** 581–594.

Steeves, T. A., and Sussex, I. M. (1989). *Patterns in Plant Development,* 2nd ed. Cambridge University Press, Cambridge.

Sussex, I. M. (1989). Developmental programming of the shoot meristem. *Cell* **56,** 225–229.

## Specific References

Bassiri, A., Irish, E. E., and Poethig, R. S. (1992). Heterochronic effects of *Teopod* 2 on the growth and photosensitivity of the maize shoot. *Plant Cell* **4,** 497–504.

Bowman, J. L., Smyth, D. R., and Meyerowitz, E. M. (1991). Genetic interactions among floral homeotic genes of *Arabidopsis. Development* **112,** 1–20.

Carpenter, R., and Coen, E. S. (1990). Floral homeotic mutations produced by transposon-mutagenesis in *Antirrhinum majus. Genes and Dev.* **4,** 1483–1493.

Church, A. H. (1904). *On the Relation of Phyllotaxis to Mechanical Laws.* Williams and Norgate, London.

Clowes, F. A. L. (1964). The quiescent center in meristems and its behavior after irradiation. *Brookhaven Symp. Biol.* **16,** 46–57.

Coen, E. S., Doyle, S., Romero, J.M., Elliott, R., Magrath, R., and Carpenter, R. (1991). Homeotic genes controlling flower development in *Antirrhinum. Development* **112,** Suppl. 1, 149–155.

Coen, E. S., Romero, J. M., Doyle, S., Elliott, R., Murphy, G., and Carpenter, R. (1990). *floricaula:* A homeotic gene required for flower development in *Antirrhinum majus. Cell* **63,** 1311–1322.

Feldman, L. J., and Torrey, J. G. (1976). The isolation and culture in vitro of the quiescent center of *Zea mays. Am. J. Bot.* **63,** 345–355.

Freeling, M. (1992). A conceptual framework for maize leaf development. *Dev. Biol.* **153,** 44–58.

Gunning, B. E. S. (1982). The root of the water fern *Azolla:* Cellular basis of development and multiple roles for cortical microtubules. In *Developmental Order: Its Origin and Regulation* (S. Subtelny and P. B. Green, Eds.), pp. 379–421. Alan R. Liss, New York.

Irish, E. E., and Nelson, T.M. (1988). Development of maize plants from cultured cells. *Planta* **175,** 9–12.

Irish, V. F., and Sussex, I. M. (1990). Function of the *apetala*-1 gene during *Arabidopsis* floral development. *Plant Cell* **2,** 741–753.

Jegla, D. E., and Sussex, I. M. (1989). Cell lineage patterns in the shoot meristem of the sunflower embryo in the dry seed. *Dev. Biol.* **131,** 215–225.

Kelly, A. J., Zagotta, M. T., White, R. A., Chang, C., and Meeks-Wagner, D. R. (1990). Identification of genes expressed in the tobacco shoot apex during the floral transition. *Plant Cell* **2,** 963–972.

Klekowski, E. J., Jr., and Kazarinova-Fukshansky, N. (1984). Shoot apical meristems and mutation: Fixation of selectively neutral cell genotypes. *Am. J. Bot.* **71,** 22–27.

Krishnamoorthy, H. N., and Nanda, K. K. (1968). Floral bud reversion in *Impatiens balsamina* under non-inductive photoperiods. *Planta* **80,** 43–51.

Leyser, H. M. O., and Furner, I. J. (1992). Characterisation of three shoot apical meristem mutants of *Arabidopsis thaliana*. *Development* **116**, 397–403.

Ma, H., Yanofsky, M. F., and Meyerowitz, E.M. (1991). *AGL1–AGL6*, an *Arabidopsis* gene family with similarity to floral homeotic and transcription factor genes. *Genes Dev.* **5**, 484–495.

Mandel, M. A., Bowman, J. L., Kempin, S. A., Ma, H., Meyerowitz, E. M., and Yanofsky, M. F. (1992). Manipulation of flower structure in transgenic tobacco. *Cell* **71**, 133–143.

Medford, J. I., Behringer, F. J., Callos, J. D., and Feldmann, K. A. (1992). Normal and abnormal development in the *Arabidopsis* vegetative shoot apex. *Plant Cell* **4**, 631/643.

Medford, J. I., Elmer, S., and Klee, H. J. (1991). Molecular cloning and characterization of genes expressed in shoot apical meristems. *Plant Cell* **3**, 359–370.

Melzer, S., Majewski, D. M., and Apel, K. (1990). Early changes in gene expression during the transition from vegetative to generative growth in the long-day plant *Sinapis alba*. *Plant Cell* **2**, 953–961.

Meyerowitz, E. M., Smyth, D. R., and Bowman, J. L. (1989). Abnormal flowers and pattern formation in floral development. *Development* **106**, 209–217.

Mitchison, G. J. (1977). Phyllotaxis and the Fibonacci series. *Science* **196**, 270–275.

Mizukami, Y. and Ma, H. (1992). Ectopic expression of the floral homeotic gene *AGAMOUS* in transgenic *Arabidopsis* plants alters floral organ identity. *Cell* **71**, 119–131.

Phillips, H. L., Jr., and Torrey, J. G. (1972). Duration of cell cycles in cultured roots of *Convolvulus*. *Am. J. Bot.* **59**, 183–188.

Poethig, R. S. (1988). Heterochronic mutations affecting shoot development in maize. *Genetics* **119**, 959–973.

Poethig, R. S. (1990). Phase change and the regulation of shoot morphogenesis in plants. *Science* **250**, 923–930.

Poethig, R. S., and Sussex, I. M. (1985). The developmental morphology and growth dynamics of the tobacco leaf. *Planta* **165**, 158–169.

Rogers, S. O., and Bonnett, H. T. (1989). Evidence for apical initial cells in the vegetative shoot apex of *Hedera helix* cv. Goldheart. *Am. J. Bot.* **76**, 539–545.

Satina, S., Blakeslee, A. F., and Avery, A. G. (1940). Demonstrations of the three germ layers in the shoot apex of *Datura* by means of induced polyploidy in periclinal chimeras. *Am. J. Bot.* **27**, 895–905.

Schultz, E. A., and Haughn, G. W. (1991). *LEAFY*, a homeotic gene that regulates inflorescence development in *Arabidopsis*. *Plant Cell* **3**, 771–781.

Schultz, E. A., Pickett, F. B., and Haughn, G. W. (1991). The *FLO10* gene product regulates the expression domain of homeotic genes *AP3* and *PI* in *Arabidopsis* flowers. *Plant Cell* **3**, 1221–1237.

Schwarz-Sommer, Z., Huijser, P., Nacken, W., Saedler, H., and Sommer, H. (1990). Genetic control of flower development by homeotic genes in *Antirrhinum majus*. *Science* **250**, 931–936.

Smith, L. G., Greene, B., Veit, B., and Hake, S. (1992). A dominant mutation in the maize homeobox gene, *Knotted*-1, causes its ectopic expression in leaf cells with altered fates. *Development* **116**, 21–30.

Smith, R. H., and Murashige, T. (1970). *In vitro* development of the isolated shoot apical meristem of angiosperms. *Am. J. Bot.* **57**, 562–568.

Sommer, H., Beltrán, J.-P., Huijser, P., Pape, H., Lönnig, W.-E., Saedler, H., and Schwarz-Sommer, Z. (1990). *Deficiens*, a homeotic gene involved in the control of flower morphogenesis in *Antirrhinum majus:* The protein shows homology to transcription factors. *EMBO J* **9**, 605–613.

Steeves, T. A. (1966). On the determination of leaf primordia in ferns. In *Trends in Plant Morphogenesis* (E. G. Cutter, Ed.), pp. 200–219. Longmans, London.

Steeves, T. A., Hicks, M. A., Naylor, J. M., and Rennie, P. (1969). Analytical studies on the shoot apex of *Helianthus annuus. Can. J. Bot.* **47,** 1367–1375.

Steeves, T. A., and Sussex, I. M. (1957). Studies on the development of excised leaves in sterile culture. *Am. J. Bot.* **44,** 665–673.

Stein, O. L., and Fosket, E. B. (1969). Comparative developmental anatomy of shoots of juvenile and adult *Hedera helix. Am. J. Bot.* **56,** 546–551.

Sussex, I. M. (1955). Morphogenesis in *Solanum tuberosum* L: Experimental investigation of leaf dorsiventrality and orientation in the juvenile shoot. *Phytomorphology* **5,** 286–300.

Vollbrecht, E., Veit, B., Sinha, N., and Hake, S. (1991). The developmental gene *Knotted*-1 is a number of a maize homeobox gene family. *Nature* **350,** 241–243.

# 10

# Biotic Factors Regulate Some Aspects of Plant Development

P lants live in intimate contact with many different kinds of microorganisms. Epiphytic bacteria and fungi grow on the surface of plant organs, whereas many kinds of soilborne microorganisms are present in the environment around plant roots (known as the rhizosphere). In some cases, a plant and a specific microorganism establish a complex relationship that involves mutual changes in gene expression, leading to developmental changes. Pathogenic relationships between fungi or bacteria and plants result in disease in the host plant, whereas symbiotic relationships have significant benefits for both the plant and the microorganism. Both symbiotic and pathogenic interactions involve changes in the plant that are similar, if not identical, to developmental changes. Here we examine some of the mechanisms by which microorganisms cause disease in plants and

fungi include the rusts (*Puccinia graminis* f.sp. *tritici*, wheat rust), powdery mildews (*Erysiphe polygoni*, a parasite of over 300 plant species), and downy mildews (*Plasmopara viticola*, downy mildew of grape). These fungi have been termed "con men" as they are highly specialized and have found a way to get around the usual plant defense mechanisms. Their success depends on their ability to invade the host plant without triggering, at least initially, the plant's defense responses.

Necrotrophic pathogens derive nutrients from dead cells   These pathogens secrete toxins that kill plant cells and then they digest the components of the dead cells. They are similar to saprophytes in their nutrition, except that they kill cells of their host first. The fungus *Monilinia fructicola* is an example of a necrotrophic pathogen. It causes the brown rot of stone fruits such as peaches and pears. The bacterium *Erwinia amylovora* causes a disease known as fireblight, which devastates many fruit trees. It produces a toxin that rapidly kills whole branches and even entire trees. These pathogens have been called "thugs" because their methods are relatively crude. They kill cells with a relatively nonselective toxin and then feed on the dead remains.

## Resistance to Disease

Most plants are resistant to most pathogens. As mentioned earlier, the reasons for this broad resistance may be rather nonspecific, such as the presence of a thick cuticle on epidermal surfaces. This cuticle can act as a barrier to pathogens. Alternatively, resistance may be the result of specific mechanisms designed to stop the invading microorganism. Even in cases where a particular fungus is highly pathogenic on a particular plant species, such as *Phytophthora infestans* on potato, many varieties of the plant are resistant to the fungus. In fact, the most common measure a farmer takes to control the late blight of potato disease is to plant a resistant cultivar or variety of potato. Plants have evolved several different mechanisms of resistance to avoid or resist invasion by a pathogen. Some of these are described in the next section.

The composition of the plant cell wall changes   Some plants alter the composition of the cell wall in the path of the invading pathogen. The deposition of lignin or suberin (a wax) in primary cell walls may block the advance of the pathogen. Some fungi are able to degrade plant cell wall carbohydrates. They synthesize and secrete enzymes that hydrolyze cellulose, hemicelluloses, and/or pectins, using the sugars released for their metabolism. If, however, the cell wall polysacccharides become encrusted with lignin and/or suberin, the wall polysaccharides cannot be degraded and the microorganism is denied its food source. Thus, plant cells sometimes synthesize and secrete lignin precursors and the enzyme peroxidase in response to fungal invasion, to lignify cell walls in the path of the invasion. The enzyme phenylalanine ammonia lyase plays a key role in the synthesis of lignin precursors and the activity of this enzyme often is markedly elevated during pathogenesis (Fig. 10.2). Plants also may modify their

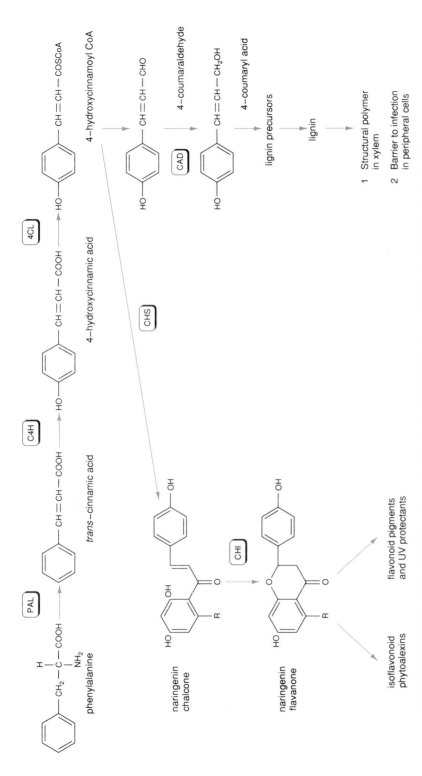

**Figure 10.2 Pathway for the biosynthesis of various molecules from phenylalanine that participate in the plant's response to pathogens**

C4H, cinnamate 4-hydroxylase; CHS, chalcone synthase; CHI, chalcone isomerase; PAL, phenylalanine ammonia lyase. Reprinted with permission from Lamb (1989).

cell walls by synthesizing and depositing hydroxyproline-rich glycoproteins (HRGPs) in the walls of cells in the vicinity of the invading pathogen. These may prevent the penetration of fungal haustoria into the cell.

Pathogenic organisms may induce a hypersensitivity response In most cases where the interaction of the pathogen and its host does not lead to disease (i.e., the interaction is incompatible, the microorganism is avirulent, and the host is resistant), it is because the microorganism induces what is known as the hypersensitivity response in the plant. The hypersensitivity response protects plants against diseases caused by bacteria, fungi, nematodes, and viruses. The hypersensitivity response is a localized response to the pathogen. The infected plant cells synthesize phenolic compounds, which are deadly to itself. The plant cells in the vicinity of the pathogen invasion, adjacent to the infected cells, also are killed by these phenolic compounds. If the pathogen is biotrophic, it will be unable to establish an infection and it also will die. Thus, the localized death of plant cells walls the infection off, preventing the invasion of the pathogen into adjacent living cells. In addition, some of these phenolic compounds may act as phytoalexins. These are chemical compounds that kill the invading pathogen.

Plants may produce phytoalexins in response to pathogens   Phytoalexins are small molecules produced by the plant that have the properties of antibiotics. Frequently, phytoalexins are also toxic to the host and their production is associated with the hypersensitivity response. These compounds accumulate in dead or dying plant cells. Resistance to the pathogen occurs when these compounds accumulate rapidly and to high concentrations at the site of infection, resulting in the death of the pathogen. The enzymes phenylalanine ammonia lyase and chalcone synthase play key roles in the synthesis of these compounds (See Fig. 10.2).

Plants may synthesize enzymes that attack the pathogen   Some plants can synthesize chitinase, an enzyme that degrades the fungal cell wall, in response to fungal attack. The destruction or weakening of the fungal cell wall renders it very sensitive to lysis as a result of unrestricted osmotic water uptake. Without a cell wall, the fungus will lyse and die.

Pathogen avirulence and host disease resistance are genetically determined   One of the most unusual aspects of the interaction between plants and their pathogens is that both compatibility and incompatibility are the result of the mutual expression of a complex set of genes when the pathogen and its host interact. This is known as the **gene-for-gene hypothesis**. It proposes that for every gene that tends to make the host plant resistant to the disease, there exists a comparable gene in the pathogen that tends to make it avirulent. A set of **resistance genes** exist in the plant and a comparable set of **avirulence genes** reside in the pathogen (Table 10.4) This relationship comes about because the product of the resistance gene is believed to interact with the product of the avirulence gene to block the formation of the pathogenic condition. This means that a given resistance gene will confer resistance on a given cultivar only if the pathogen expresses a complimentary avirulence gene. Both the avirulence genes and the resistance genes are dominant genes.

At present, the molecular mechanisms of these avirulence–resistance systems are not known. Although avirulence genes have been cloned, the mechanism by which their expression results in avirulence is not known.

Pathogens invade a host by expressing pathogenicity genes    The steps leading to a compatible reaction (disease) are shown in Table 10.5. A pathogenic organism is able to invade its host, in part, because it possesses certain **pathogenicity genes**, which encode pathogenicity functions. Pathogenicity functions may include some or all of the following:

- Attachment to the plant surface
- Formation of fungal penetration structures
- Degradation of host cuticle and cell wall
- Production of toxins

The microorganism must adhere to the plant if it is to initiate infection. It is possible that adherence involves a relatively nonspecific molecular interaction between some components of the plant cell surface and the cell surface of the pathogen. In some cases, however, the pathogen synthesizes an adhesive and secretes it when it comes in contact with a potential host. Genes encoding the enzymes required to synthesize the adhesive may be pathogenicity genes for this organism. The pathogen next must penetrate the epidermis of the plant if it cannot enter through a wound or through the stomates. Some fungi synthesize

**Table 10.4   Pattern of interactions between a pathogen and its host**[a]

| Genotype of pathogen | Host cultivar genotype | | | |
|---|---|---|---|---|
| | R1R1 R2R2 | R1R1 r2r2 | r1r1 R2R2 | r1r1 r2r2 |
| P1P1P2P2 | − | − | − | + |
| P1P1p2p2 | − | − | + | + |
| p1p1P2P2 | − | + | − | + |
| p1p1p2p2 | + | + | + | + |

[a] In this case there are two resistance genes (R) in the host plant, which are complemented by two avirulence genes (P) in the pathogen. A minus indicates an incompatible reaction; a plus indicates that the interaction was compatible.

**Table 10.5    Early events in the infection of a root by a fungus**

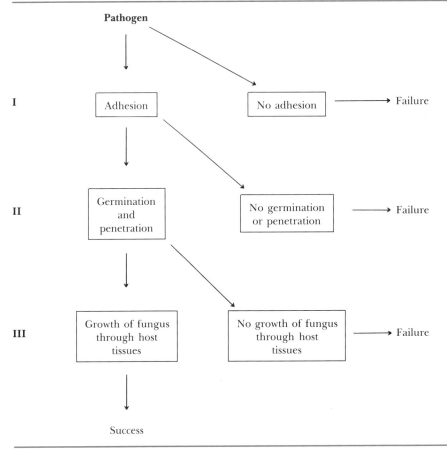

Reprinted with permission from Ralton *et al.* (1986).

and secrete a cutinase enzyme that will destroy waxy cuticle covering the epidermis, enabling it to penetrate into the tissues or the leaf or stem. Once inside the plant organ, invading fungi may grow intercellularly, without invading the cytoplasm of living cells. In these cases, fungal hyphae grow within the cell wall and do not enter the cytoplasm. They secrete enzymes, such as cellulases or hemicellulases, that enable them to digest the cellulose and/or noncellulosic polysaccharides of the wall and they use the sugars released in this digestion for their metabolism. In these cases, the genes encoding the cutinase, cellulase, or hemicellulase could be considered pathogenicity genes.

Close to 100 genes have been found to be involved in pathogenicity. *Erwinia amylovora*, the bacterium that causes fireblight of apple, pear, and other members of the rose family, synthesizes and secretes a toxic protein known as harpin. Harpin is a 44-kDa glycine-rich hydrophobic protein that kills plant cells by

**Table 10.6**   **Examples of some molecules that can act as elicitors to initiate the hypersensitivity reaction in higher plants**

| Elicitor | Source | Experimental plant |
| --- | --- | --- |
| **Nonspecific Elicitors** | | |
| *Abiotic elicitors* | | |
| Actinomycin D | Commercial | Pea endocarp |
| Psoralens | | |
| Ultraviolet light | | |
| CdCl$_2$ | | |
| Autoclaved ribonuclease A | | French bean suspension-cultured cells and hypocotyl sections |
| *Fungal components* | | |
| Hepta-$\beta$-glucoside containing 1,6 and 1,3 linkages | *Phytophthora megasperma* var. *sojae* | Soybean cotyledons, hypocotyls, and suspension-cultured cells |
| Chitosan heptamer of $\beta$-1,4-glucosamine | Commercial (but a known component of fungal cell walls) | Pea endocarp Soybean suspension-cultured cells |
| Glucan | *Uromyces phaseoli* | French bean |
| Polysaccharide containing Glc, Man, Ara | *Colletotrichum lindemuthianum* | French bean suspension-cultured cells |
| Eicosapentaenoic acid and arachidonic acid; $\beta$-glucan and arachidonic acid | *Phytophthora infestans* | Potato tuber tissue |
| | *Phytophthora infestans* | Potato tuber tissue |
| *Plant components* | | |
| Trideca-$\alpha$-1,4-D-galacturonide | Commercial polygalacturonic acid or castor bean cell walls digested by endopolygalacturonase from *Rhizopus stolonifer* | Castor bean |
| Dodeca-$\alpha$-4-D-galacturonide ("endogenous elicitor") | Plant cell walls (soybean, tobacco, sycamore, wheat, citrus pectin) | Soybean cotyledon |
| Proteinases inhibitor-inducing factor (PIIF) (pectic polysaccharides) | Tomato and potato leaves | Pea endocarp |
| | | Castor bean seeding |
| **Race-Specific Elicitors** | | |
| High-molecular-weight glycan with 1,3 and 1,4 $\beta$-linkages + Rha–Man–Gal and protein | *Colletotrichum* spp. | Red kidney bean |
| Glucomannan | *Phytophthora megasperma* f. sp. *glycinea* | Soybean |
| Glycoprotein | *Phytophthora megasperma* f. sp. *glycinea* | Soybean |
| Peptidogalactoglucomannan | *Cladosporium fulvum* | Tomato |

Reprinted with permission from Ralton *et al.* (1986).

disrupting the ion flux across the plasma membrane. Virulent strains of this pathogen all have an *hrp* gene encoding the harpin protein; however, the harpin protein also is an elicitor of the hypersensitivity response. Harpin both causes the disease symptoms and initiates the plant's defense against the disease.

### Cell–Cell Communication between the Host and Pathogen

Molecular communication occurs between the pathogen and its host. This communication may be inadvertent, and molecules produced for a completely different purpose may be used as a signal to trigger a new pattern of gene expression in both the host and the pathogen. The nature of the genes expressed will determine whether the interaction is compatible or incompatible. Signals emanating from the pathogen may trigger the hypersensitivity response in the host, whereas signals from the host may induce the expression of pathogenicity genes in the pathogen.

Host-derived molecules induce the expression of pathogenicity genes    Establishment of infection usually requires a specific signal from the plant. For example, the outer surface of the plant is coated with cutin, a waxy polymer produced and secreted by the epidermal cells. Cutin monomers constitute the signal that induces the expression of the cutinase gene in the fungus *Fusarium solani*, whereas fragments of the plant cell wall induce fungal enzymes that further degrade the wall in many other pathogens. In both cases, the fungus produces a very low constitutive level of the cutinase or wall-degrading enzymes. The activity of these enzymes produces the fragments that lead to the massive expression of the genes, with wide-scale invasion of the pathogen and cellular destruction.

Elicitors induce the hypersensitivity response    An elicitor is a molecule that triggers the hypersensitivity response in the plant. Elicitors are very diverse molecules without any chemical similarity, except that they trigger the hypersensitivity response (Table 10.6). In some cases these elicitors are formed when plant-derived enzymes degrade fungal cell walls. For example, a beta-linked heptaglucoside has been shown to be a highly effective elicitor of the hypersensitivity response in soybean roots infected with the root-rotting fungus *Phytophthora megasperma*. In other cases, elicitors are derived from the degradation of the host plant cell wall. In fact, a variety of chemical and physical agents that damage cells act as elicitors. For example, ultraviolet light and heavy metals can act as elicitors. The elicitors do not appear to be highly specific, although they are exceedingly potent. Nanomolar concentrations of some elicitors will evoke the hypersensitivity reaction. Possibly, the specificity of these plant–pathogen interactions stems from the mechanism by which the elicitors are produced.

The elicitor, whatever its source or chemical nature, triggers the hypersensitivity response by first binding to a receptor in or on the surface of the host plant

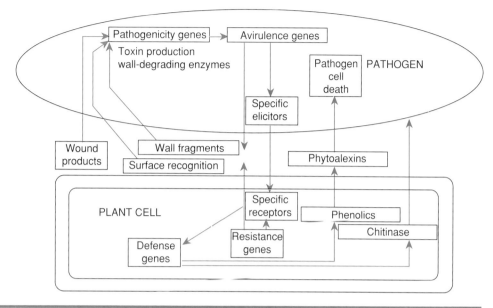

**Figure 10.3 Summary of the interactions between pathogenic microorganisms and their plant hosts**
Reprinted with permission from Lamb *et al.* (1989).

cell. We know that plant cells have an elicitor receptor because of binding studies that have been conducted with labeled elicitor. The specific nature of these receptors is, however, not known. Also, the mechanism by which the binding of the elicitor to the receptor triggers the hypersensitivity response has yet to be determined. Presumably there is a signal transduction mechanism that is activated by elicitor–receptor binding. This signal transduction pathway may involve calcium ions and it is possible that it is similar to the signal transduction pathways shown to be involved in some hormonal responses.

**Elicitor–receptor binding induces expression of plant defense genes** Elicitors bind to their receptors and this, in turn, induces the expression of a group of plant defense genes. These genes encode enzymes required for the synthesis of phytoalexins, which may kill the pathogen, and similar phenolic compounds that bring about the hypersensitivity response. The enzymes include phenylalanine ammonia lyase, chalcone synthase, chalcone isomerase, cinnamyl-alcohol dehydrogenase, and other enzymes (see Fig. 10.2). Other defense genes encode enzymes, such as chitinase and glucanase, that hydrolyze the bacterial or fungal cell wall.

Box 10.1 and Fig. 10.3 summarize pathogenic interactions between plants and microbes.

## Crown Gall and Plant Transformation

### Crown Gall, a Plant Cancer

Plants are susceptible to many different kinds of disease, including some types of cancer or tumors. A plant tumor is called a gall. There are many different kinds of galls, and many different agents can induce gall formation, including

6. Plants may synthesize and secrete enzymes that attack the pathogen. Some plants produce a chitinase, an enzyme that degrades the fungal cell wall, in response to fungal attack.

F. Pathogenic organisms may produce elicitors when they attack their plant host.

G. An elicitor is a molecule that triggers the hypersensitivity reaction in the host cells.

H. Many different molecules have been shown to act as elicitors:
1. Fungal wall fragments formed when plant enzymes degrade the fungal cell wall.
2. Plant cell wall fragments formed when fungal pathogen enzymes degrade the host plant cell wall.

I. Elicitors have the following characteristics:
1. The specificity of the host–pathogen interaction does not lie in the chemical nature of the elicitor.
2. Elicitors often are active at very low concentrations (nanomolar).
3. Elicitors bind to receptors in the plant cell membrane.
4. Elicitor–receptor binding triggers the expression of plant defense genes.

J. Plant defense genes encode one of several classes of proteins:
1. Enzymes involved in phenylpropanoid metabolism, which synthesize molecules that act as phytoalexins or trigger the hypersensitivity response.
2. Phenylalanine ammonia lyase.
3. Chalcone synthetase.
4. Chalcone isomerase.
5. Enzymes that degrade fungal cell walls such as chitinase and glucanase.
6. Proteins involved in strengthening the plant wall (HPRGs).

wounding, insects, viruses, a certain genetic composition, and bacteria. Insect galls are self-limiting, but in many other cases, the gall tissue may grow to the point that it kills the plant. Of these various types of gall diseases, the most interesting from a biological point of view is **crown gall**. The causal agent of this disease is the soil-dwelling bacterium *Agrobacterium tumefaciens*, a member of a group of bacteria known as Rhizobiaceae, which also includes the symbiotic

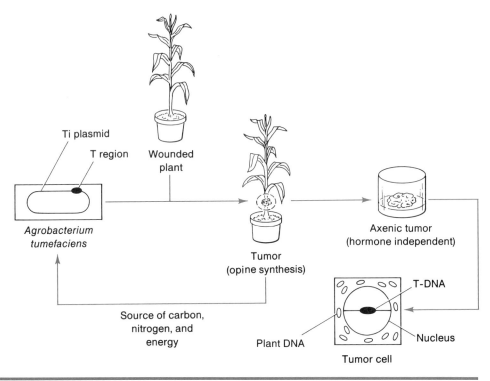

Ti plasmid

T region   Wounded
plant

Agrobacterium
tumefaciens

Tumor
(opine synthesis)

Axenic tumor
(hormone independent)

T-DNA

Nucleus

Plant DNA

Tumor cell

Source of carbon,
nitrogen, and
energy

**Figure 10.4 Plant transformation by *Agrobacterium tumefaciens***
Virulent strains of the soil-dwelling bacterium *A. tumefaciens* contain a large plasmid
known as the Ti plasmid. The T-DNA region of the Ti plasmid contains genes
encoding enzymes required for the synthesis of the plant hormones auxin and
cytokinin. These T-DNA genes are phytooncogenes. When the bacterium infects a
wound on a dicot, it transfers TDNA from its Ti plasmid into the plant cell, which
becomes incorporated into the nuclear genome of the host plant. Plant cells that have
incorporated foreign DNA into their genome are said to be transformed. Expression of
these T-DNA phytooncogenes in the plant genome results in the synthesis of high
levels of auxin and the cytokinin zeatin. These hormones stimulate the abnormal
proliferation of the transformed cells, resulting in the formation of the gall, or tumor.
Reprinted with permission from Nester *et al.* (1984).

nitrogen-fixing *Rhizobium. A. tumefaciens* enters the plant through a wound and
genetically transforms the plant cell. It induces tumors by transferring some of
its genetic material into the plant cell, where the expression of genes encoded
in this bacterial DNA in the plant cell stimulates plant cell division. As the cells
proliferate, they form a tumor whose growth cannot be regulated by the plant
(Fig. 10.4).

Bacteria-free gall tissues will grow in culture on simple medium
lacking hormones   The unique aspects of this disease can be demonstrated

by exposing wounded plant tissue to a virulent strain of *A. tumefaciens* for several days and then heating the infected tissue to about 45°C. Plant cells are able to survive this treatment, but the bacteria are not; however, even though the bacteria have been killed, the plant cells still form a tumor. Furthermore, the tumor tissues can divide and grow indefinitely when they are cultured on a simple medium lacking the plant hormones, auxin and a cytokinin. Normal tissue cannot grow under these conditions. Normal plant tissues require plant hormones to grow in tissue culture. The ability of *A. tumefaciens* to transform plant cells is dependent on a large plasmid, known as the Ti (for tumor-inducing) plasmid. All strains of *A. tumefaciens* that are able to induce the formation of crown galls have one of these plasmids, whereas strains of the bacterium that have lost the Ti plasmid no longer cause galls. Some of the general features of the biology of crown gall are illustrated in Fig. 10.4.

Crown gall tissues synthesize unusual amino acids    An unusual characteristic of crown gall tissue, in addition to the fact that it is an unregulated tumor, is that it produces one or more of a class of rare amino acids known as opines. Two of the most common opines are nopaline and octopine. These are not found in normal plant tissue, but only in plant tissue that has been transformed by *Agrobacterium*. Furthermore, there is a close relationship between the ability of the plant gall tissue to make a particular type of opine and the type of Ti plasmid that induced the gall. There are many different strains of *A. tumefaciens* that can incite galls on plant tissues, all of which have a Ti plasmid; however, there are several different types of Ti plasmids,. The different Ti plasmids are classified according to the type of opine made by the gall tissue they incite. The nopaline-type Ti plasmids and the octopine-type Ti plasmids are the most common, but there are other types of Ti plasmids that incite tumors that manufacture different opines. Opines are produced and secreted by the plant cells of gall tissues, and they can be used by *A. tumefaciens* as a source of carbon, nitrogen, and energy. The ability of the bacterium to use the opine depends on genes that are found on the Ti plasmid, but the bacterium itself does not synthesize the opine it is capable of using. The genes for opine synthesis also are carried by the Ti plasmid, but they are expressed only after they are inserted into the plant genome. Structures of some of the opines are illustrated in Fig. 10.5. This is a type of genetic colonialism. *Agrobacterium* genetically transforms plant cells with genes the plant cannot regulate, forcing the plant cell to synthesize and secrete nitrogen-rich compounds that only the bacterium can use.

## T-DNA

How is the T-DNA identified?    How much of the Ti plasmid DNA is transferred to the plant cell during the infection process? Ti plasmids are relatively large plasmids, containing 100,000 to 200,000 bp of DNA as a circular, extrachromosomal genetic element (Fig. 10.6). When investigators began comparing the sequences of these different Ti plasmids, they noted that the sequences exhibited

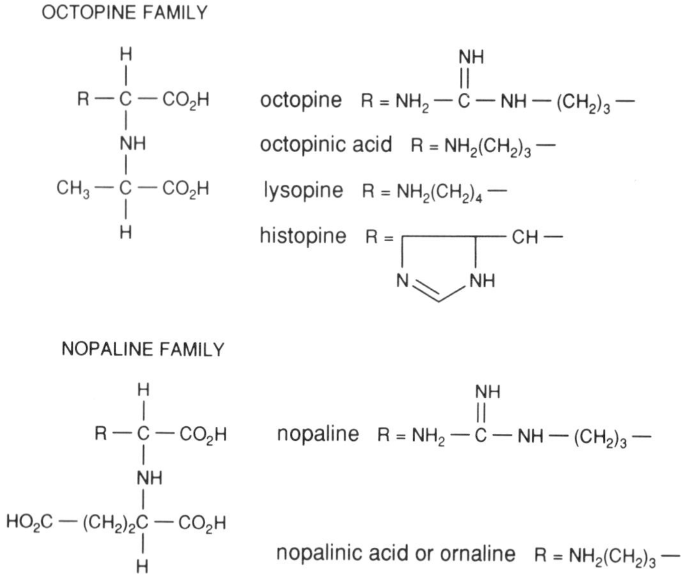

OCTOPINE FAMILY

octopine  $R = NH_2 - \overset{\overset{NH}{\|}}{C} - NH - (CH_2)_3 -$

octopinic acid  $R = NH_2(CH_2)_3 -$

lysopine  $R = NH_2(CH_2)_4 -$

histopine  R =

NOPALINE FAMILY

nopaline  $R = NH_2 - \overset{\overset{NH}{\|}}{C} - NH - (CH_2)_3 -$

nopalinic acid or ornaline  $R = NH_2(CH_2)_3 -$

**Figure 10.5 Structures of some opines**
Reprinted with permission from Guyon *et al.* (1980).

relatively little homology. There are, however, two regions that are smilar in sequence in the different Ti plasmids. One of these is known as the *vir* region because it contains genes that are known to be required for transformation, or genes that enhance the process. These are the genes responsible for the virulence of the disease, so they are known as the virulence or *vir* genes. In addition, the region known as T-DNA has a great deal of sequence similarity among the different Ti plasmids. Investigators used Southern analysis of genomic DNA extracted from untransformed and transformed plant cells to determine the extent of the Ti DNA that the bacterium inserted into the plant cells. The only portion of the Ti plasmid found in the genomic DNA of the transformed plants is that corresponding to the T-DNA. In fact, it was called T-DNA because it is the transforming DNA of the Ti plasmid. Usually, only one or at the most a few copies of the T-DNA are found per diploid genome in the transformed plant cells.

How is the T-DNA organized?   Nopaline-type T-DNA usually exists as one copy per transformed plant cell where it is found as one continuous piece of DNA, about 15 kb in length. The structure of nopaline T-DNA in the plant cell is very similar to its structure in the Ti plasmid. In other words, it is integrated into the host genome without major rearrangement. In the bacterial plasmid and occasionally in plant genomes transformed with this DNA, it is flanked at either end by a 23-bp directly repeated DNA sequence. These repeats are known as the right and left T-DNA borders. The right border, and most of the T-DNA,

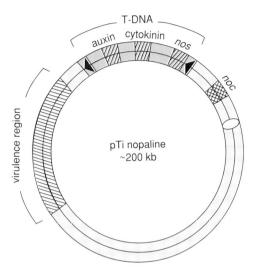

**Figure 10.6 Map of the Ti plasmid showing the location of some genes important for crown gall induction**

The Ti plasmid is a circular DNA molecule that ranges from 100,000 to 200,000 bp. Shown here is a simplified diagram of a nopaline Ti plasmid. The genes important for crown gall induction are those in the virulence region, which contains seven or more different operons known as the *vir* genes. The T-DNA is the region of the plasmid that is transferred to the plant cell. It is flanked by a 23-bp direct repeat and contains genes encoding enzymes that can synthesize plant hormones acting as an auxin and a cytokinin. These are the oncogenes whose expression results in the overproduction of growth hormones and unrestricted, tumorous growth of the transformed cells. The *nos* gene carried by the T-DNA encodes the enzyme nopaline synthase, which will enable the transformed plant cells to synthesize the opine nopaline. The *noc* gene encodes an enzyme that enables the bacterium to use nopaline for its metabolism. Redrawn with permission from Singer and Berg (1991).

is found intact in the plant genomic DNA where it is inserted apparently at random; however, the left border usually undergoes rearrangement during the transfer or integration process. A number of experiments have shown that any DNA sequence between the left and right borders in the Ti plasmid will be integrated into the plant genome, usually intact and with its base sequence unchanged. The T-DNA in octopine-type Ti plasmids also is flanked by direct repeats on its right and left borders.

What genetic information does T-DNA contain?    Clearly, T-DNA contains genes that transform the plant cell, giving it the ability to proliferate uncontrollably. These genes are known as oncogenic or *onc* genes. In addition, T-DNA contains genes that enable the plant cell to synthesize an opine. Northern analysis of RNA extracted from crown gall tissues demonstrates that transformed plant cells contain transcripts from the T-DNA. Typically, a tumor incited by

*A. tumefaciens* carrying a nopaline Ti plasmid will contain about 13 different mRNAs derived from the T-DNA, whereas a tumor incited by an octopine-type Ti plasmid will have about 7 different transcripts from the T-DNA. These are typical plant mRNA molecules. They are transcribed by the plant RNA polymerase II and are capped at their 5′ ends and polyadenylated at their 3′ ends. Thus, we can conclude that T-DNA contains genes that have eukaryotic promoters rather than prokaryotic promoters. This explains why these genes are not expressed in the bacterium, but rather are expressed only in the plant cells after transformation.

Mutation analysis of T-DNA has enabled investigators to determine the function of some of the T-DNA genes. When the T-DNA of an octopine Ti plasmid is mutated in gene 1 or 2 before infecting plant cells, it induces a type of abnormal growth known as a teratoma, in which there is an uncontrolled proliferation of shoots from the transformed cells. This is known as the shooty phenotype (Fig. 10.7). The shooty phenotype can be mimicked in cultured tobacco tissues by treating the tissues with an excess of cytokinin and an abnormally low level of auxin. Similarly, when plants are transformed with Ti plasmids containing an insertional mutation in gene 4, a different type of teratoma is formed in which there is an abnormal proliferation of roots (see Fig. 10.8). This is known as the rooty phenotype. The rooty phenotype can be mimicked in cultured normal tobacco tissue by reducing the level of cytokinin while maintaining a high level of auxin. When genes 1, 2, and 4 are all mutated at the same time, the plasmid cannot incite the formation of either tumors or teratomas, although the T-DNA can still be detected in the cells, and the transformed cells will synthesize octopine. This demonstrates that DNA transfer can be separated from the oncogenicity of the T-DNA. Genes 1, 2, and 4 are the oncogenes. It is the transfer and expression of these genes that induce the formation of the tumor. The tumor forms as a result of overproduction of plant growth regulators.

As many of the phenotypic characteristics of crown gall tumors and teratomas can be induced in normal tissues by manipulating the ratio and absolute levels of the two plant hormones auxin and cytokinin, it was not surprising to discover that the T-DNA genes encode enzymes in the biosynthesis of these plant hormones. Genes 1 and 2 have been shown to encode two different essential enzymes in the biosynthetic pathway of the auxin indole-3-acetic acid. Gene 1, known as *iaaM*, encodes the enzyme tryptophan-2-monooxygenase. It converts tryptophan into indole-3-acetamide. Gene 2, is *iaaH*. It encodes the enzyme indolylacetamide hydrolase. It converts indole-3-acetamide into indole-3-acetic acid (Fig. 10.8). Gene 4 is *iptZ*; it encodes the enzyme isopentenyl transferase, a key enzyme in the biosynthetic pathway of the plant growth regulator cytokinin (Fig. 10.9). Mutations in gene 3 result in plasmids that induce tumors on susceptible plants, but these tumors do not form octopine. Gene 3 encodes the enzyme octopine synthase.

Why can the plant cell not regulate the activity of these genes or the enzymes encoded by these genes?  The plant regulates its growth and

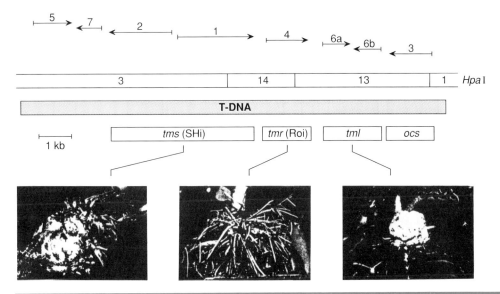

**Figure 10.7 Map of the T-DNA from an *Agrobacterium* Ti plasmid, showing the effects of T-DNA mutations on crown gall tumor morphology**

The shaded bar represents the T-DNA. The numbered arrows above the T-DNA represent the locations of genes expressed in plant cells transformed with the T-DNA. Genes 1 and 2 encode two different enzymes involved in auxin biosynthesis; gene 4 encodes a cytokinin biosynthetic enzyme. Mutations that inactivate genes 1 and 2 (*tms*) result in a T-DNA that produces "shooty" tumors (Shi). Mutations that delete gene 4 (*tmr*) result in a T-DNA that causes "rooty" tumors (Roi) to form on infected plants. Mutations in gene 6 (*tml*) change the size of the tumors formed. Gene 3 encodes octopine synthase, the enzyme responsible for the formation of octopine by the transformed cells. Octopine is an "opine," one of the unusual amino acids produced only by crown gall tissues as a result of their transformation by *Agrobacterium*-derived T-DNA. The bacteria are able to use these opines as an energy source. Reprinted with permission from Morris (1986) and the *Annual Review of Plant Physiology*, **37,** ©1986 Annual Reviews Inc.

development, in part, by regulating the levels and distribution of the plant growth regulators. Why can the plant not regulate the activity of these genes as it does the endogenous genes for growth regulator biosynthesis? We cannot answer this question at present, but the answer undoubtedly will be found in the nature of the promoters for these genes. It is possible that the promoters for these genes on T-DNA lack the sequences that enable them to be developmentally regulated. Certainly the nopaline synthase gene, which has been studied in considerable detail, lacks the signals for any significant developmental regulation. It tends to be expressed at fairly high levels in most tissues of the plant, regardless of their composition or developmental state.

Box 10.2

## Summary of Crown Gall and Plant Transformation

A. Crown gall is a form of plant cancer caused by *Agrobacterium tumefaciens*.
 1. *A. tumefaciens* enters the plant through a wound.
 2. The bacterium genetically transforms the plant cell.
B. The transformed plant tissue differs from normal plant tissues in at least three ways:
 1. It continues to proliferate even after the inciting bacteria have been killed.
 2. Bacteria-free galls will grow in culture on simple medium lacking hormones.
 3. The gall tissues produce and secrete unusual amino acids known as opines. *A. tumefaciens* can use opines for metabolism.
C. *A. tumefaciens* contains a large plasmid, known as the Ti plasmid, necessary for plant cell transformation.
 1. There are several different types of Ti plasmids and each type incites tumors with specific characteristics.
 2. Ti plasmids are designated by the type of opine they regulate: nopaline-type Ti plasmids, octopine-type Ti plasmids, and so forth.
 3. Ti plasmids contain genes for both the synthesis and the utilization of opines, but genes for synthesis are expressed only in the plant and genes for utilization are expressed only in the bacterium.
 4. Only a small part of the Ti plasmid is transferred to the plant cell during infection.
 5. The transferred DNA is known as T-DNA.
 6. T-DNA is integrated into the host plant cell genome.
 7. T-DNA carries the tumor-inducing (*onc*) genes as well as genes for opine synthesis.
 8. In addition to T-DNA, two other regions of Ti plasmids are important for the biology of crown gall.
     a. A 30-kb region known as the *vir* region contains genes required for the integration of T-DNA into the host genome.
     b. Genes for opine utilization are found elsewhere on the Ti plasmid.
D. The following are five important characteristics of T-DNA genes and their expression:
 1. Nopaline T-DNA is 15 kb long and is flanked at either end by a 23-bp direct repeat border sequence.
 2. Thirteen different mRNA transcripts are transcribed from nopaline T-DNA, and seven are transcribed from octopine T-DNA.
 3. Six of these mRNA transcripts are common to these two groups of T-DNAs.
 4. All T-DNA genes are transcribed by the plant's RNA polymerase II and they are not transcribed in the bacterium.
 5. All these mRNAs are polyadenylated at their 3′ end and, therefore, are similar to other cellular mRNAs.

E. Analysis of T-DNA gene function by transposon mutagenesis has led to the following conclusions:
   1. Ti plasmids with mutations in T-DNA gene 1 or 2 induce the formation of shoots instead of tumors.
   2. Ti plasmids with mutations in T-DNA gene 4 induce the formation of roots instead of tumors.
   3. When genes 1, 2, and 4 are deleted, the transformed plant cells do not divide, although they will synthesize an opine.
   4. Genes 1, 2, and 4 are the oncogenes.
   5. Genes 1 and 2 encode enzymes involved in the synthesis of the plant growth regulator auxin.
   6. Gene 4 encodes an enzyme required for synthesis of the plant growth regulator cytokinin.
   7. Genes 1, 2, and 4 differ from the endogenous genes involved in plant growth regulator synthesis and their expression is not regulated by the plant.
   8. Overproduction of these plant growth regulators results in the tumor phenotype.
F. The steps in plant transformation by *A. tumefaciens* are as follows:
   1. Bacteria must first attach to the plant cell surface. Bacterial genes affecting attachment are on chromosomes, not the Ti plasmid.
   2. Induction of the *vir* genes on the Ti plasmid is necessary for T-DNA transfer.
   3. *vir* gene expression is induced by a plant wound phenolic (e.g., acetosyringone).
      a. There are seven *vir* operons on most Ti plasmids.
      b. *vir* operons A and G are expressed constitutively.
      c. VirA protein is a receptor for plant phenolic inducers of *vir* gene expression.
      d. VirA–phenolic complex is an active kinase, phosphorylating VirG protein.
      e. Phosphorylated VirG is a DNA-binding protein that recognizes a 12-bp sequence in the other *vir* gene promoters.
      f. VirG binding to *vir* gene promoters permits *vir* gene transcription.
   4. The T-DNA must be processed by *vir* gene products.
      a. One VirD protein is an endonuclease that recognizes and nicks the T-DNA border sequences.
      b. The T strand of T-DNA is released from the Ti plasmid.
   5. The T strand is then transferred to the plant cell, a process that requires protein products of the *virB* operon.
   6. The T strand is integrated into the plant genomic DNA.
   7. The T-DNA genes are expressed.

on other strains of Ti. In this connection, it is interesting to note that the wide-host-range strains of *Agrobacterium* elicit a hypersensitivity reaction from grape vines instead of tumor formation; however, some wide-host-range strains with mutations in their *virC* genes are tumorigenic on grape vines.

### The Ti Plasmid and Plant Genetic Engineering

The Ti plasmid has been developed as a vehicle for introducing foreign genes into plants. For this purpose it was necessary to disarm the plasmid so that it does not cause tumors. This was done by deleting those genes in the T-DNA that encode enzymes controlling auxin and cytokinin synthesis, the phytoonco-genes. In addition, it was necessary to introduce a gene into the T-DNA that will enable the investigator to select the transformed cells. A gene for antibiotic resistance frequently has been used for this purpose. A desirable cloned gene can then be inserted into the T-DNA of the engineered Ti plasmid and used to infect either cultured cells or leaf disks. The infected cells are placed on a culture medium containing auxin and cytokinin, as well as the antibiotic, to induce growth. Only the transformed cells can grow in the presence of the antibiotic. These transformed cells have received the T-DNA containing not only the gene for antibiotic resistance, but also the foreign gene. To obtain a plant containing the foreign gene, it is necessary to regenerate plants from the cultured, trans-formed cells. Fortunately, methods to accomplish this have been developed for many plants, although not yet all important crop species. At present, both trans-formation of the cereal grains with *Agrobacterium* and their regeneration from cultured cells are very difficult. Nevertheless, some remarkable successes have been achieved with this approach. Investigators have introduced a number of foreign genes into plants, including storage protein genes in soybean and herbi-cide resistant genes in tobacco. In the future, genes for such desirable character-istics as disease resistance and salt tolerance will be transferred to crop plants via this technique. This approach also is very useful for the study of DNA sequences involved in the control of gene expression.

Box 10.2 summarizes the crown gall and plant transformation.

## Symbiotic Interactions: Legumes and *Rhizobium*

All organisms require a source of reduced nitrogen for their metabolism. Al-though there is an abundance of nitrogen in the atmosphere, eukaryotic organ-isms are unable to use this source of nitrogen directly, in part because it is chemically very inert and a large input of energy is required to reduce molecular nitrogen to a form that can be used. Various bacteria are able to reduce atmospheric nitrogen and, in so doing, make these reduced nitrogen compounds available to plants and other organisms. There are two types of nitrogen-reducing (or nitrogen-fixing, as this process is called) bacteria: free-living nitrogen-fixing bacteria and symbiotic nitrogen-fixing bacteria. In this chapter we do not deal with the chemical reactions of nitrogen fixation in any detail; rather, we are concerned with the impact of these nitrogen-fixing organisms on the development of plants that form a symbiotic relationship with bacteria.

Most symbiotic nitrogen-fixing bacteria belong to a group of soil-dwelling bacteria known as the Rhizobiaceae, and many of these belong to the genus *Rhizobium*. *Rhizobium* bacteria form symbiotic nitrogen-fixing relationships only with members of the plant family known as Leguminosae. This family includes peas, beans, soybeans, clover, alfalfa, and many other useful crop plants. In fact, Leguminosae is one of the most successful groups of angiosperms and includes many plants that can act as pioneers, growing in nutrient-poor soils where other plants cannot grow. The importance of this nitrogen-fixing relationship to ecological success is dramatically illustrated when legumes and nonlegumes are grown in a nutrient solution lacking fixed nitrogen. If the legume establishes a nitrogen-fixing symbiosis, it will thrive, whereas the nonlegume, or the legume that is not inoculated with the right strain of *Rhizobium*, will struggle to survive.

## Chemistry of Atmospheric Nitrogen Fixation

The reduction of atmospheric nitrogen requires a great deal of energy. The overall reaction for the process is

$$N_2 + 6e^- + 12ATP + 12H_2O \xrightarrow{\text{nitrogenase}} 2NH_4^+ + 12ADP + 12P_i + 4H^+$$

The reaction is catalyzed by the bacterial enzyme complex nitrogenase. *Rhizobium* bacteria are able to carry out this reaction only when they are in a symbiotic relationship with plant roots and after they have differentiated to become **bacteroids** inside plant root tissues. (In becoming a bacteroid, the bacteria increase in size up to 40-fold and become club-shaped. They also develop an internal membrane system not usually found in bacteria). The chemical reaction not only has a very high energy requirement, but is also very sensitive to oxygen and to one of its products, ammonia. The nitrogenase reaction is inhibited by both oxygen and ammonia.

## Important Contributions by the Plant to This Symbiosis

As it is clear that the bacterium actually does the nitrogen fixation, what, we might ask, does the plant contribute to the symbiosis? Why does the bacterium go to the trouble of actively seeking out and infecting susceptible plants? In fact, these bacteria can fix nitrogen only when they are intimately associated with the plant and encased in special structures constructed by the plant called nodules. Figure 10.10 shows the appearance of bacteroids inside the nodules of a soybean root.

The plant makes the following contributions to the symbiosis:

1. The plant provides metabolizable substrates for the respiration necessary to provide the energy for the process. Ultimately the energy is supplied by photosynthesis. Sugar, mostly sucrose, derived from photosynthesis is transported to the roots, where it is available for respiration by the bacteria to provide the ATP and electrons necessary for nitrogen fixation.
2. The plant provides protection against oxygen poisoning of the nitrogenase through a protein pigment it synthesizes, leghemoglobin. Actually the leghemoglobin itself is a symbiotic product. It consists of a protein with a molecular weight of about 15,000 Da, to which a heme prosthetic group is attached. The protein is a product of the plant, whereas the heme is a product of the bacteria.
3. The plant removes the ammonium ions that competitively inhibit the nitrogenase reaction. In many legumes, the plant does so by synthesizing an enzyme, glutamine synthase, that uses the ammonium ions produced by the nitrogenase to make glutamine from glutamic acid.

### Establishment of the Symbiosis

Figure 10.11 illustrates the process by which nitrogen-fixing bacteria infect plant roots and induce the formation of nodules. There are three phases in the establishment of an effective symbiosis between the bacterium and the plant: recognition, infection, and differentiation.

The bacteria must be able to find the plant (recognition)    Associations between plants and bacteria are highly specific. For every species of legume, there is a particular species of *Rhizobium* that infects it. For example, clover is infected with *R. trifolii*, the common bean with *R. phaseoli*, alfalfa with *R. meliloti*, lotus with *R. loti*, and so forth. There are even varieties of the bacterium that infect particular cultivars of a species.

The recognition phenomenon at the molecular level appears to be effected by the interaction between a component of the bacterial cell wall and a minor component of the root hair cell wall, known as a **lectin.** Lectins are glycoproteins with a high degree of affinity for particular sugars. The specific binding of the bacterium to the lectin attaches the bacterium to root hairs protruding from the surface of the root.

The bacteria must be able to penetrate the plant (infection)   The binding of the bacterium to the root hair sets the stage for the next step, the infection process, which is somewhat unusual. Bacterial binding causes the root hairs to curl so that the tip bends back on the hair shaft, trapping some of the bacteria. The hair stops growing in length, but cellulose deposition continues inside the hair to form the **infection thread.** It is not clear how the bacteria get across the root hair cell wall or how they come to be encased inside the infection thread, but there is some evidence that they secrete enzymes that digest the plant cell wall. The bacteria become embedded inside the infection thread, and as it grows, the infection thread carries the bacteria through the epidermal cells and into the cells of the root cortex. The bacteria are then discharged into the cortical cells. In the course of their passage into the cortical cells they become enveloped by a membrane produced by the cortical cells, the **peribacteroid membrane.** The bacteria are inside the plant cells, but at the same time are topologically outside the cytoplasm. The peribacteroid membrane regulates the transport of materials into the bacteroids from the plant cell, as well as the transport of ammonium ions and heme from the bacteroids into the plant cytoplasm.

The bacteria must be able to differentiate within the plant (differentiation)   Once the bacteria are in the cells, they differentiate into structures known as bacteroids. Before this differentiation they cannot fix nitrogen and lack the nitrogenase enzyme. After differentiation they become competent to fix nitrogen and now synthesize the nitrogenase necessary for this process. While this is happening, the plant cells also grow and differentiate. The plant cells are first stimulated to divide to form the nodule, which looks somewhat like a small tumor. In this sense, establishment of the nitrogen fixation symbiosis is somewhat analogous to formation of a gall in response to *Agrobacterium,* except that the plant cells are not transformed, in the genetic sense, by *Rhizobium* infection. The infected cortical cells form a meristem, the nodule meristem, which will produce the nodule. In some ways nodules are similar to lateral roots, although they have a different origin.

## Molecular–Genetic Analysis of Nitrogen Fixation in *Rhizobium*

Expression of *nif* and *nod* genes in the bacterium is necessary for the formation of nitrogen-fixing nodules   Two groups of bacterial genes are necessary for the formation of a productive symbiotic relationship with the plant: the *nod* genes, which are involved in the formation of the nodule, and the *nif* genes, which are involved in nitrogen fixation. Both groups of genes are

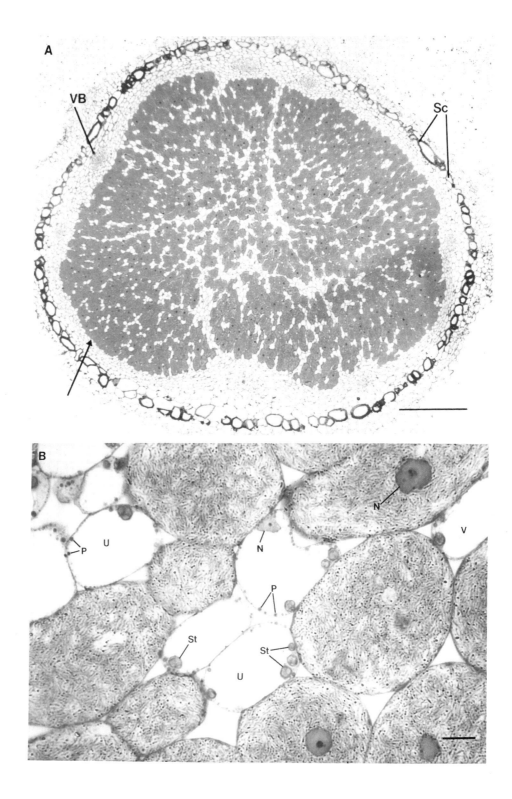

located on a large plasmid, about 200 kb in size, that is carried by all strains of *Rhizobium* able to form nitrogen-fixing nodules. Various mutations of these plasmids have been made by transposon-induced mutagenesis. The total numbers of *nif* and *nod* genes involved are not known, but they are clustered into a 45-kb region of the plasmid. The early *nod* genes are involved in the infection process. These bacterial genes are expressed only during nodulation and their expression is induced by specific plant-derived molecules that are members of a group of chemicals known as **flavonoids,** which include daidzein, produced by soybean roots, and luteolin, produced by clover roots (Fig. 10.12).

The *nod* genes A, B, and C are required for root hair curling, for formation of the infection thread, and for the cortical cell divisions necessary for nodule formation. These genes are involved in the synthesis of a lipooligosaccharide that acts as a signaling molecule for the plant cell. Secretion of the lipooligosaccharide onto the root hairs not only induces hair curling, but initiates changes in the cortical cells that lead to the formation of the nodule primordium. Before infection, all root cortical cells have a large central vacuole and have ceased cell division. After the *Rhizobium* bacteria bind to the root hair and initiate infection thread formation, the nuclei of the cortical cells enlarge and are transported to the center of the cell. The nucleus maintains this central position, supported by the cytoskeleton. To some extent the behavior of these cortical cells is similar to that of other large vacuolate cells that have been induced to divide by hormones or wounding. In the case of these legumes, however, only some of the cells of the inner cortex actually divide and give rise to the nodule primordium. The cells of the outer cortex, whose nuclei also have enlarged and been transported into the center of the cell, do not divide. Instead, they form aligned cytoplasmic

**Figure 10.10  Structure of nitrogen-fixing root nodules**
(A) Section through a 4-week-old soybean nodule showing the localization of the bacteroid-containing cells (stained darkly), which are interspersed with uninfected cells. The central region containing the infected cells is surrounded by the nodule cortex, made up of a layer of thick-walled sclerenchyma (Sc) and parenchyma with interspersed vascular bundles (VB). The arrow indicates that boundary between the central, infected region and the uninfected cortical region. Arrowheads point to rays of uninfected parenchyma. Bar = 0.5 mm. (B) Infected and uninfected cells in a soybean nitrogen-fixing nodule. Infected cells (I) contain hundreds of bacteroids that obscure most cytoplasmic organelles, although a large, probably polyploid nucleus (N) often is visible. Uninfected cells (U) lack bacteroids, have a much smaller nucleus (N), and contain numerous peroxisomes (P), starch grains (St), and a large central vacuole (V). × 1300. Bar = 10 μm. Reprinted with permission from Selker and Newcomb (1985).

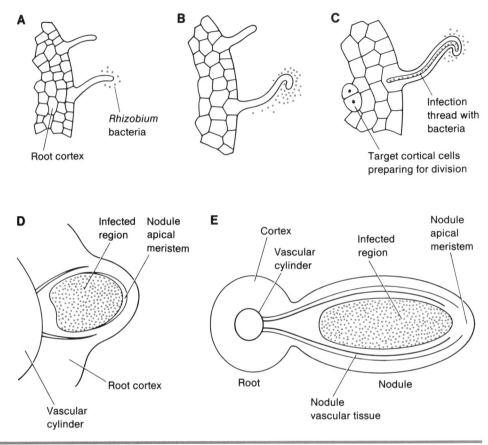

**Figure 10.11 Root nodule development**

(A) Rhizobium bacteria in the soil are attracted to root hairs, to which they bind. (B) This induces the root hair to curl. (C) An infection thread carries the dividing bacterium down the hair and into the cortical cells. Chemical signals produced by the bacterium induce cell divisions in target inner root cortical cells. (D) Derivatives of these cortical cells form an apical meristem, which grows out through the cortex as it produces the tissues of the nodule. (E) Diagram illustrating the different regions of a nodule and their relation to the root. The nodule is connected to the vascular cylinder of the root by vascular bundles that surround it, and the cells in the central region of the nodule are infected with bacteroids. Modified from Bergersen (1982).

bridges through which the infection thread will grow, on its way to the dividing cells of the inner cortex.

The *nod*D gene encodes a DNA-binding protein that activates the expression of the other *nod* genes. This is the master switch. *nod*D gene is turned on by the inducer, and its protein product in turn binds to the promoters of the other *nod* genes, turning them on. Some of the late *nod* genes encode functions necessary

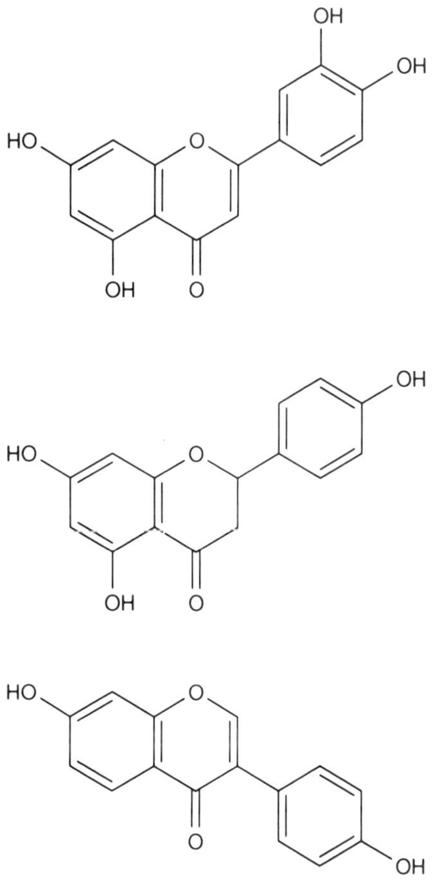

luteolin
(flavone)

naringenin
(flavanone)

daldzein
(isoflavone)

**Figure 10.12 Three compounds produced by plants that induce *nod* gene expression in *Rhizobium* bacteria**

for the differentiation of bacteria into bacteroids. The *nif* genes encode proteins involved directly in the chemistry of nitrogen fixation and include the genes encoding nitrogenase.

Specific plant gene expression also is required for the formation of nitrogen-fixing nodules The genome of legumes contains a group of genes, nodulin genes, that are expressed more or less exclusively during the establishment of the nitrogen-fixing symbiosis. The transcription of these genes is triggered by *Rhizobium* infection. Nodulins include genes encoding enzymes of nitrogen and carbon metabolism. The soybean gene encoding uricase is a nodulin and the gene encoding glutamine synthetase is a nodulin in many other species.

**Box 10.3**

### Summary of Symbiotic Nitrogen Fixation

A. Nitrogen is an essential element for all organisms. Although nitrogen is one of the most abundant elements and constitutes 80% of the atmosphere, atmospheric $N_2$ is unavailable to most organisms.

B. Certain bacteria, mostly members of the genus *Rhizobium*, are able to fix atmospheric nitrogen when they are in a symbiotic relationship with plants, most of which are members of the plant family known as Leguminosae.

C. Symbiotic nitrogen fixation requires metabolic cooperation between the plant and the bacterium and it takes place in special structures called nodules.

D. Nitrogen fixation (reduction) is a highly energetic process carried out with a bacterial enzyme complex, known as nitrogenase, and energy supplied by the plant.

$$N_2 + 6e^- + 12ATP + 12H_2O \xrightarrow{\text{nitrogenase}} 2NH_4^+ + 12ADP + 12P_i + 4H^+$$

E. The plant makes several important contributions to the fixation:
  1. The plant supplies the reduced carbon (from photosynthesis) used for bacterial metabolism.
  2. The plant provides the protein leghemoglobin, which protects the nitrogenase against oxygen poisoning of the reaction.
  3. The plant uses the ammonium ions produced by the nitrogenase reaction, which are competitive inhibitors of the reaction.

F. A cell–cell recognition event must occur between the bacterium and the plant to establish the symbiosis.
  1. The bacterium must be able to find the plant.
  2. The association between the bacterium and the plant is highly specific.
  3. Bacteria are attracted to the roots of susceptible plants and bind to the root hairs.
  4. This recognition phenomenon involves components of the bacterial cell surface that interact with components of the plant cell wall.

The gene encoding the leghemoglobin apoprotein also is a nodulin and is expressed only in nodules. Leghemoglobin is an oxygen-carrying protein, somewhat similar to the hemoglobin of blood, that helps maintain the low oxygen tension necessary for effective nitrogen fixation.

Characteristics of symbiotic nitrogen fixation relationships are summarized in Box 10.3.

G. Binding of the bacterium to the root hair induces the formation of an infection thread.
   1. The infection thread forms from the tip of the root hair.
   2. The infection thread is formed by the root hair cell by altering the mechanism by which the cell deposits its cell wall.
   3. The infection thread carries the bacterium through the epidermal cells and into the cortical cells.
H. Once inside the cortical cells, the nodule begins to differentiate.
   1. Both the bacteria and the cortical cells differentiate to form the nodule.
   2. The bacteria become bacteroids and begin to synthesize nitrogenase. The cortical cells divide and begin to produce leghemoglobin.
I. Specific bacterial genes, called *nod* genes, are expressed only during and after infection of the plant.
   1. *nod* genes regulate the infection and nodulation process.
      a. *nod* gene expression is induced by flavonoids produced by the plant.
      b. *nod* genes A, B, and C are involved in synthesis of a lipooligosaccharide signaling molecule that induces the formation of the infection thread and root cortical cell divisions.
      c. *nod* D gene encodes a DNA-binding protein that activates the expression of the other *nod* genes.
   2. *nif* genes encode enzymes required for nitrogen fixation, including the bacterial genes for nitrogenase.
J. The plant has an equivalent set of genes, *nodulin* genes, which are expressed only during *Rhizobium* infection.
   1. Nodulin genes include genes encoding enzymes of nitrogen and carbon metabolism. A gene encoding uricase is a soybean nodulin, and a gene encoding glutamine synthetase is a nodulin in many other species.
   2. The gene encoding the leghemoglobin apoprotein also is a nodulin and is expressed only in nodules.

## Questions for Study and Review

1. What is meant by the term *saprophyte?* The organism *Pseudomonas syringae* causes a disease known as "bacterial canker" on almond trees. Is this consistent with its role as a saprophyte on strawberries and many other plants? Although *P. syringae* is a saprophyte on most

plants, it can damage even these in special circumstances. How can this damage be brought about?

2. What is a compatible reaction between a host and a plant pathogen? What are the consequences of a compatible reaction? What characteristics must the pathogen and host have for a compatible reaction to occur?

3. What is a necrotrophic pathogen? How does a necrotrophic pathogen kill its host? Give some examples of necrotrophic pathogens.

4. What are biotrophic pathogens? What are haustoria and what function do they serve?

5. How does lignification help make plants resistant to a particular disease? What are lignin precursors and what are the key steps in their biosynthesis?

6. What is the hypersensitivity response? How does an effective hypersensitivity response render a plant resistant to a particular pathogen?

7. What are phytoalexins? What are some of the key steps in the biosynthesis of phytoalexins? What role do they play in the plant's defense response to disease organisms?

8. What are the different pathways a pathogen might use to enter a host plant? If a fungus grows through the cell walls of a plant, digesting the components of the primary walls and using these molecules for its nutrition, what measures might the plant take to block the progress of the fungus through its tissues?

9. What are resistance and avirulence genes? What role do they play in plant disease?

10. What are pathogenicity genes? What types of functions do pathogenicity genes encode? What triggers the expression of pathogenicity genes?

11. What is an elicitor? Give several examples of molecules that act as elicitors. How are elicitors produced? What function do they have in plant disease?

12. What is the class of molecules known as "opines"? What role do opines play in crown gall disease? Where is the genetic information necessary for the synthesis and utilization of opines?

13. What are the Ti plasmids carried by the crown gall bacterium *Agrobacterium tumefaciens*? What features do all Ti plasmids that cause galls have in common?

14. What is the T-DNA of Ti plasmids? What role does it play in crown gall disease? What is the significance of the 23-bp direct repeat sequence found at either end of T-DNA?

15. What genetic information does the T-DNA carry? Explain the effect of mutations in T-DNA genes 1, 2, 3, and 4 on the characteristics of tissues transformed by Ti plasmids with these mutations, both singly and when a given plasmid has mutations in all these genes.

16. What role do plant hormones play in crown gall disease?

17. What are the *vir* genes of the Ti plasmid? What role do they play in crown gall disease? What induces the expression of the *vir* genes? Where do these compounds come from?

18. The overall reaction for the reduction of atmospheric $N_2$ shows that it takes 12 moles of ATP and 6 electrons to reduce 1 mole of $N_2$. Where does this ATP come from? This reaction is possible only because the legume and the bacterium are in a symbiotic relationship. Explain what each partner contributes to the reaction.

19. How do *Rhizobium* bacteria find and recognize the plant species they colonize in nitrogen-fixing nodules?

20. How do *Rhizobium* bacteria reach the cortical cells of the root? What is the infection

thread and of what is it made? What is the significance of the peribacteroid membrane for nitrogen fixation and where does it come from?

21. How is a bacteroid different from a free-living *Rhizobium* bacterium in terms of the genes it expresses? What is the difference between *nif* and *nod* genes in terms of the processes in which their products function?

22. What are nodulins? What role do the nodulins play in nitrogen fixation?

## Further Reading

### General References

Agrios, G. N. (1988). *Plant Pathology,* 3rd ed. Academic Press, San Diego, CA.

Carcfoot, G. L., and Sprott, E. R. (1969). *Famine on the Wind: Plant Diseases and Human History.* Angus Robertson, London.

Dixon, R. O. D., and Wheeler, C. T. (1986). *Nitrogen Fixation in Plants.* Chapman and Hall, New York.

Giles, K. L., and Atherly, A. G. (1981). *Biology of the Rhizobiaceae.* Academic Press, New York.

Hartman, H. T., Kofranek, A. M., Rubatzky, V. E., and Flocker, W. J. (1988). *Plant Science,* 2nd ed. Prentice-Hall, Englewood Cliffs, NJ.

Matthews, R. E. F. (1991). *Plant Virology,* 3rd ed. Academic Press, San Diego, CA.

Smith, C. J. (1991). *Biochemistry and Molecular Biology of Plant–Pathogen Interactions.* Clarendon Press Oxford.

### Pathogenic Interactions

Baulcombe, D. (1989). Strategies for virus resistance in plants. *Trends Genet.* **5,** 56–60.

Bradley, D. J., Kjellbom, P., and Lamb, C. J. (1992). Elicitor- and wound-induced oxidative cross-linking of a proline-rich plant cell wall protein: A novel, rapid defense response. *Cell* **70,** 21–30.

Castresana, C., DeCarvalho, F., Gheysen, G., Habets, M., Inzé, D., and Van Montagu, M. (1990). Tissue-specific and pathogen-induced regulation of a *Nicotiana plumbaginifolia* β-1,3-glucanase gene. *Plant Cell* **2,** 1131–1143.

Dong, X., Mindrinos, M , Davis, K. R., and Ausubel, F. M. (1991). Induction of *Arabidopsis* defense genes by virulent and avirulent *Pseudomonas syringae* strains and by a cloned avirulence gene. *Plant Cell* **3,** 61–72.

Ecker, J. R., and Davis, R. W. (1987). Plant defense genes are regulated by ethylene. *Proc. Natl. Acad. Sci. USA* **84,** 5202–5206.

Frank, S. A. (1992). Models of plant–pathogen coevolution. *Trends Genet.* **8,** 213–219.

Lamb, C. J., Lawton, M. A., Dron, M., and Dixon, R. A. (1989). Signals and transduction mechanisms for activation of plant defenses against microbial attack. *Cell* **56,** 215–224.

Lynn, D. G., and Chang, M. (1990). Phenolic signals in cohabitation: Implications for plant development. *Annu. Rev. Plant Physiol. Plant Mol. Biol.* **41,** 497–526.

Ralton, J. E., Howlett, B. J., and Clarke, A. E. (1986). Receptors in host–pathogen interactions. In *Hormones, Receptors and Cellular Interactions in Plants* (C. M. Chadwick and D. R. Garrod, Eds.), pp. 281–318. Cambridge University Press, London.

Roby, D., Broglie, K., Cressman, R., Biddle, P., Chet, I., and Broglie, R. (1990). Activation of a bean chitinase promoter in transgenic tobacco plants by phytopathogenic fungi. *Plant Cell* **2,** 999–1007.

Rumeau, D., Maher, E. A., Kelman, A., and Showalter, A. M. (1990). Extensin and phenylalanine ammonia-lyase gene expression altered in potato tubers in response to wounding, hypoxia, and *Erwinia carotovora* infection. *Plant Physiol.* **93,** 1134–1139.

Vandenackerueken, G. F. J. M., Vankar, J. A. L., and DeWit, P. J. G. M. (1992). Molecular analysis of the avirulence gene AVR9 of the fungus tomato pathogen *Cladosporum fulvum* fully supports the gene-for-gene hypothesis. *Plant J.* **2,** 359–366.

Yalpani, N., Silverman, P., Wilson, T. M. A., Kleier, D. A., and Raskin, I. (1991). Salicylic acid is a systemic signal and an inducer of pathogenesis-related proteins in virus-infected tobacco. *Plant Cell* **3,** 809–818.

## Agrobacterium and Crown Gall

An, G., Watson, B. D., Stachel, S., Gordon, M. P., and Nester, E. W. (1985). New cloning vehicles for transformation of higher plants. *EMBO J.* **4,** 277–284.

Feldman, K. A., and Marks, M. D. (1987). *Agrobacterium*-mediated transformation of germinating seeds of *Arabidopsis thaliana:* A non-tissue culture approach. *Mol. Gen. Genet.* **208,** 1–9.

Gasser, C. S., and Fraley, R. T. (1989). Genetically engineering plants for crop improvement. *Science* **244,** 1293–1299.

Guyon, P., Chilton, M.-D., Petit, A., and Tempe, J. (1980). Agropine in "null-type" crown gall tumors: Evidence for generality of the opine concept. *Proc. Natl. Acad. Sci. USA* **77,** 2693–2697.

Herrera-Estrella, A., Chen, Z.-M., Van Montagu, M., ,and Wang, K. (1988). VirD proteins of *Agrobacterium tumefaciens* are required for the formation of a covalent DNA–protein complex at the 5′ terminus of T-strand molecules. *EMBO J.* **7,** 4055–4062.

Herrera-Estrella, A., VanMontagu, M., and Wang, K. (1990). A bacterial peptide acting as a plant nuclear targeting signal: The amino-terminal portion of *Agrobacterium* VirD2 protein directs a β-galactosidase fusion protein into tobacco nuclei. *Proc. Natl. Acad. Sci. USA* **87,** 9534–9537.

Howard, E., and Citovsky, V. (1990). The emerging structure of the *Agrobacterium* T-DNA transfer complex. *BioEssays* **12,** 103–108.

Howard, E. A., Winsor, B. A., DeVos, G., and Zambryski, P. (1989). Activation of the T-DNA transfer process in *Agrobacterium* results in the generation of a T-strand–protein complex: Tight association of VirD2 with the 5′ ends of T-strands. *Proc. Natl. Acad. Sci. USA* **86,** 4017–4021.

Jin, S., Roitsch, T., Christie, P. J., and Nester, E. W. (1990). The regulatory VirG protein specifically binds to a *cis*-acting regulatory sequence involved in transcriptional activation of *Agrobacterium tumefaciens* virulence genes. *J. Bacteriol.* **172,** 531–537.

Stachel, S. E., Messens, E., VanMontagu, M., and Zambryski, P. (1985) Identification of the signal molecules produced by wounded plant cells that activate T-DNA transfer in *Agrobacterium tumefaciens. Nature* **318,** 624–629.

Stachel, S. E., and Zambryski, P. C. (1986). *Agrobacterium tumefaciens* and the susceptible plant cell: A novel adaptation of extracellular recognition and DNA conjugation. *Cell* **47,** 155–157.

Weiler, E. W., and Schroder, J. (1987). Hormone genes and crown gall disease. *Trends Biochem. Sci.* **12,** 271–275.

Young, C., and Nester, E. W. (1988). Association of the VirD2 protein with the 5′ end of T strands in *Agrobacterium tumefaciens. J. Bacteriol.* **170,** 3367–3374.

Zambryski, P. (1988). Basic processes underlying *Agrobacterium*-mediated DNA transfer to plant cells. *Annu. Rev. Genet.* **22,** 1–30.

Zambryski, P., Tempe, J., and Schell, J. (1989). Transfer and function of T-DNA genes from *Agrobacterium* Ti and Ri plasmids in plants. *Cell* **56,** 193–201.

## Nitrogen Fixation: Rhizobium and Root Nodule Formation

Bergersen, F. J. (1982). *Root Nodules of Legumes: Structure and Function.* Research Studies Press, New York.

Caetano-Anollés, G., and Gresshoff, P. M. (1991). Alfalfa controls nodulation during the onset of *Rhizobium*-induced cortical cell division. *Plant Physiol.* **95,** 366–373.

Díaz, C. L., Melchers, L. S., Hooykaas, P. J. J., Lugtenberg, B. J. J., and Kijne, J. W. (1989). Root lectin as a determinant of host-plant specificity in the *Rhizobium*–legume symbiosis. *Nature* **338,** 579–581.

Dudley, M. E., Jacobs, T. W., and Long, S. R. (1987). Microscopic studies of cell divisions induced in alfalfa roots by *Rhizobium meliloti. Planta* **171,** 289–301.

Gloudemans, T., and Bisseling, T. (1989). Plant gene expression in early stages of *Rhizobium*–legume symbiosis. *Plant Sci.* **65,** 1–14.

Györgypal, Z., Botond Kiss, G., and Kondorosi, A. (1991). Transduction of plant signal molecules by the *Rhizobium* NodD proteins. *BioEssays* **13,** 575–581.

Long, S. R. (1989). *Rhizobium*–legume nodulation: Life together in the underground. *Cell* **56,** 203–214.

Long, S. R., and Atkinson, E. M. (1990). Nitrogen fixation: *Rhizobium* sweet-talking. *Nature* **344,** 712–713.

Maxwell, C. A., Hartwig, U. A., Joseph, C. M., and Phillips, D. A. (1989). A chalcone and two related flavonoids released from alfalfa roots induce *nod* genes of *Rhizobium melilot. Plant Physiol.* **91,** 842–847.

Nap, J.-P., and Bisseling, T. (1990). Developmental biology of a plant–prokaryote symbiosis: The legume root nodule. *Science* **250,** 948–954.

Roche, P., Debelle, F., Lerouge, P., Vasse, J., Truchet, G., Prome, J. C., and Denarie, J. (1992) The lipo-oligosaccharidic symbiotic signals of *Rhizobium meliloti. Biochem. Soc. Trans.* **20,** 288–291.

Selker, J. M. L., and Newcomb, E. H. (1985). Spatial relationships between uninfected and infected cells in root nodules of soybean. *Planta* **165,** 446–454.

Spaink, H. P., Sheeley, D. M., Van Brussel, A. A. N., Glushka, J., York, W. S., Tak, T., Geiger, O., Kennedy, E. P., Reinhold, V. N., and Lugtenberg, B. J. J. (1991). A novel highly unsaturated fatty acid moiety of lipo-oligosaccharide signals determines host specificity of *Rhizobium. Nature* **354,** 125–130.

Verma, D. P. S. (1992). Signals in root nodule organogenesis and endocytosis of *Rhizobium. Plant Cell* **4,** 373–382.

# Index

Page locators in italics denote references to material in figures and tables.